Gendered Touch

Nuncius Series

Studies and Sources in the Material and Visual History of Science

VOLUME 9

The titles published in this series are listed at *brill.com/nuns*

Gendered Touch

Women, Men, and Knowledge-Making in Early Modern Europe

Edited by

Francesca Antonelli
Antonella Romano
Paolo Savoia

BRILL

LEIDEN | BOSTON

Cover illustration: Edmé-Gilles Guyot, *Neue physikalische und mathematische Belustigungen*. Bey Eberhard Kletts sel. Wittwe: Augsburg, 1772 (Frontispiece).

The Library of Congress Cataloging-in-Publication Data is available online at https://catalog.loc.gov
LC record available at https://lccn.loc.gov/2022012126

Typeface for the Latin, Greek, and Cyrillic scripts: "Brill". See and download: brill.com/brill-typeface.

ISSN 2405-5077
ISBN 978-90-04-51260-3 (hardback)
ISBN 978-90-04-51261-0 (e-book)

This book is printed on acid-free paper and produced in a sustainable manner.

Contents

PART 3
Eighteenth-century Spaces of Gendered Knowledge

Acknowledgements

The essays presented here are the result of a collective discussion, which much benefited from the presence of scholars who kindly accepted to comment on our papers. We thus wish to thank those who were present at the conference and shared with us their thoughts, in particular Elisa Andretta, Monica Azzolini, Marco Beretta, Paola Bertucci, Simon Dagenais, Paola Govoni, Cynthia Klestinec, Anne Rasmussen, Silvia Sebastiani and Stéphane van Damme.

Figures

Notes on Contributors

Francesca Antonelli
is currently Postdoctoral Fellow at the University of Bologna. In May 2021 she received her PhD from the University of Bologna and the École des Hautes Études en Sciences Sociales, with a thesis on Marie-Anne Paulze-Lavoisier (1758–1836) and Lavoisier's laboratory notebooks, on which she is now preparing a book. Her main research interests are the history of scientific practices, material culture, and gender in the long eighteenth century.

Juliet Claxton
is an independent lecturer and scholar with experience of teaching and research roles at both Queen Mary, University of London and King's College, London. She is the Features Editor for *Jewellery History Today* (Society of Jewellery Historians) and is currently researching the material culture of bio-mineral gemstones, particularly with reference to their inclusion in early modern medicinal recipes.

Paula Findlen
is Ubaldo Pierotti Professor in Italian History. Her publications include *Possessing Nature: Museums, Collecting, and Scientific Culture in Early Modern Italy* (University of California Press, 1994), *Athanasius Kircher: The Last Man Who Knew Everything* (Routledge, 2004) and most recently *Leonardo's Library: The World of a Renaissance Reader* (Stanford University Library, 2019), *Empires of Knowledge: Scientific Networks in the Early Modern World* (Routledge, 2019), *Early Modern Things: Objects and Their Histories, 1500–1800*, 2nd edition (Routledge, 2021), and Camilla Erculiani, *Letters on Natural Philosophy*, ed. Eleonora Carinci, trans. Hannah Marcus, foreword by Paula Findlen (ITER, 2021).

Amparo Fontaine
is Lecturer in Early Modern History at the Pontificia Universidad Católica de Valparaíso in Chile. After receiving her M.Phil and Ph.D. from the University of Cambridge in 2019, she was a Max Weber postdoctoral fellow in the European University Institute. Her research focuses on the cultural history of music and knowledge in the long eighteenth century, especially in France. She is currently working on a book project that explores the notion of musical harmony in French culture, combining scientific practices, material culture, performance,

and the French Revolution. Additionally, she is also working on two research projects on cultural encounters between early-modern Europe and the Americas. From 2022, she will be Marie Skłodowska-Curie Fellow at the École des Hautes Études in Paris.

Paola Govoni

is Associate Professor at the University of Bologna. She is the author of books and articles on science and society in modern and contemporary times. Her latest publications in the area of this book are: *Drawing Nature, Building Knowledge: Between Beauty and 'Gut Feelings' in the Sciences*, in *Non-Fiction Picturebooks: Sharing Knowledge as an Aesthetic Experience* (ed. by G. Grilli, Florence, 2020); *Hearsay, Not-So-Big Data, and Choice: On Understanding Science and Maths by Looking at Men Who Supported Women*, in *Against all Odds: Women's Ways to Mathematical Research Since 1800* (ed. by E. Kaufholz-Soldat and N. Oswald, Springer, 2020).

Matteo Martelli

(PhD Greek Philology, 2007; PhD History of Science, 2012) is professor of History of Science at the University of Bologna. His research focuses on Graeco-Roman and Byzantine science – with particular attention to alchemy and medicine (pharmacology) – and its reception in the Syro-Arabic tradition. His publications include *The Four Books of Pseudo-Democritus* (2014), *Collecting Recipes. Byzantine and Jewish Pharmacology in Dialogue* (2017; edited with L. Lehmhaus), and *L'alchimista antico. Dall'Egitto greco-romano a Bisanzio* (2019). He is the principal investigator of the ERC project *AlchemEast*, and he is currently working on a critical edition of the Syriac alchemical books ascribed to Zosimos of Panopolis.

Sabrina Minuzzi

is a Marie Skłodowska-Curie Global Fellow in Early Modern History at Ca' Foscari and at Brown University, Providence (H2020 [2018], G. A. 844886). She is developing her interdisciplinary project *MAT-MED in Transit. The Transforming Knowledge of Healing Plants*, which interweaves the study of the circulation of *naturalia* with that of the reception of related medical-scientific knowledge through the object book. Among her works *Sul filo dei segreti. Farmacopea, libri e pratiche terapeutiche a Venezia in età moderna* (Milano: Unicopli, 2016), and "Quick to say quack. Medicinal secrets from the household to the apothecary's shop in early eighteenth-century Venice," *Social History of Medicine*, 2019/1.

Lucia Raggetti
is Assistant Professor of the History of Ancient Sciences in the University of Bologna. After receiving her PhD in Arabo-Islamic studies in Naples, she held a DAAD Fellowship in Hamburg and then worked as research assistant at Freie Unversität Berlin, in the research group on *Wissensgeschichte*. Her main research interests are Arabic philology and the history of natural sciences and medicine in the Arabo-Islamic milieu, on which she has published a variety of articles. She is author of *ʿĪsā ibn ʿAlī's Book on the Useful Properties of Animal Parts: Edition, Translation and Study of a Fluid Tradition* (de Gruyter, 2018) and *Un coniglio nel turbante. Intrattenimento e Inganno nella scienza arabo-islamica* (Bibliografica 2021).

Antonella Romano
has been researcher at the CNRS since 1997 and professor of the history of science at the European University Institute (Florence) between 2005 and 2013. She is currently full professor at the EHESS, member and former director of the Alexandre-Koyré Centre. Her work, mainly focused on the 16th century, is based on the study of religious orders and missionaries as institutions and agents of the recomposition of the European grammar of knowledge about the world. Among her latest pubblications: *Impressions de Chine. L'Europe et l'englobement du monde (16e–17e siècles)* (Paris: Fayard, 2016, trans. into Spanish and Italian); with Bert de Munck (ed.), *Knowledge and the Early Modern City* (London: Routledge, 2019); with Elisa Andretta and Romain Descendre (ed.), *Un mondo di Relazioni. Giovanni Botero e i saperi nella Roma del Cinquecento* (Rome: Viella, 2021).

Paolo Savoia
is Assistant Professor of the History of Science at the University of Bologna. He studied Philosophy at the Universities of Bologna and Pisa, and History of science at Harvard University. His publications concern the history of early modern medicine, the history of practical knowledge related to food and science, gender and history, and the historiography of science. Recently, he published the book *Gaspare Tagliacozzi and Early Modern Surgery: Faces, Men and Pain* (Routledge, 2020).

INTRODUCTION

Gender, History, and Science in Early Modern Europe

Francesca Antonelli and Paolo Savoia

1. The idea of this book comes from a two-day conference co-organized by CIS – The International Centre for the History of Universities and Science of the University of Bologna, and the Centre Alexandre Koyré (Ehess-Paris) which took place at the University of Bologna on June 28–29, 2019, developing a suggestion coming from one of the series editors, Marco Beretta.[1] The conference took place during a heatwave (the first day temperatures in Bologna reached 40° C), the essays then written during the first outbreak of a global pandemic, and as we sit writing this introduction, we are at the beginning of a record-breaking – if focused on the richest areas of the world – vaccination campaign against Covid-19. As it is almost a truism that the problems of the present guide the questions historians ask the past, it is worth noting that writing about the history of science and knowledge means dealing with very specific historical and scientific conditions these days. In our case, pandemics and climate change certainly sharpen the effects of worldwide differences and inequalities based on class, race, and gender.

The essays collected here aim at exploring how practical expertise, or embodied knowledge, of the gendered bodies intersected with the production of knowledge in early modern Europe. *Gendered touch* looks at both how representations of gendered bodies contributed to the production of knowledge, and at how practice itself was gendered. These two intertwined research dimensions will be pursued in parallel to inquiry about how knowledge was produced, translated, appropriated, and transmitted among different kinds of actors – both women and men – such as craftspeople, physicians, alchemists, apothecaries, music theorists, natural philosophers, and natural historians. In

1 This introduction is the outcome of shared approach and fruitful discussions between the two authors. Francesca Antonelli authored sections 1, 3, 6; Paolo Savoia authored sections 2, 4, 5. We both wish to thank Antonella Romano for the generous exchange of ideas and comments throughout the writing of this text.

 | DOI:10.1163/9789004512610_002

pursuing these goals, one of the major efforts of this book is to keep a balance between the history of women and men, and reflections on gender.[2]

2. This search for a balance between the social history of women (and men) and the significance of gender roles for historical inquiry is certainly not new. In the 1980s, building on intellectual and political work of at least two decades of feminist work, and expanding enquiries about social history, historians constructed powerful narratives and shaped new analytical tools about the history of women, science, and gender.

In 1980, Carolyn Merchant published her book *The Death of Nature*, subtitled *Women, Ecology and the Scientific Revolution*, famously claiming that the transformations of the idea of nature that happened during the seventeenth century brought about a new era of joint exploitation of nature – conceived as a passive resource – and of women – conceived as passive, domestic, beings.[3] Merchant's book was much subtler than it appeared to its first readers, and painted a complex picture of the bright and the dark sides of the so-called "scientific revolution" in light of the history of women. It is true that the book painted too narrow a picture of both nature's and women's predicaments in early modern Europe, as the period was mostly productive in relation to the new claims of gender studies.[4] On the other hand, a historiographical category as crucial to Merchant's analysis as that of "scientific revolution" has by then been the object of profound revisions by historians.[5] It was still a time where history of science was not interested by the social and gendered dimension of science making, except when women emerged as instruments: a new audience

2 See Gianna Pomata, "Histoire des femmes et 'gender history'," *Annales. Économies, Sociétés, Civilisations*, 48, 4 (1993): 1019–1026.

3 Carolyn Merchant, *The Death of Nature: Women, Ecology, and the Scientific Revolution* (New York: Harper, 1980).

4 Gianna Pomata, "Donne e rivoluzione scientifica: verso un nuovo bilancio," in Nadia Maria Filippini, Anna Scattigno, and Tiziana Plebani, ed., *Corpi e storia* (Roma: Viella, 2011), pp. 165–191; Charis Thompson, "Back to Nature? Resurrecting Ecofeminism after Poststructuralist and Third-Wave Feminisms," *Isis*, 97 (2006): 505–519. For a different but equally insightful reaction to Merchant's book, see also Sylvana Tomaselli, "The Enlightenment Debate on Women", *History Workshop Journal*, 20 (1985): 101–124. In the field of anthropology of science, *A Cyborg Manifesto* by Donna Haraway, first published in 1985, opened a new line of investigation where gender and science were deeply re-elaborated (the *Manifesto* was originally published as "Manifesto for Cyborgs: Science, Technology, and Socialist Feminism in the 1980s," *Socialist Review*, 80 (1985): 65–108). See also Donna Haraway, "Situated Knowledges: The Science Question in Feminism and the Privilege of Partial Perspective," *Feminist Studies*, 14, 3 (1988): 575–599.

5 See Antonella Romano, "Fabriquer l'histoire des sciences modernes. Réflexions sur une discipline à l'ère de la mondialisation", *Annales. Histoire, Sciences Sociales*, 70, 2 (2015): 381–408 and its bibliography.

for enlightened scientists, or a new instrument of its diffusion through the typically feminine space of sociability, the "salon".[6] Merchant's work remains important, however, in that it opened up a space to think together the history of women, the history of science, and the history of gender roles as they mattered not only to social history but also to the very process of the production of scientific knowledge.

A few years later, Joan Scott famously invited historians of women to pause and reflect on the vast mass of empirical material that had been put together in the past decades, and to do so by experimenting with the category of gender. Writing in 1986, Scott recalled that while the grammar term 'gender' had been sporadically used by 19th-century thinkers to describe sexual characters, it was only recently that feminists begun using the word to refer to "the social organization of the relationship between the sexes."[7] Feminists used gender mainly to underline and expose the implicit biological reductionism implied by the insistence on "sexual difference", and to underline the fundamentally social character of distinctions based on sex. At a more academic level, 'gender' seemed to allow scholars the possibility of studying not only women but also men, or masculinity, from a social and historical point of view. In other words, 'gender' opened up a view on a world where men and women were reciprocally defining social and cultural constructs.[8] The stakes were higher than simply doing 'women's studies': it was a matter of writing a new kind of history altogether. Historians like Joan Kelly had already claimed that a feminist history of women should not be only a new sub-discipline, but, on the contrary, that the history of women could bring about within mainstream history new topics such as the history of experience and the history of the body. 'Gender', often combined with class and race, answered these new needs. Using gender as a historiographical tool meant to claim that a separated sphere inhabited by women never existed, but that men and women constantly entertained complex social, cultural and symbolic relations. Scott, writing in a period of mounting success of post-structuralist theory, also underlined that an excessive

6 See for exemple Daniel Roche, *Les républicains de lettres. Gens de culture et Lumières au XVIIIe siècle* (Paris: Fayard, 1998); Giuliana Gemelli, Maria Malatesta, "Le avventure della sociabilità," in Maria Malatesta, Giuliana Gemelli, ed., *Forme di sociabilità nella storiografia francese contemporanea* (Milano: Feltrinelli, 1982) pp. 9–120; Antoine Lilti, *Le monde des salons. Sociabilité et mondanité à Paris au XVIII*e siècle (Paris: Fayard, 2005). For an historiographical overview on the notion of "sociability", see also Stéphane Van Damme, "La sociabilité intellectuelle. Les usages historiographiques d'une notion", *Hypothèses*, 1, 1 (1998): 121–132.

7 Joan W. Scott, "Gender: A Useful Category of Historical Analysis," *The American Historical Review*, 91, 5 (1986): 1053–1075, quotation p. 1053.

8 Natalie Zemon Davis, "Women's History in Transition: The European Case," *Feminist Studies*, 3, 3/4, (1976): 83–103.

reliance on Foucault's philosophical work or Lacanian psychoanalysis lead historians to select their sources too partially and narrowly, to find what they already expected to find, and therefore to lose the empirical, experimental dimension of the historian's craft. It was a matter of going back to the archives with new questions, not to abandon them in the name of post-modern irony or the power of discourse. For Scott, 'gender' was not only about gender identities, but rather a new way to look at historical questions: "gender is a constitutive element of social relationships based on perceived differences between the sexes, and gender is a primary way of' signifying relationships of power."[9]

Londa Schiebinger's *The Mind has No Sex?* can be considered one the best embodiments of Scott's hopes. Published in 1989, the book makes a very conscious use of gender ("gender shaped sex, rather than vice versa"[10]) in the history of science, and is at the same time very attentive in reconstructing the life and work of several women artisans in early modern Europe as well as those of several women doing science through aristocratic sociability and patronage practices. Schiebinger's book described not only the history of women but also the history of the socially constructed gender roles in the process of the formation of scientific knowledge – as an eloquent example, she focused on Linnaeus' classification system based on the concept of mammals, showing that it emerged in parallel with new campaign for breastfeeding that swept eighteenth-century Europe. Schiebinger distanced her work from essentialism more vigorously than Merchant, and argued that the point of feminist historiography was not just to unearth the history of important women, but also to describe – in view of disrupting it – the space of scientific knowledge as marked by a distribution of power along gendered lines.[11]

These US approaches have been crucial in reshaping the theoretical and empirical landscape for gender and history of science. At the same time, in

9 Scott, "Gender," p. 1067. Since then, the relationships between gender and women's history have inspired many complex and stimulating debates. Besides the already mentioned article by Pomata, see also Laura Lee Downs, "If 'Woman' Is Just an Empty Category, Then Why Am I Afraid to Walk Alone at Night? Identity Politics Meets the Postmodern Subject." *Comparative Studies in Society and History*, 35, 2 (1993): 414–437; the *AHR Forum: Revisiting "Gender: A Useful Category of Historical Analysis"*, in *American Historical Review*, 113, 5 (2008); Ida Fazio, "Introduzione. Genere, politica, storia. A 25 anni dalla prima traduzione italiana de Il "genere": un'utile categoria di analisi storica" in Joan W. Scott, *Genere, politica, storia* (Rome: Viella, 2013), pp. 6–30.

10 Londa Schiebinger, *The Mind Has No Sex? Women in the Origins of Modern Science* (Cambridge, Ma: Harvard University Press, 1989), p. 161.

11 Schiebinger, *The Mind Has No Sex?*, pp. 265–277. Some of these issues were later developed by Schiebinger in her *Nature's Body. Gender in the Making of Modern Science* (London: Routledge, 1993). See also Londa Schiebinger, ed., *Women and Gender in Science and Technology*, 4 vols. (London: Routledge, 2014).

Europe, similar issues were developed along the lines of a new social history which was putting women at the center. In this perspective, the debate about women in history developed in the framework of the "History Workshop Journal" in the 1970s, and it culminated in important works on the 18th century.[12] Later, the editorial project launched in France, *Histoire des femmes en occident*, became a model all around the world.[13] Published in French and Italian in 1991, and shortly after translated into six other languages, this five-volume collection stimulated a wider debate on the role of women in history, while raising methodological and historiographical questions that challenged historical practice *tout court*. It is obviously not possible here to summarize the variety of orientations and historiographic fields that were defined, also thanks to those early debates, in the following years.[14] However, in the context of the present volume, it is worth noting that many of the issues discussed since then touch on key problems of doing history, such as – just to mention one – that of how to combine the study of cultural representations with an investigation of social practices and individual trajectories. Especially in Italy, the adoption of a micro-historical approach and the use of a biographical focus in historical analysis proved quite crucial in this sense, highlighting the tensions between the individual and the collective and encouraging further inquiries on how historical actors re-appropriated social and gender norms.[15] In this line, and also in the wake of the interest in "subaltern" subjects encouraged once again by microhistory, women in/and science gradually became a legitimate field

12 Sally Alexander, Anna Davin, "Feminist History," *History Workshop Journal*, 1, 1 (1976): 4–6; *Women's History, History Workshop Journal*, 35, 1 (1993): 1–116; *Sex and Gender, History Workshop Journal*, 41, 1 (1996): 1–90.

13 Georges Duby, Michelle Perrot, ed., *Histoire des femmes en Occident*, 5 vols. (Paris: Plon, 1990–1991); Sylvie Steinberg, "1991: Duby et Perrot rendent l'histoire impensable sans les femmes", http://40ans.ehess.fr/2015/11/08/1991-une-histoire-sans-les-femmes-nest-plus-possible/. For some translations: *Storia delle donne in Occidente* (Roma-Bari: Laterza, 1991–1992), and *Historia de las mujeres en Occidente* (Madrid: Taurus, 1992). For a general review of these discussions, see Mónica Bolufer, *Mujeres y hombres en la historia. Una propuesta historiográfica y docente* (Granada: Comares, 2018), esp. chap. 2; on the French context, Françoise Thébaud, *Écrire l'histoire des femmes et du genre* (Lyon: ENS Éditions, 2007).

14 For a synthesis, see once again Bolufer, *Mujeres y hombres en la historia*, chap. 2. For some recent developments see also Sylvie Steinberg, Mélanie Traversier, Camille Noûs, "Aperçus sur les développements récents de l'histoire des femmes et du genre à l'époque moderne", in *Combats, débats, transmission: les 20 ans de Mnemosyne*, ed. Julie Verlaine, Patrick Farges, *Genre & Histoire*, 26, special issue, 2020, https://journals.openedition.org/genrehistoire/5933.

15 Ida Fazio, "Storia delle donne e microstoria," in Marina Caffiero, Maria Pia Donato, and Giovanna Fiume, ed., *Donne potere religione. Studi per Sara Cabibbo* (Milano: FrancoAngeli, 2017), pp. 81–94.

of investigation. Significantly, the first studies devoted to this topic within the Italian historiographical landscape dealt with the individual trajectories of some 17th- and 18th-century women who creatively reinterpreted gender norms of the time to carve out a space for themselves in contemporary scientific debates, circles and institutions.[16]

These different strands of analysis and methodological issues concentrated in the works of Marta Cavazza who, around the mid-1990s, began working on the history of early modern "women graduate and lecturers (*dottrici e lettrici*)" reconstructing the careers of 17th and 18th century women who exceptionally made it through the gates of university and scientific academies.[17] Cavazza's historical scholarship has been particularly inspiring as an example of the perfect balance between social history and gender history, since her approach always "naturally" mixed empirical description and meta-reflections on such descriptions. For example, Cavazza described the dynamic of pope Benedict XIV's patronage of women doing natural philosophy, mathematics, and anatomy (Laura Bassi, Maria Gaetana Agnesi, and Anna Morandi Manzolini) by asking what role did the patronage of learned women played in the context of the so-called Catholic Enlightenment. She argued that these projects of support of learned women were not so much part of a politics of enhancing women's education (although strains of Catholic reformist thought advocated for that), but rather of a strategy of renovation of Catholic religious and lay institutions, connected to papal support of experimental sciences and the ideal of public utility of applied sciences. The pope saw in learned women an opportunity to attract the attention of the European republic of Letters to Bologna again, after its 17th century decline, also exploiting a largely mythical tradition of learned women at the university of Bologna, in a historically complex knot

---

16 In addition to the works by Marta Cavazza, briefly discussed in the next paragraph, see the series of case-studies collected in Raffaella Simili, ed., *Scienza a due voci* (Florence: Olschki, 2006). For a more comprehensive review of Italian historiography on women, gender, and science, see Marta Cavazza, "Dalla rimozione alla riscoperta. Gli studi sul contributo femminile alla scienza nell'Italia del Settecento", in Dario Generali, ed., *Clelia Borromeo Arese. Un salotto letterario settecentesco tra arte, scienza e politica*, 2 vols., vol. 1 (Florence: Olschki, 2011), pp. 149–164, now in Id., *Laura Bassi: Donne, genere e scienza nell'Italia del Settecento* (Milan: Bibliografica, 2020), pp. 13–24. A microhistorical approach is also influential in the recent Federica Favino, *Donne e scienza nella Roma dell'Ottocento* (Rome: Viella, 2020).

17 Marta Cavazza, "'Dottrici' e 'lettrici' dell'Università di Bologna nel Settecento," *Annali di storia delle università italiane*, 1 (1997): 109–126, now in Ead., *Laura Bassi: Donne, genere e scienza nell'Italia del Settecento* (Milan: Bibliografica, 2020), pp. 37–72. The first article Cavazza devoted to the Bassi case was published in 1995: Ead., "Laura Bassi e il suo gabinetto di fisica sperimentale: realtà e mito", *Nuncius*, 10 (1995): 715–753, now also in Ead., *Donne, genere e scienza*, pp. 131–161.

of different strategies involving both sensationalism and commitment to the new sciences by these women.[18] Cavazza's work outlined the interest of the Enlightenment as a period where gender relations could take new avenues, as other investigations demonstrated.[19] At the same time, Cavazza moved in the territory of gender roles (male and female) with great ease, thus teaching historians that it is not possible to write a history of science, or of women in science, without at the same time writing a history of men in science (and vice versa). Paving the path for other research both in Italy and elsewhere, in the 18th century and other periods, her work remains a turning point to which this volume aims at being an homage.

3. The essays collected here make no sharp distinction between the social history of men, women, and gender, simply because during the very process of historical research and writing, these things are not separable. As Evelyn Fox Keller wrote in 1985: "The widespread assumption that a study of gender and science could only be a study of women still amazes me: if women are made rather than born, the surely the same is true of men. It is also true of science."[20] In any case, this is not some nostalgic 1980s revival. If historians in the 1980s were afraid to confine the history of women and science in a separated space, disconnected from the mainstream of history, today – even taking into account various different national traditions and institutional arrangements – it is becoming largely impossible to teach and write history of science and knowledge without including the history of women and gender.[21] The articles collected in this volume take full advantage of more recent studies of women, gender, men, and science in early modern Europe, a field so rich, pluralistic, and multifaceted that it is even difficult to call it a "field."[22]

18 Marta Cavazza, "Between Modesty and Spectacle: Women and Science in Eighteenth-Century Italy," in Paula Findlen, Wendy Wassyng Roworth, and Catherine Sama, ed., *Italy's Eighteenth Century: Gender and Culture in the Age of the Grand Tour* (Stanford: Stanford University Press, 2009), pp. 275–302.

19 Barbara Taylor, Sarah Knott, ed., *Women, Gender and Enlightenment* (London: Palgrave MacMillan, 2005); Silvia Sebastiani, *The Scottish Enlightenment: Race, Gender, and the Limits of Progress* (London: Palgrave MacMillan, 2013).

20 Evelyn Fox Keller, *Reflections on Gender and Science* (New Haven: Yale University Press, 1996), pp. 3–4.

21 Karen Offen, Chen Yan, ed., *Women's History at the Cutting Edge* (London: Routledge, 2020).

22 It is largely beside the scope of this introduction to compile a bibliography on gender and early modern science, but see, among others, Katharine Park, *Secrets of Women: Gender, Generation, and the Origins of Human Dissection* (New York: Zone Books, 2006); Sandra Cavallo, *Artisans of the Body in Early Modern Italy: Identities, Families and Masculinities* (Manchester: Manchester University Press, 2010); Rebecca Messbarger, *The Lady Anatomist: The Life and Work of Anna Morandi Manzolini* (Chicago: University of Chicago Press); Mary E. Fissell, "Introduction: Women, Health, and Healing in Early Modern Europe,"

In different ways, the essays collected in this volume pay close attention to materiality and spaces, as both perspectives are themselves indebted with gender approaches.[23] This volume also aims at connecting in circular dynamics the expertise of those who have been traditionally considered "subaltern" knowledge makers – including women – with the theoretical activities of the natural philosophers. In this respect, we have also been inspired, besides feminist historians, by historians as different from each other as Edgard Zilsel and Piero Camporesi, who significantly enlarged the number of actors who can legitimately be counted as part of early modern knowledge-making.[24] The essays composing this volume therefore explore a fruitful tension – displaying itself along gendered lines – between the textual transmission and appropriation of knowledge, and the practical and performance-like aspects of producing knowledge. The cases of alchemical recipes, food production, adapting mechanical bodies to musical instruments, and taking notes in the new experimental spaces are all pervaded by such tension.

While these historiographical references served as the background for our discussions, each author was left to personally reinterpret the editors' broad

---

Bullettin of the History of Medicine, 82 (2008): 1–17; Elaine Leong *Recipes and Everyday Knowledge: Medicine, Science, and the Household in Early Modern England* (Chicago: The University of Chicago Press, 2018); Sharon Strocchia, *Forgotten Healers: Women and the Pursuit of Health in Late Renaissance Italy* (Cambridge: Harvard University Press, 2019); Adeline Gargam, *Les Femmes savantes, lettrées et cultivées dans la littérature française des Lumières ou la conquête d'une légitimité*, 2 vols. (Paris: Honoré Champion, 2013). For a survey of the recent literature, see the introduction by Adeline Gargam to *Femmes de sciences de l'Antiquité au XIXe siècle. Réalités et représentations* (Dijon: Éditions Universitaire de Dijon, 2014) and its rich bibliography, and Marta Cavazza, "Introduzione: dalla rimozione alla riscoperta" in Ead., *Laura Bassi. Donne, genere e scienza nell'Italia del Settecento*, pp. 13–34.

23 Among others, see Marco Beretta, *Storia materiale della scienza* (Rome: Carocci, 2020); David N. Livingstone, *Putting Science in Its Place. Geographies of Scientific Knowledge* (Chicago: University of Chicago Press, 2003). For a critical review of recent historiography on the so-called "material turn" of the history of science see, among others, Stéphane Van Damme, *La prose des savoirs. Pragmatiques des mondes intellectuels* (Strasbourg: Presses universitaires de Strasbourg, 2020), chap. 3. For an evaluation of the rich historiography on the spaces of knowledge production, see among others, the *Premessa* by Sabina Brevaglieri e Antonella Romano to *Produzione di saperi. Costruzione di spazi*, special issue of *Quaderni storici*, 158, 1 (2013): 3–19. A fundamental contribution to this topic is also Christian Jacob, ed., *Lieux de savoir*, 2 vols., (Paris: Albin Michel, 2007–2011).

24 Edgar Zilsel, "The Sociological Roots of Science," in Id., *The Social Origins of Modern Science*, ed. D. Raven, W. Krohn, and R.S. Cohen (Dordrecht: Springer, 2003), pp. 7–21; Piero Camporesi, *La miniera del mondo: artieri inventori impostori* (Milan: Il Saggiatore, 1990).

initial insights. The idea was to compare our respective research questions and methodologies through the presentation of case studies on which each of us was working and which involved, to a different extent and in various ways, women as well as men. Accordingly, the articles gathered in this book cover different settings, from domestic experimentation with preparation and preservation of food items to the humanists' libraries, from the alchemists' working spaces to eighteenth-century laboratories. They also cover different kinds of actors, the men and women cooperating and fighting for the production of knowledge, and deal with different chronologies and geographies. However, the variety of perspectives that can be found in what follows stems more from the intent to combine the analytical tools offered by different historiographies than from the intent of proposing another big narrative on early modern science and knowledge.

4. The first part of this book is titled *The Gendered Construction of Textual Traditions: The Case of Maria the Alchemist*. The two essays by Lucia Raggetti and Matteo Martelli form a dyad which must be read together. These essays complement each other, as they examine the role of Maria The Alchemist – an ambiguous and mythical figure, also known as Maria The Jewess or Maria *Prophetissa* – in different textual traditions. Raggetti opens her analysis by noting that, before being celebrated by early modern European alchemists as the inventor of the heated bath – a heating technique eponymously named "bain-marie" – the figure of Maria "had come a long way across more than one millennium, and through the eyes and the idioms of different cultures and traditions." Her focus is on the Arabo-Islamic tradition, in which Maria entered through the translations of Greek texts, in particular the Graeco-Egyptian alchemical corpus. Following the circulation of the myth in the new context of reception, Raggetti explores the multiple identities of Maria, reconstructing an "elusive portrait" which, as she shows throughout her essay, was to a great extent the result of various cultural reappropriations. Maria's name – in the Arabic version of Maryam – was itself a source of speculation and could alternatively be associated with the mother of Jesus, the sister of Moses and Aaron, a Coptic slave offered to the prophet Muhammad, as well as and with other obscure figures of an ancient and often mythical past. Her origins, that now shifted from Jewish to Copt and Egyptian, were also the object of attention in several sources. As the author highlights, many complex issues were at stake in these adaptations such as, for instance, the relationship between Islam and the Jewish tradition that the myth of Maria involved. It is not until the seventeenth century, however, that we find an Arabic source connecting Maria to the famous heated bath. It is a treatise on chemical medicine by Ibn Sallūm,

the chief-physician of the Ottoman sultan Mehemet IV, based in Constantinople, who probably read about the bain-marie in Latin alchemical works that were available in the Ottoman capital. Several passages of this text – which appears as a significant example of how women's alchemical knowledge was re-appropriated and represented by men in different cultural contexts – are translated by Raggetti at the end of the essay.

Martelli inquiries on how the mythical figure of Maria the Alchemist and the practical alchemical knowledge attributed to her in many languages and cultures were interpreted, used, and authenticated by early modern alchemists, humanists, and natural philosophers in their search for authority. Such different attempts at reading Maria's work form what Martelli defines "a complex and polyphonic discussion on Maria and her alchemical achievements".

Martelli tracks down the readings of Maria's *Practica* – and in particular commentaries about her procedures for "whiting the stone" and the famous specific heating method – in a backward process beginning from Netwon back to Orthelius and to Parcelsus, in Latin as well as in several European vernaculars. The case of Netwon's notes on Maria's special heating technique shows that the name of Maria and the alchemical equipment associated to her were inherited by early modern alchemists from an earlier tradition, which was deeply rooted into Arabo-Latin alchemical treatises. Renaissance and early modern humanists – such as Isaac Casaubon and Justus Scaliger – played an important role in the identification of the figure, knowledge, and techniques related to Maria, as Greek alchemical literature, either in translation or through manuscripts, provided new pieces of information, which could complement and, sometimes, rectify what had been already extracted from the Arabo-Latin tradition.

In this intricate textual history, Maria is one of the three or four women who were deemed able to perform alchemical knowledge in the Byzantine tradition. Ultimately, alchemists', humanists', and philosophers' historical and textual inquiries on Maria the Alchemist allowed them to redesign the chronological and geographical boundaries of their art and to root their own practices and theories into this convenient past, a past which produced the 'ancient' alchemical literature that they never stopped to reinterpret and re-contextualize. Women's alchemical knowledge is therefore placed by Martelli at the center of a process of rewriting the past in search for origins and authority.

5. The second part is titled *Domestic and Apothecary Workshops: Food and Pharmacy in the Seventeenth Century*. Paolo Savoia's article begins by noting that early modern cheesemaking was accompanied by a specific scientific interest to understand such a peculiar natural process of transformation of

natural matter by collecting information from those who mastered it in their everyday life. It turns out that in early modern Europe the art of transforming milk into cheese and butter was almost exclusively the domain of women. By showing connections between ideas about transformation of natural matter, gender, and techniques of manipulation of nature, Savoia argues that these observations and artisanal practices contributed to changing the way knowledge was made and nature was known by physicians, natural philosophers, and natural historians around the time of the "scientific revolution." This contribution was a process of transformation of expertise into explanation. Savoia focuses on the issues of the observation, translation and eventually appropriation of both artisanal practical skills embodied in women's cheese-making, and on a supposedly female intimate knowledge with milk and dairy products by learned natural historians and natural philosophers from the late fifteenth century to the seventeenth century. Savoia examines significant examples of the dynamics of expertise/explanation: the late fifteenth-century book on dairy products written by Piedmontese physician Pantaleone of Confienza; the reports and experiments on cheese-making at the Royal Society in London and the Académie des Sciences in Paris in the 1660s; the 1664 booklet on people disgusted by cheese by the Dutch theologian and natural philosopher Martin Schook.

Savoia argues that women's expertise with cheesemaking, which early modern sources take for granted, involved women's experience not only with medicaments and recipes, but also with touching natural matter, mastering complex natural processes of transformation, and devising/utilizing technology. Savoia claims that the polarity between expertise and explanation, or practical skills and their scientific appropriation and translation, was reinforced by notions of gender and by the social work gender roles played in acquiring expertise, disseminating practical knowledge, and establishing new knowledge. Far from being only a matter of "representations," gender roles mediated, on the one hand, the observation and recording of experimental and practical knowledge; and, on the other hand, some of the processes of theory-making and induction which are typical of the scientific revolution.

Juliet Claxton explores the culinary and medicinal experimentation that has been associated with Alathea Talbot, Countess of Arundel (*c.*1585–1654) by discussing both her literary output, and household inventories. Claxton specifically focuses on the knowledge procedures taking place in particular spaces within this elite household, as well as the coordinated activity between the lady and her domestic staff. Claxton links Talbot's recipe collections to the 1641 inventory of the Countess' Westminster home Tart Hall, and therefore she sites her experimental activities within contemporary courtly architecture and illustrates where and how the Countess' own 'touch' can be identified within different domestic locations.

Claxton focuses on the gendering of the preparation of marmalade and other preservation techniques in seventeenth-century English households, where "household collectives"[25] performed cooking and preserving experiments with foodstuff. Such preparations aroused the attention of canonical natural philosophers and naturalists such as Robert Boyle, John Beale, and Samuel Pepys. In elite establishments where the management of a large and complex household required both hands-on knowledge and lived experience, most domestic food preparation was performed by servants, but the lady of the house was expected to have the knowledge and ability to manage and oversee the everyday running of her home. In fact, making marmalade and conserving fruit was a complex task that required specific knowledge, an ability to judge temperature, and familiarity with visual signals to achieve the correct setting point. Observing an expert was the chief method of learning, and women initiated each other into the techniques.

Claxton argues that while the knowledge transfer generated by seventeenth-century culinary and pharmacological literature have been extensively studied, much less critical discussion has been concerned with the siting of knowledge production or the equipment specified within recipe collections. Preserving and distilling was an experimental process, classified as 'natural philosophy' and often termed 'kitchen physic' that needed its own space and technology. Claxton inquiries about the space where this type of culinary experimentation and knowledge transfer happened, what equipment was used, and who were the female and male actors actually performing the procedures. Claxton argues that while Alathea Talbot's Dutch 'Pranketing Room' – a space devoted to collective food experimentation directed by her – remained primarily a space for entertainment, the room also included a performative aspect where the final elements of food preservation became part of courtly 'scientific' entertainment.

The case of the Countess of Arendel is interesting if read against Minuzzi's article on women experimenting with pharmacology in early modern Venice. Minuzzi eloquently begins her article – which analyzes women's activities in a variety of spaces, including households, apothecary's guilds workshops – by acknowledging that "the concept of marginality of women in the history of medicine can only be considered true on the surface, yet rather superseded in substance". By providing an overview of recent trends in medical history,

25 Leong, *Recipes and Everyday Knowledge*, p. 4.

Minuzzi recalls that through methodological tools such as "medical agent" and "bodywork" the field of early modern male and female healthcare providers – and, we add, knowledge-carriers – has dramatically expanded. Minuzzi argues that by now we are able to see that women's marginality in respect to the organised work in arts and colleges and to the medicine taught at the universities cohabits with a dense multiplicity of women active in many contexts of medical practice and in the exchange of knowledge even with the academic and professional, lay and religious, as well as commercial universe. Therefore, Minuzzi argues that in the history of early modern medicine, women, as well as other non-officially sanctioned healthcare providers, form a series of "connective tissues" with respect to medical care of the body in general.

Minuzzi's article starts from a reflection on its sources, namely the authorisations for the handling of medicinal secrets in early modern Venice. While this kind of source privileges male representation, in reviewing the few female figures that emerge from it Minuzzi intersects other biographical information available, and in a couple of cases print books and manuscripts, which are capable of bridging in part the silence or the omissions that characterise the main serial source. Minuzzi therefore puts together clues to reconstruct the socio-cultural profiles of the women engaged in the invention and/or handling of medicinal secrets within a normative system that allowed them to start up or continue a trade in the drug on a small scale.

Minuzzi shows that the "women in secrets" reviewed in her article are characterised by their belonging to a middling-sort, in which the lowest extreme does not appear and the highest one is represented, as in the case studied by Juliet Claxton in this volume, by a English lady, Margaret Paston. Moreover, the women in secrets active in Venice were all regularly authorised and were eager to be so. Unlike the noblewomen at the Italian courts, these middle-class, artisan women in secrets were more keen in handling on their knowledge to the public and to posterity. Minuzzi argues that it was in the spaces between the apothecary and the home, between shop laboratory and urban kitchen more or less equipped that the women-artisans of secrets lived. Between the domestic walls the trade labels and the gender differences in knowing and doing fell, the medicinal secrets continued to be handled and sold in the same way.

6. The third part – titled *Eighteenth-century Spaces of Gendered Knowledge* – begins with Paula Findlen's chapter analysing the context that led the Academy of the Sciences of Bologna to admit Neapolitan noblewoman Faustina Pignatelli (1705–1769), as its second female member in 1732, after the much more famous and celebrated Laura Bassi. Building on Marta Cavazza's studies, Findlen discusses the Bolognese academicians' debates not only around the relative merits of these two women, but also about the desirability of admitting more than one woman to the Bologna Academy of Sciences, as many did not want their recognition of Pignatelli, a foreigner, to set a worrisome precedent. Bassi had been a singular instance of a city publicly embracing the demonstrable talent of a local prodigy, a young woman well versed in natural philosophy; but the prospect of Pignatelli's admission raised concerns that the Bologna Academy of Sciences might now admit any and every learned woman who demonstrated aptitude in the sciences. Ultimately, Pignatelli's admission was the outcome of a successful campaign mounted by Neapolitan political and scientific authorities to establish close links with the Bologna Academy of the Sciences. Findlen argues that by looking at correspondence as a material object it appears that the exchange of letters was the most powerful tool for Pignatelli' success in gaining reputation as a skilled mathematician. Findlen places this episode in the intellectual and political context of early modern Naples, and argues that Pignatelli was the expression of a noble world and a court society that was about to end in the long eighteenth century, thus illuminating how gendered knowledge intersects with different sorts of social and political powers: for both her detractors and her defenders, she was the exception that prevented the fearsome rule of women becoming part of elite scientific societies.

The question of spaces, and of how historical actors put them to use in order to access scientific knowledge and build their own reputation, is discussed also in Francesca Antonelli's essay, which focuses on the case of Marie-Anne Paulze-Lavoisier (1758–1836). Known as the wife and associate of the French chemist and tax farmer Antoine-Laurent Lavoisier – and in particular as a translator and illustrator of chemical texts – Paulze-Lavoisier is here presented as a "secrétaire" that is, following the meaning of the term in eighteenth-century French, a cultural and political agent whose main activity is that of writing on behalf of someone. The aim of this form of writing was not necessarily publication, but rather the storing of information: a task that Paulze-Lavoisier took on since the very beginning of her collaboration with her husband in the early 1770s. Antonelli works on Lavoisier's laboratory notebooks (the *Registres de laboratoire*) and especially on some notes relating to a series of experiments on diamonds and precious stones that were performed by the Lavoisiers in different settings – a famous Parisian public garden, the Jardin de l'Infante, as well as domestic laboratories – between the early 1770s and the early 1780s. Following Paulze-Lavoisier's hand, the author develops a two-fold analysis. On the one hand, she examines the material, spatial, and social dimensions of her work as note-taker, putting the problem of information management at the core of her collaboration with Lavoisier. On the other hand, she investigates how her trajectories as a *secrétaire* intersected with other kind of practices, mostly connected with polite sociability. The question underlying the analysis is in fact to what extent Paulze-Lavoisier's involvement in note-taking practices went beyond Lavoisier's need for a vast and well-organized archive and became a mean of self-promotion, through which she could negotiate her position within scientific circles of the time. From a broader point of view, through the case of the Lavoisiers the author intends to highlight how a set of social and gender norms – such as, just to name one, modesty – could be reinterpreted by historical actors in creative ways. Through the same lens, she argues, it is possible not only to relook at some well-known sources – such as Paulze-Lavoisier's illustrations for her husband's *Mémoires de physique et de chimie*, dating to the early 1790s – from a different angle, but also to cast some new light on Lavoisier himself. In contrast with the image of the solitary genius, compiling the whole of his laboratory notebooks by himself, the French chemist represented here is a man of his time, needing the constant support of his associates and eager to take part, together with his wife, in the social life of eighteenth-century Paris.

With Amparo Fontaine's essay we are still in eighteenth-century France, but this time the focus is on musical performance as a gendered practice, involving contemporary physiological and medical knowledge as well as moral, social and political values. Basing on a wide set of sources – from didactic music manuals to other prescriptive literature such as acting and dancing methods, but also musical instruments, literary texts and visual representations of the performers – Fontaine mantains that "perfomers of musical instruments expressed and mobilised different understanding of the physical body". As it is known, musical instruments had been associated since the Renaissance with human physiology, offering various metaphors for the functioning of the body. However, Fontaine goes further in her analysis, stating that musical instruments, invested as they were with "gender, social status, character, morality, and anatomy", were "translated into the espressive requirements to the performer which", she argues, "relied on the performer's physical body". She thus established three models aiming at describing how musical bodies were adopted by eighteenth-century musical performers: the "courtly body", the "mechanical body", and the "expressive body". Each of these models emerges in a specific historical conjucture – although the chronologies sometimes overlapped, so that the three models end up coexisting in the second half of the eighteenth century – and stems from different conceptions of the body but also from the gendered and social expectations directed to the performers. This does not mean, however, that these models were passively adopted by the musicians: on the contrary, they met frequent oppositions and personal reinterpretations through, for instance, the introduction of new material technologies and instrumental techniques. They were also part of wider intellectual debates which, as the author shows, were related not only to musical performance itself but also to the role of women in musical practice.

In the postface, Paola Govoni offers some thought-provoking remarks on women, men, gender, and science that exceed the chronological boundaries of the early modern period. In particular, she focuses on the reciprocal influence of gender on scientific practice, and on how the sciences have molded gender ideas. Govoni's postface provides therefore the best possible end for a book: an invitation to further reserch, inquiry, and reflection.

References

AHR Forum: Revisiting "Gender: A Useful Category of Historical Analysis", in *American Historical Review*, 113, 5 (2008).

Alexander, Sally, Anna Davin, *Sex and Gender, History Workshop Journal*, 41, 1 (1996): 1–90.

Alexander, Sally, Anna Davin, "Feminist History," *History Workshop Journal*, 1, 1 (1976): 4–6.

Alexander, Sally, Anna Davin, *Women's History, History Workshop Journal*, 35, 1, (1993): 1–116.

Beretta, Marco, *Storia materiale della scienza* (Rome: Carocci, 2020).

Bolufer, Mónica, *Mujeres y hombres en la historia. Una propuesta historiográfica y docente* (Granada: Comares, 2018).

Brevaglieri, Sabina, Antonella Romano, "Premessa," in Sabina Brevaglieri, Antonella Romano, ed., *Produzione di saperi. Costruzione di spazi*, special issue of *Quaderni storici*, 158, 1 (2013): 3–19.

Camporesi, Piero, *La miniera del mondo: artieri inventori impostori* (Milan: Il Saggiatore, 1990).

Cavallo, Sandra, *Artisans of the Body in Early Modern Italy: Identities, Families and Masculinities* (Manchester: Manchester University Press, 2010).

Cavazza, Marta, "Between Modesty and Spectacle: Women and Science in Eighteenth-Century Italy," in Paula Findlen, Wendy Wassyng Roworth, and Catherine Sam, ed., *Italy's Eighteenth Century: Gender and Culture in the Age of the Grand Tour* (Stanford: Stanford University Press, 2009), 275–302.

Cavazza, Marta, *Laura Bassi: Donne, genere e scienza nell'Italia del Settecento* (Milan: Bibliografica, 2020).

Downs, Laura Lee, "If 'Woman' Is Just an Empty Category, Then Why Am I Afraid to Walk Alone at Night? Identity Politics Meets the Postmodern Subject," *Comparative Studies in Society and History*, 35, 2, (1993): 414–437.

Duby, Georges, Michelle Perrot, ed., *Histoire des femmes en Occident*, 5 vols. (Paris: Plon, 1990–1991).

Favino, Federica, *Donne e scienza nella Roma dell'Ottocento* (Rome: Viella, 2020).

Fazio, Ida, "Introduzione. Genere, politica, storia. A 25 anni dalla prima traduzione italiana de Il "genere": un'utile categoria di analisi storica" in Joan W. Scott, *Genere, politica, storia* (Rome: Viella, 2013), 6–30.

Fazio, Ida, "Storia delle donne e microstoria," in Marina Caffiero, Maria Pia Donato, Giovanna Fiume, ed., *Donne potere religione. Studi per Sara Cabibbo* (Milano: FrancoAngeli, 2017), 81–94.

Fissell, Mary E., "Introduction: Women, Health, and Healing in Early Modern Europe," *Bullettin of the History of Medicine*, 82 (2008): 1–17.

Fox Keller, Evelyn, *Reflections on Gender and Science* (New Haven: Yale University Press, 1996).

Gargam, Adeline, ed., *Femmes de sciences de l'Antiquité au XIXe siècle. Réalités et représentations* (Dijon: Éditions Universitaire de Dijon, 2014).

Gargam, Adeline, *Les Femmes savantes, lettrées et cultivées dans la littérature française des Lumières ou la conquête d'une légitimité*, 2 vols. (Paris: Honoré Champion, 2013).

Jacob, Christian, ed., *Lieux de savoir*, 2 vols. (Paris: Albin Michel, 2007–2011).

Leong, Elaine, *Recipes and Everyday Knowledge: Medicine, Science, and the Household in Early Modern England* (Chicago: The University of Chicago Press, 2018).

Lilti, Antoine, *Le monde des salons. Sociabilité et mondanité à Paris au XVIIIe siècle* (Paris: Fayard, 2005).

Livingstone, David N., *Putting Science in Its Place. Geographies of Scientific Knowledge* (Chicago: University of Chicago Press, 2003).

Merchant, Carolyn, *The Death of Nature: Women, Ecology, and the Scientific Revolution* (New York: Harper, 1980).

Messbarger, Rebecca, *The Lady Anatomist: The Life and Work of Anna Morandi Manzolini* (Chicago: University of Chicago Press).

Offen, Karen, Chen Yan, ed., *Women's History at the Cutting Edge* (London: Routledge, 2020).

Park, Katharine, *Secrets of Women: Gender, Generation, and the Origins of Human Dissection* (New York: Zone Books, 2006).

Pomata, Gianna, "Donne e rivoluzione scientifica: verso un nuovo bilancio," in Nadia Maria Filippini, Anna Scattigno and Tiziana Plebani, ed., *Corpi e storia* (Rome: Viella, 2011), 165–191.

Pomata, Gianna, "Histoire des femmes et 'gender history'," *Annales*, 48, 4 (1993): 1019–1026.

Roche, Daniel, *Les républicains de lettres. Gens de culture et Lumières au XVIIIe siècle* (Paris: Fayard, 1998).

Gemelli, Giuliana, Maria Malatesta, "Le avventure della sociabilità," in Maria Malatesta, Giuliana Gemelli, ed., *Forme di sociabilità nella storiografia francese contemporanea* (Milano: Feltrinelli, 1982), 9–120.

Romano, Antonella, "Fabriquer l'histoire des sciences modernes. Réflexions sur une discipline à l'ère de la mondialisation", *Annales. Histoire, Sciences Sociales*, 70, 2 (2015),: 381–408.

Schiebinger, Londa, *Nature's Body. Gender in the Making of Modern Science* (New Brunswick: Routledge, 1993).

Schiebinger, Londa, ed., *Women and Gender in Science and Technology*, 4 vols. (London: Routledge, 2014).

Schiebinger, Londa, *The Mind Has No Sex? Women in the Origins of Modern Science* (Cambridge, Ma: Harvard University Press, 1989).

Scott, Joan W., "Gender: A Useful Category of Historical Analysis," *The American Historical Review*, 91, 5 (1986): 1053–1075.

Simili, Raffaella, ed., *Scienza a due voci* (Florence: Olschki, 2006).

Steinberg, Sylvie, Mélanie Traversier, Camille Noûs, "Aperçus sur les développements récents de l'histoire des femmes et du genre à l'époque moderne", in Julie Verlaine, Patrick Farges, ed., *Combats, débats, transmission : les 20 ans de Mnemosyne, Genre & Histoire*, 26, special issue, 2020, https://journals.openedition.org/genrehistoire/5933.

Strocchia, Sharon, *Forgotten Healers: Women and the Pursuit of Health in Late Renaissance Italy* (Cambridge: Harvard University Press, 2019).

Taylor, Barbara, Sarah Knott, ed., *Women, Gender and Enlightenment* (London: Palgrave MacMillan, 2005).

Sebastiani, Silvia, *The Scottish Enlightenment: Race, Gender, and the Limits of Progress* (London: Palgrave MacMillan, 2013).

Thébaud, Françoise, *Écrire l'histoire des femmes et du genre* (Lyon: ENS Éditions, 2007).

Thompson, Charis, "Back to Nature? Resurrecting Ecofeminism after Poststructuralist and Third-Wave Feminisms," *Isis*, 97 (2006): 505–519.

Tomaselli, Sylvana, "The Enlightenment Debate on Women," *History Workshop Journal*, 20 (1985) :101–124.

Haraway, Donna, "Situated Knowledges: The Science Question in Feminism and the Privilege of Partial Perspective," *Feminist Studies*, 14, 3 (1988): 575–599.

Van Damme, Stéphane, "La sociabilité intellectuelle. Les usages historiographiques d'une notion", *Hypothèses*, 1, 1 (1998): 121–132.

Van Damme, Stéphane, *La prose des savoirs. Pragmatiques des mondes intellectuels* (Strasbourg: Presses universitaires de Strasbourg, 2020).

Zemon Davis, Natalie, "Women's History in Transition: The European Case," *Feminist Studies*, 3, ¾ (1976): 83–103.

Zilsel, Edgar, "The Sociological Roots of Science," in Id., *The Social Origins of Modern Science*, ed. D. Raven, W. Krohn, and R. S. Cohen (Dordrecht: Springer, 2003), pp. 7–21.

PART 1

The Gendered Construction of Textual Traditions: The Case of Maria the Alchemist

CHAPTER 1

Maria the Alchemist and Her Famous Heated Bath in the Arabo-Islamic Tradition

Lucia Raggetti

1 Maria the Alchemist: Graeco-Egyptian Origins and Arabo-Islamic Reception

Before being celebrated by early-modern alchemy as *Maria Prophetissa*,[1] the figure of Maria the Alchemist had travelled a long way across more than one millennium, and through the eyes and the idioms of a number of different cultures and traditions. Most of her original works are now lost and only known to us through the indirect tradition of those who claimed to be her pupils (such as Zosimus, 3rd–4th cent. Egypt) or through the quotations of Byzantine authors (for instance, the 7th-cent. anonymous alchemists referred to as Christianos).[2] The different aspects and narrative components affixed to Maria in the course of time and transmission – i.e. a Jewish origin, the prophecy and the prophetic connections with Moses and Jesus, her relations with fellow alchemists like the Persian Ostanes – mix and blur to form an elusive portrait. The occurrence of a woman's name among the great savants of antiquity clashes with the paucity of information about her. In fact, Maria the Alchemist is one of those fascinating yet evasive historical personalities about whom much is said but very little is known for certain.

The Arabo-Islamic tradition received and translated many Greek texts, including a large selection from the Graeco-Egyptian alchemical corpus. In the Arabo-Islamic milieu, Maria (*Māriya*) was known as 'the Copt' and 'the Egyptian'. Different explanations and hypotheses have been formulated to

1 This article is being published within the framework of the *AlchemEast* project, which has received funding from the European Research Council (ERC) under the European Union's Horizon 2020 research and innovation programme (G.A. 724914).

For the tradition of Maria the Alchemist in late mediaeval and early modern Europe, see Matteo Martelli's chapter in this volume.

2 In the Jewish tradition, Maria was identified as the sister of Moses, especially in Gnostic circles. See Raphael Patai, *The Jewish Alchemists* (Princeton, NJ: Princeton University Press, 1995); Manfred Ullmann, *Die Natur- und Geheimwissenschaften im Islam* (Leiden: Brill, 1972), pp. 181–182; Roshdi Rashed, ed., *Encyclopedia of the History of Arabic Science*, 3 vols. (London/New York: Routledge, 1996), pp. 861–862; see also Lawrence M. Principe, *The Secrets of Alchemy* (Chicago, IL/London: University of Chicago Press, 2013), pp. 15–16.

 | DOI:10.1163/9789004512610_003

account for this shift in her origins. The Arabic version of Maria's name, Maryam, is mentioned several times in the Qur'ān and Sūra 19 is named after her. There appears to be some confusion, though, between the mother of ʿĪsā (Jesus, Qur'ān 19:16–19) and a much earlier biblical figure, that of the sister of Moses and Aaron (Qur'ān 19:27–28).[3] The canonic collections of Prophetic traditions preserve traces of the moment when a companion of the Prophet, Muġīra ibn Šuʿba, asks Muḥammad about the comment he received from the Christians of Najran regarding this ambiguity: the Christians seem to have immediately spotted the incoherence in the Quranic words, and are probably mocking the Muslims "of the first hour" about it. The Prophet, however, explains the phenomenon by referring to an alleged ancient custom of naming people after apostles and pious figures.[4] Moreover, the Islamic tradition also suggests an identification with the Coptic slave bearing the same name (*Maryam*),[5] who was offered to the Prophet Muhammad and bore his only son.[6] This shared name and origin, however, seems to be more of a coincidence than a solid argument for identification.

The account provided by the *Book of Gifts and Rarities* places Maria in Egypt, at the time of the Caliph al-Ma'mūn, for whom she performed a transmutation of sand into gold.[7] In a tradition reported by the 14th-century Persian alchemist al-Ǧildakī, however, Maria remains Jewish, and is presented as the daughter of the Queen of Sheba, whose royal connection with the Jews passed

3 See Ullmann, *Natur- und Geheimwissenscahften*, p. 182 and Patai, *Jewish Alchemists*, pp. 186–187. This problematic identification became one of the arguments in Adrian Reland's *De Religione Mohammedica*, "XX. Num in Alcorano Maria ἀειπάρθεν appellatur soror Mosis", see Hadriani Relandi, *De Religione Mohammedica Libri Duo* (Trajecti ad Rhenum: Libraria Guglielmi Broedelet, 1717), pp. 211–216. The Marian argument also has space in the contemporary religious controversies, with social media as their new stage. The YouTube channel 'Apostate Prophet', for instance, mentions it in the frame of its ongoing discourse on what is considered incoherent or ethically unacceptable in Islam: "concise, critical, often satirical videos. I'm a learned ex-Muslim who focuses on factual and fair criticism of my former religion", <https://www.youtube.com/watch?v=DPD9_mHKZmA>.

4 See *Saḥīḥ* Muslim 2135 (Book 38, *ḥadīṯ* 16) and *Ǧamiʿ* al-Tirmiḏī 3155 (Book 47, *ḥadīṯ* 207).

5 To the best of my knowledge, *Māriya* and not *Maryam* is the name used for the Graeco-Egyptian alchemist in Arabic literature.

6 The Prophet's only son, named Ibrahim, did not reach adulthood. His mother is very briefly mentioned among the women of the Prophet. See Montgomery Watt, *Muhammad at Medina* (Oxford: Clarendon Press, 1956), p. 396.

7 See Ullmann, *Natur- und Geheimwissenscahften*, p. 182; Ghāda al Hijjāwī al-Qaddūnī, *Book of Gifts and Rarities* (Kitāb al-Hadāyā wa al-Tuḥaf) *Selections Compiled in the Fifteenth Century from an Eleventh-Century Manuscript on Gifts and Treasure* (Cambridge, MA: Harvard University Press, 1996), pp. 128–130. This Arabic text is an 11th-century collection of stories and anecdotes related to pearls, precious stones, metals, and jewels.

through the court and the bed of Salomon.[8] There are several other sources that omit Maria's origin entirely, and in which she is only named.[9] However, it remains hard to say which Maria or Maryam actually came to people's mind when the alchemist was mentioned.

Be that as it may, presenting her as a Copt confirmed the link between Maria and Egypt, cradle of alchemy and ancient knowledge more generally. Moreover, a Coptic connection had the advantage of avoiding the stirring up of tensions between the Islamic and the Jewish traditions, frictions that could cast a potentially negative shadow over Maria. An unequivocally positive profile for her authority was crucial for the transmission of knowledge into the new reception context.

The Arabo-Islamic stream of the tradition preserves a number of texts attributed to Maria:

- *The Book of Maria and the Wise Men.*[10]
- *Letter of Maria, Daughter of [the Queen of] Sheba, the Copt, to Aras*, also preserved in its Latin translation with the title *Practica Mariae prophetissae sororis Moysi.*[11]
- *The Epistle of the Crown and the Creation of the Newborn Baby*, a dialogue between the Mother and her sons (the five base metals) about their genesis and the possibility of their transmutation.[12]

In this latter text, the female body is at the core of all metaphors in the dialogue, almost always related to childbearing and motherhood. The epistle opens with

8 Al-Ǧildakī suggests this in his *Kitāb anwār al-durar fī īḍāḥ al-ḥaǧar* ('Book of the lights of pearls about the explanation of the stone'), see MS Berlin Landberg 606 (Ahlwardt 4187), f. 12b; see also Ullmann, *Natur- und Geheimwissenscahften*, p. 182.

9 See Ullmann, *Natur- und Geheimwissenscahften*, p. 182.

10 In this case, it is possible that the authorship and the role of protagonist shifted from Cleopatra to Maria the Alchemist, see Ullmann, *Natur- und Geheimwissenscahften*, p. 182.

11 Ullmann, *Die Natur-*, p. 181. On the contrary, the title *Risāla Māriya ilā Aras wa-su'āluhū wa-ǧawābuhā lahū*, "Letter of Mary to Aras, his question and her answer to him", is recorded by Fuat Sezgin, *Geschichte des arabischen Schrifttums, Band IV. Alchimie – Chemie – Botanik – Agrikultur bis ca. 430 H.* (Leiden: Brill, 1971), p. 72. The translation of this text was printed several times in Europe, starting with *Theatrum chemicum*. It contains a discussion on how to whiten stones in a few days.

12 In the Arabo-Islamic culture, dialogues are an important stylistic and rhetorical form largely adopted in the transmission of knowledge in connection with its teaching and learning. See Regula Forster, *Wissensvermittlung im Gespräch. Eine Studie zu klassisch-arabischen Dialogen* (Leiden: Brill, 2017). For the English translation of the text, see Eric J. Holmyard, "An Alchemical Tract Ascribed to Mary the Copt," *Archeion*, 8 (1927): 161–167. The edition of the Arabic text and a new English translation are part of a monograph dedicated to Graeco-Egyptian alchemical authorities in the Arabic tradition, currently in preparation.

a general theoretical statement about the combination of a masculine and a feminine element being the origin of everything that exists. There are two voices in the dialogue: the Mother – who is synonymous with Nature – and five of her seven children, that is the base metals. These "children" loudly complain to their mother about their lack of perfection compared to their other two noble brothers (gold and silver). As children of her prime, gold and silver are perfect whereas the others suffered from a lack of balance that equally affected her youth and her old age. The children insist that they too seek perfection, though they wonder how it is possible to restore them to their moist, fetal condition. The Mother cannot carry out the process alone, and her actions can only be completed by the Father (that is to say, the alchemist). The space of the furnace to which the alchemist adds the appropriate ingredient to bring base metals to perfection is compared to a womb in which a second gestation can take place.

2 *Balneum Mariae*?

Zosimus credited Maria with the invention of several alchemical instruments, though he mentioned no specific connection with the eponymous heated bath and its characteristic double boiler, that is the *bain-marie*. Curiously, neither the Greek nor the Arabic sources establish a direct connection between this particular instrument and Maria, and the association might well derive from an earlier description by Hippocrates and Theophrastus.[13]

Manfred Ullmann actually mentions only one Arabic source referring to the connection between Maria and her *balneum*. This text is a work composed by the 17th-century physician and alchemist Ṣāliḥ ibn Naṣr Allāh al-Ḥalabī, best known as Ibn Sallūm. Born in Aleppo, Ibn Sallūm was chief physician to the Ottoman sultan Mehmet IV (r. 1648–1687). It seems that he could read Latin and, in Istanbul, had access to European printed books. He became indirectly acquainted with iathrochemistry and the works of Paracelsus through the compositions of his followers, in particular Daniel Sennert (d. 1637) and Oswald Croll (d. 1609). Ibn Sallūm's translations introduced iatrochemistry

13 Edmund von Lippmann, "Zur Geschichte des Wasserbades", in *Abhandlungen* 2: 185; Id., *Enstehung und Ausbreitung der Alchemie* (Berlin: Julius Springer, 1919), vol. 1, p. 50. Berthelot, in the second volume of *La chimie au moyen âge*, reproduced the illustration from a Syriac manuscript that he labelled as '*bain marie*', whereas the manuscript itself gives the caption 'hot water' (*l'eau chaude*), see Marcellin Berthelot, *La chimie au moyen âge* 2 (Paris: Imprimerie Nationale, 1893), p. 113.

to the Ottoman cultural milieu.[14] His *Ġāyat al-itqān fī tadbīr badan al-insān* ('Book of the Culmination of Perfection in the Treatment of the Human Body') is a crucial source for understanding this exchange between East and West.[15] The rather frequent choice of Greek or Latin names for plants and the direct mention of Paracelsus are perhaps the most striking clues of this exchange.[16] In his encyclopaedia of chemical medicine, Ibn Sallūm not only mentioned *bain-marie* rather frequently, but also described the instrument and the process, specifying that it is particularly suitable for the distillation of thin and delicate substances. The use of this distillation technique and its related device is most often mentioned with regard to the cold and hot extraction of "waters" (distillates).[17]

Ibn Sallūm also mentions a number of other instruments and techniques for performing the procedures described in his treatise. Along with the *bain-marie*, the Ottoman physician names the alembic, the gourd, and the *Iflāṭūnī* ('the vessel of Plato') as other possible distillation devices.

---

14 For the life and works of Ibn Sallūm, see Manfred Ullmann, *Medizin im Islam* (Leiden: Brill, 1970), p. 182; Emilie Savage-Smith, "Drug Therapy of Eye Diseases in Seventeenth-Century Islamic Medicine", *Pharmacy in History*, 29 (1987): 3–28; see also the synthetic biography of Ibn Sallūm on the website of the National Library of Medicine (Washington DC) dedicated to the Islamic medical manuscripts in its collections, <https://www.nlm.nih.gov/hmd/arabic/bioI.html#sallum>. For the influence of Paracelsusian theories on Ottoman medicine, see Nil Sarı and M. Bedizel Zülfikar, "The Paracelsusian Influence on Ottoman Medicine in the Seventeenth and Eighteenth Centuries", in Ekmeleddin İhsanoğlu, ed., *Transfer of Modern Science & Technology to the Muslim World. Proceeding of the International Symposium on "Modern Sciences and the Muslim World" Science and Technology Transfer from the West to the Muslim World from the Renaissance to the Beginning of the XXth Century* (*Istanbul, 2–4 September 1987*) (Istanbul: Research Centre for Islamic History, Art and Culture IRCICA, 1992), pp. 157–179.

15 The *Ġāyat al-itqān* is divided into four sections: the first gives the general notions necessary for the preparation of compound substances; the second one is devoted to the preparation and use of specific compound drugs; the third deals with diseases that affect specific parts of the body listed from head to toe; and the fourth and last section is on the diseases that affect the whole body. The Arabic text has been edited, see Ibn Sallūm, *Ġāyat al-itqān fī tadbīr badan al-insān*, taḥqīq wa-dirāsa Muḥammad Yāsar ibn Muḥammad Ǧamīl Zakkūr (Beirut: Dar al-Kotob al-Ilmiyah, 1971). Some pages and a detailed description of the manuscript copy are kept in the collection of the National Library of Medicine, see <https://www.nlm.nih.gov/hmd/arabic/chemical_medicine4.html>; a complete manuscript copy from the British Library collection has been digitized thanks to the Qatar Digital Library Project and can be perused online, see <https://www.qdl.qa/%D8%A7%D9%84%D8%B9%D8%B1%D8%A8%D9%8A%D8%A9/archive/81055/vdc_100059288113.0x000001>.

16 For instance, see 'sage water', 'cinquefoil water', and 'tartar water' in the Appendix.

17 The technical lexicon, in particular plant names and other ingredients, are heavily influenced by modern Greek and Ottoman Turkish. For the English translation of the relevant passages in Ibn Sallūm's *Ġāyat al-itqān*, see the Appendix.

Among them, the *bain-marie* is clearly meant to distill at relatively low temperatures. In some cases, hot ashes or sand may represent an alternative to water as a conducting substance between the two vessels. Ibn Sallūm describes a composite apparatus that includes an outer recipient made of copper, a flat lid with openings, a gourd that serves as an inner receptacle, an alembic connected to the gourd, and one last receptacle for collecting the distillate dripping from the alembic's pipe. If the capacity of the outer receptacle allows it, more than one gourd – and related alembic – can be placed in the heated bath. Ibn Sallūm indicated two different ways of performing distillations using a *bain-marie*: the first involves immersing the inner receptacle deeply into water; the second one uses a stand to support the inner receptacle, which does not come into direct contact with the water.

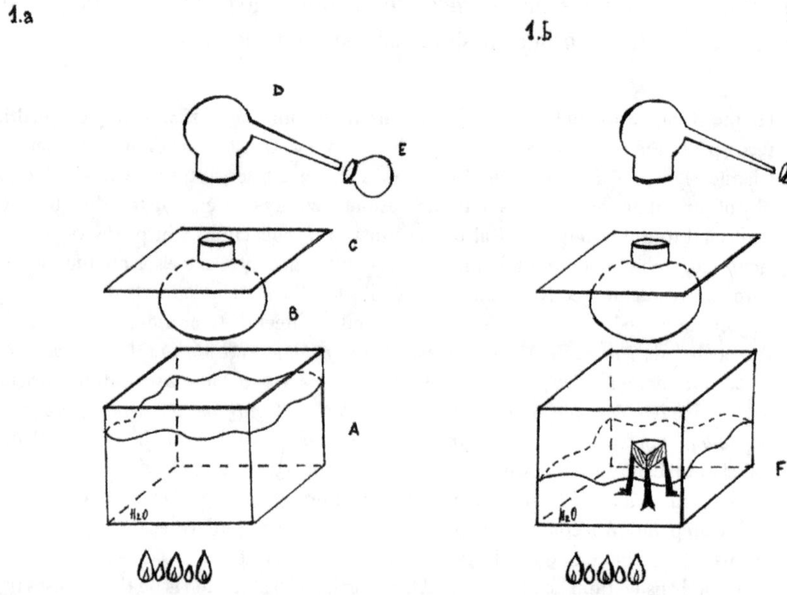

FIGURE 1.1 Ibn Sallūm's two ways of distilling through the technique of *bain-marie*. 1.1.a. First method of distillation using a *bain-marie*. The inner receptacle is deeply immersed in the water that is contained in the outer receptacle. From bottom to top: copper vessel filled with water (A, *inā' min al-nuḥās*), the gourd (B, *qar'a*) is inserted into an opening on the flat lid (C, *ġiṭā' musaṭṭaḥ*), the alembic is attached to it (D, *inbīq*), and this, in turn, is attached to the receptacle that collects the distillate (E, *qābila*). 1.1.b. Second method of distillation using a *bain-marie*. The inner receptacle is suspended above the water by means of a stand (F, *manṣib*) that is placed inside the outer receptacle. The upper part of the distillation device remains the same.

DRAWING BY THE AUTHOR

3 Concluding Notes

Entering the Arabo-Islamic tradition, Maria the Alchemist acquired a number of additional features and new attributes. All these peculiar elements shed some light on the kind of adjustments made in the course of transmission from the Greek tradition into the new cultural and linguistic context of reception (e.g. relinquishing the Jewish connection). As for her eponymous heated bath, the *bain marie*, there seems to be no clear and explicit mention of such an expression before the Ottoman period. It was probably Ibn Sallūm's fascination for iatrochemistry and the Paracelsusian tradition that introduced the expression *bain-marie* to Arabic technical literature. In fact, Ibn Sallūm's *Ġāyat al-itqān* remains, to the best of our knowledge, the only Arabic source that clearly relates this device to Maria the Alchemist. A Western mediaeval tradition of the *bain-marie* later received by Paracelsus and his followers presents itself as a possibility for further research.

Appendix

This appendix contains the translation of those passages in Ibn Sallūm's *Ġāyat al-itqān* that mention the *bain marie* as one of the different methods of distillation. The mention of the *bain-marie* is confined to two specific sections of the work. The first deals with the basic knowledge that a seller of medical syrups must acquire, and explains how distillation in a heated bath works and which instruments to use. The second is a detailed description of the procedures for obtaining compound substances. Those passages from these two sections that do not mention the *bain-marie*, but simply refer to distillation more generally are indicated by paragraph title only – the title is given in translation between squared brackets.

Second Part of the First Section about What Is Necessary for the Syrup-Seller to Know about the Instruments and the Procedure[18]

[*What is necessary to know for those who prepare liquid extracts and syrups: warming up, cooling, exposure to the sun, sinking and staying afloat as a test for substances, moistening, soaking, melting and softening, dissolving, drying, roasting, burning, addition*

18 Ibn Sallūm's *Ġāyat al-itqān*, pp. 131–146, *Faṣl fī al-taqṭīr bi-l-qarʿa wa-l-inbīq*. Introducing this chapter, Ibn Sallūm gives an ideal portrait of the producer and merchant of medical syrups and various extracts. He states that the condition of the syrup-seller is not

of perfumes, grinding, cleansing, washing, cooking, fermentation, sifting, squeezing, purification]

Section on Distillation with the Gourd and the Alembic[19]

This is a separation of the thin and watery parts, which rise [as vapours] from the heavy and earthy parts. Distillation has many methods and its instruments are numerous and venerable.[20]

Among these methods, there is distillation in a *bain-marie*. This consists of making a copper vessel – either square, or elongated, or circular – its thinness should be well proportioned. It should have a flat lid, the lid should have a place for the gourd, sometimes on this flat lid there are numerous openings for the gourds, in proportion to the vessel's capacity. Then, water must be poured into this vessel made of copper; then, the gourds must be fixed to the openings on the flat surface of the lid, so that the gourds are in the middle of the vessel and most of [the gourd] is immersed in the water. Then, the alembics must be attached to the gourds and connected to them using lute and egg white. The receptacle that collects the distillate (*qābila*) must be connected to the pipe of the alembic. Then, a fire must be lit under them until it starts dripping. All the water that reduces in the above-mentioned vessel must be topped up with other water, and this additional water should be hot too.

The distillation thus performed is mostly used in distillation processes that involve thin herbs, flowers, juices, fruits, and the distillation of some kinds of meat.[21]

Another method of distillation using a *bain-marie*: water is placed in the vessel; a stand must be placed in the middle of the water inside this vessel, the gourd must be placed on top of the stand so that it does not touch the water. Then, a lid with a number of openings, relative to the capacity of the selected gourds, must be placed on top of it. Then, the gourds must be placed on the stand or stands, and the alembics must, in

---

dissimilar to that of the physician; indeed, he is the physician's right hand. Along with possessing a refined knowledge of simple medicaments and the ways to compound them, he also ought to be trustworthy and modest, and he ought to know how to write properly and how to calculate weights.

19 Ibn Sallūm's *Ġāyat al-itqān*, pp. 142–143.

20 Here the expressions 'watery' and 'earthy' refer to the elemental composition of a substance.

21 The Arabic text of the edition mentions an illustration; the editor specifies that this was actually never realized in the manuscript, and proposes the illustration of a distillation device from a Jabirian manuscript. Despite representing and labelling very clearly the different components of the distillation device (gourd, alembic, collection vase), this illustration, does not appear to depict the same apparatus described in Ibn Sallūm's text.

turn, be placed on top of and connected to [the gourds] in the way described before. And the operation will be perfect thanks to this illustration.[22]

[*Dry distillation: suitable for plants with much moistness (to be completed in six hours); separation of oils from water; another distillation placing the gourd on ashes, sand, or iron wastes ground in a dry bath* (ḥammām yābis)*; about lute* (lutum sapientiae)*; clay to seal glass instruments; a clay to glue broken glass; a paste to seal cracks from which vapours leak out; another kind of distillation done with fire but without covering, either moist or dry, suitable for minerals and oils; method of distillation that alchemists* (aṣḥāb al-kīmiyāʾ) *call* tankīs (*inversion?*) *used in particular for oils extracted from different kinds of wood*]

Second Section on All the Composite Medicaments (aqrābaḏīn) *for the Preparation of Compounds*

First Chapter: On the Extraction of Distilled Waters and Thin Spirits (*arwāḥ laṭīfa*)[23]

Section on the Extraction of Plain Water

Firstly, you must know that when you want to extract the water from herbs (*ḥašāʾiš*), it is necessary to grind these herbs in a stone mortar. In fact, it is necessary to achieve the fermentation that takes place in the 'horse belly'[24] for a few days; then remove the contents and distil them in a *bain-marie*.

Section on the Extraction of Cold Waters

EXTRACTION OF LETTUCE WATER: this cools the blood and hotness in the liver and in the brain, induces sleep, quenches thirst, and produces milk; lettuce must be taken in its season, ground in a stone mortar, then it must be distilled in a *bain-marie* as explained; the distilled water must be placed in the sun for a few days, the opening of the flask must be closed with a perforated leaf so that the vapours can escape, its faculty can be preserved for use for one year in a cool and dry place.

[EXTRACTION OF] PURSLANE WATER: this cools the blood and the liver, quenches thirst, is useful against the expectoration of blood and hot cough, it is used in times of plague as well; purslane must be taken, ground, and placed in the gourd on hot ashes until it is distilled, then must be placed in the sun, like the first water.

22 The copyist usually left a blank space for the illustrations, which were added later on. It is not uncommon for such projects not to be completed and, as in this case and in the previous one, an illustration promised by the text is not actually attested in the manuscript.

23 Ibn Sallūm, *Ġāyat al-itqān*, pp. 159–177.

24 Hole in the ground or buried vessel filled with horse manure.

[EXTRACTION OF] NENUPHAR (EUROPEAN WHITE WATER LILY) WATER: this cools the blood and all the hot parts, is useful against jaundice, pleurisy, hot cough, and against the headache that follows various fevers and the epidemic fever, nothing is more useful than this against hectic fever; the flower of nenuphar must be taken, distilled in a *bain-marie*, and placed in the sun as described before.

[EXTRACTION OF] SWEET VIOLET WATER: this cools the blood, heart, and the liver, it is useful against the hotness of lungs and the roughness of the chest due to hotness; it must be distilled like the first [water] in a *bain-marie*, and placed in the sun for a few days.

[EXTRACTION OF] SORREL WATER: this cools the blood, liver, and the spleen, is useful against skin disease such as itch and scabies, is given to drink against plague together with the theriac; sorrel leaves must be taken, ground in a stone mortar, distilled in a *bain-marie*, and placed in the sun for a few days.

[EXTRACTION OF] ENDIVE WATER: this calms the hotness of stomach and liver when drunk or applied externally; fresh endive roots must be taken, ground, distilled in a *bain-marie*, and placed in the sun for a few days.

[EXTRACTION OF] FUMITORY WATER: this is useful against jaundice and scabies, cleanses the blood from burning humours (*aḫlāṭ*), and produces an abundant flow in the veins; it is made in the same way as endive water.

[EXTRACTION OF] GROUND CHERRY WATER (*ALKEKENGI*): this is useful against hot headache and makes kidney stones crumble; it is distilled in a *bain-marie*, and placed in the sun for a few days.

[EXTRACTION OF] WHITE HERBS WATER: this induces sleep, calms pains and hot headache; the leaves of the herbs must be taken fresh, ground, distilled in a *bain-marie*, exposed to the sun, and preserved.

[EXTRACTION OF] QUINCE WATER: this strengthens the heart and the stomach, helps the digestion, calms vomit and diarrhea; take as much quince as you want, cut it and throw it [in a vessel] with its peelings, turn it into grains, grind [the grains] in a stone mortar, distil it in a moist bath, store it for the moment of need, its faculty can be preserved for use for two years.

[EXTRACTION OF] CENTIFOLIA ROSE WATER: this strengthens the heart, is useful against fainting and palpitations; it must be distilled in a *bain-marie*, exposed to the sun, and then can be preserved for up to three years.

[EXTRACTION OF] FRESH GREEN NUTS WATER: this is useful against plague and dropsy; nuts must be taken fresh, ground, and distilled in a *bain-marie*.

[EXTRACTION OF] GOURD WATER: this calms the hotness of fevers and thirst, cleanses kidneys and bladder, thanks to copious urination; the gourd must be taken, peeled, ground, and distilled in a *bain-marie*.

[EXTRACTION OF] SEMPERVIVUM WATER: this is cooling when used, either when drunk or applied externally; this one must be distilled in a *bain-marie* too.

[EXTRACTION OF] COMMON PLANTAIN WATER: this is useful for those affected by consumption, dropsy, dysentery, it is also useful against quartan fever, in particular the one that produces ulcers on the kidneys; it must be distilled in a *bain-marie*, again following the description above.

[EXTRACTION OF] YELLOW MELON WATER: this makes the urine flow copiously, crumbles the stones, and expels the sand from kidneys and bladder; the way of extracting it is the same as that applied to gourd water.

[EXTRACTION OF] COUCH GRASS WATER: this stops diarrhea and makes the urine flow copiously, cleanses the kidneys and the bladder, opens up occlusions, and kills worms; couch grass must be taken fresh along with its roots, ground in a stone mortar, water must be gradually poured over it while grinding, so that it remains moist since it naturally contains little water, and then distilled in a *bain-marie*.

[EXTRACTION OF] SOUR CHERRY WATER: this is useful against dropsy, apoplexy, and laxity of the bladder; sour cherry must be taken, ground, squeezed, and distilled in a *bain-marie*.

Section on the Extraction of Hot Waters

[EXTRACTION OF] TONKA BEANS (*BATŪNKĀ*) WATER: this is useful against dropsy, opens up occlusions, warms up the body parts, eases childbirth; tonka beans must be taken in April, ground while spraying water over them, and distilled in a *bain marie*.

[EXTRACTION OF] SAGE (*SĀLWIYĀ*) WATER: this strengthens the brain, is useful against apoplexy, hemiplegia and spasms, causes menstruation; it is prepared like the tonka beans, its faculty can be preserved for use for two years.

[EXTRACTION OF] MARJORAM WATER: this warms up and strengthens the brain, is useful against spasms and hemiplegia, makes forgetfulness cease, causes menstruation, and strengthens the womb; it, too, is distilled like the tonka beans.

[EXTRACTION OF] CHAMOMILE (*BĀBŪNAǦ*) WATER: this makes the pain of the abdomen and colic cease, crumbles stones, causes menstruation, expels the dead foetus, is given to drink at the beginning of an attack of recurrent fever; fresh flowers of chamomile must be distilled in a *bain-marie* and then preserved.

[EXTRACTION OF] GIANT FENNEL WATER: this strengthens the brain and the heart, expectorates the humours (*aḫlāṭ*), is useful against the cough, makes the milk flow, strengthens the eyesight either by drinking it or by applying it in drops in the eye; it is extracted in a *bain-marie* by distilling ground fresh giant fennel.

[EXTRACTION OF] PENNYROYAL WATER: this warms up the womb, causes menstruation, and is useful against chronic headache; it is prepared in the same way as the tonka beans.

[EXTRACTION OF] ROSEMARY WATER: this strengthens the stomach, cleanses blood, opens up occlusions, is useful against jaundice, asthma, and breathing difficulties; rosemary must be taken, ground, moistened with water or wine – and this is even better – distilled in a *bain-marie*, and its faculty can be preserved for use for four years.

[EXTRACTION OF] *IFRĀNǦIYYĀ* (?) WATER: this strengthens the eyesight either by drinking it or by applying it in drops; its moist part must be distilled in a *bain-marie* after grinding.

[EXTRACTION OF] LINDEN (*IḪLĀMŪR*) FLOWER[25] WATER: this is useful against apoplexy, hemiplegia, facial palsy, and the cold disease of the brain; it is prepared like rose water.

[EXTRACTION OF] MAIDENHAIR FERN WATER: this cleanses the lungs and the liver, and crumbles stones; its extraction is made by distillation in a *bain-marie* after grinding.

[EXTRACTION OF] HYSSOP WATER: this is useful against asthma and chronic cough, helps expectorate thick humours; hyssop must be taken in August when its flowers start to appear, must be ground and moistened with warm water while grinding, and distilled in a *bain-marie*.

[EXTRACTION OF] WHITE HOREHOUND WATER: this is useful against chronic cough and breathing difficulties, strengthens the heart and the spleen, is useful to those severely affected by dropsy; its extraction is like that for the extraction of hyssop water.

[EXTRACTION OF] HOLY THISTLE (*KĀRDŪSANṬŪ*) WATER: this is useful against headache, strengthens the 'preservative faculty',[26] induces perspiration, and is a good therapy for those affected by quartan fever; it is prepared as previously described.

[EXTRACTION OF] LEMON BALM WATER: this strengthens the 'preservative faculty' and strengthens all the senses, slows the appearance of grey hair, strengthens the heart and the stomach, cheers up, is very useful against hemiplegia, calms toothache, removes the foul smell that comes from the gums and the mouth, when used as a mouth rinse; the desired amount of fresh lemon balm must be taken, ground, macerated in basil liquor or honey wine, for a few days, then distilled and preserved, it retains its faculty for four years.

[EXTRACTION OF] *AL-SANFĪṬIS AL-KABĪR* (?) WATER: this accelerates the setting of broken bones, is useful against stroke, fall, and the expectoration of blood; take the desired amount of *al-sanfīṭis al-kabīr* (?), it must be ground, moistened with water while grinding, and distilled in a *bain-marie*.

[EXTRACTION OF] CINNAMON WATER: this strengthens the parts of the head and warms up the stomach; good cinnamon must be taken, crushed, soaked, either in

25 Alternative form for *iflāmūr*, Ottoman Turkish from the Greek φλαμούρι, in turn from Byzantine Greek φλάμμουρον or φλάμπουρον. See Gustav Meyer, "Türkische Studien. I. Die griechischen und romanischen Bestandtheile im Wortschatze des Osmanisch-Türkischen", in *Sitzungsberichte der philosophisch-historischen Classe der Kaiserlichen Akademie der Wissenschaften*, vol. 128 (Wien: In Commission bei F. Tempsky, 1892), p. 33.

26 This expression may refer to one of the faculties that alchemy associated with the different substances already in authors such as the Pseudo-Democritus. Their denomination closely recalls medical lexicon.

quince water or warm rosewater, for one day and one night, distilled in a *bain-marie* and preserved, it retains its faculty for four years.

[EXTRACTION OF] ABSINTHE WATER: this warms up the stomach and the digestion, kills worms, opens up occlusions, expels yellow gall thanks to copious urination during fevers, calms toothache and vomit without damaging the brain as absinthe otherwise does; fresh absinthe must be taken, ground, soaked in water or wine for four days, then distilled in a *bain-marie*.

[EXTRACTION OF] MINT WATER: this strengthens stomach, liver, and the spleen, improves digestion too; it is prepared like the absinthe water.

[EXTRACTION OF] THYME WATER: this is useful against moist cough, dropsy, and quartan fever, and it releases trapped winds; thyme must be taken, ground, moistened by spraying water, then distilled in a *bain-marie*.

[EXTRACTION OF] COMMON CENTAURY WATER: this kills worms, is useful against fevers, strengthens the stomach and improves digestion; it is prepared like thyme water.

[EXTRACTION OF] COMMON AGRIMONY WATER: this opens up occlusions, strengthens the liver, is useful against dropsy, and improves the digestion; its preparation is as described before.

[EXTRACTION OF] ARISTOLOCHIA WATER: this is useful against spasms, joints pain, epilepsy, and gripes, it also improves digestion; aristolochia must be taken fresh along with its roots, ground in a stone mortar, macerated in a bit of warm water, and distilled in a *bain-marie*.

[EXTRACTION OF] GENTIAN WATER: this is useful against incontinence of urine, causes menstruation, strengthens the stomach, is useful against all fevers; gentian must be taken as it is, including its roots, stems, and leaves, ground, and distilled in a *bain-marie*.

[EXTRACTION OF] ROCKET WATER: this warms up the kidneys and the bladder, increases sperm, strengthens the stomach; it is prepared in a way already described.

[EXTRACTION OF] BANYAN WATER: this stops gripes and crumbles stones, is useful against asthma and breathing difficulties; it is prepared in the way already described.

[EXTRACTION OF] 'GLASS HERB' WATER: this opens up occlusions of the spleen and of the liver, cleanses the kidneys and the bladder, causes menstruation and urine to flow; it is prepared in the way already described.

[EXTRACTION OF] CINQUEFOIL (*BANTĀFILON*) WATER: this opens up occlusions, crumbles stones, eliminates decay, cuts the root of fevers; it is prepared like aristolochia water.

[EXTRACTION OF] RADISH WATER: this makes urine flow, crumbles stones in the kidneys and in the bladder, is useful against jaundice; radish must be taken as it is, crushed, and distilled in a *bain-marie*.

[EXTRACTION OF] ELECAMPANE WATER: this strengthens the stomach and the brain, crushes stones in the kidneys and the bladder; elecampane must be taken as it is, crushed, and distilled in a *bain-marie*.

[EXTRACTION OF] *ṬARĀĠIYŪN* (?) WATER: this crumbles stones in the kidneys and in the bladder, is useful against the plague and malignant fevers; it is prepared in the way already described.

[EXTRACTION OF] MUGWORT WATER: this warms up the stomach, is useful against dropsy and cough, induces sleep, helps the expectoration of thick humours; it is prepared in the way already described.

[EXTRACTION OF] *HAMĪL* – THAT IS A VARIETY OF IVY – WATER: this cleanses the blood and the black burning humours, opens up occlusions of the spleen and of the liver, brings humours to ripeness and gets them ready before their expulsion with the feces; it is necessary to extract its water in April, so that the extremities of the plant are fresh, then must be distilled in a *bain-marie*, following the procedure.

[EXTRACTION OF] 'BULL TONGUE' WATER: this strengthens heart and brain, cleanses the blood, is useful against fainting and palpitations, cheers up the heart; 'bull tongue' must be taken as it is, ground, and distilled in a *bain-marie*.

[EXTRACTION OF] ASARABACCA (LIT. *ASARUM*) WATER: this is useful against fevers, expels the foetus; it is prepared like the 'bull tongue' water.

[EXTRACTION OF] *SCOLOPENDRIUM* (?) WATER: this opens up occlusions of the liver and the spleen, cleanses the blood, is useful against quartan fever; it is prepared in the way already described.

[EXTRACTION OF] SPANISH OREGANO WATER: this opens up occlusions, strengthens stomach and the brain, is useful against phlegmatic fevers; it is prepared in the way already described.

Section on the Extraction of Waters That Act through Their Peculiar Property

[EXTRACTION OF] HUMAN BLOOD WATER: this is useful against the malignant diseases, in particular against malignant and epidemic fevers, and there is nothing more useful against hectic fever, if given to drink with julep.

The blood of a healthy man must be taken, his age must range from thirty to thirty-three, [the blood] must be distilled with an *iflāṭūnī* ['vessel of Plato'][27] over a moderate fire. Some people mix it with the same quantity of wine sublimate (*ṣāʿid al-šurāb*) and distil it; some other people let the blood dry, crush it finely, mix it with an abundance of 'bull tongue' water or sorrel water, keep it in the 'horse belly' for some days, then take it out and distil it.

27 Arabic alchemy presents Plato as one of the ancient authorities in the field, and an instrument for distillation was named after him.

[EXTRACTION OF] HUMAN EXCREMENT WATER: you should know that the masters of alchemy (*arbāb al-kīmiyā'*) used to distil human excrement believing that this would have made silver yellow and heavier. As for what we are concerned with, we have already seen that it has peculiar properties [that act] on the external parts of the body; in fact, this is useful against malignant swellings and vicious sores, stops glaucoma in the eye; if given to drink, it is useful against epilepsy and dropsy, crumbles kidney and bladder stones, and there is nothing more useful than this against the bite of a rabid dog.
[EXTRACTION OF] MISTLETOE WATER: it is useful against epilepsy and specifically cures it, it crumbles kidney and bladder stones; mistletoe has to be taken as it is, with its leaves and stems, then distilled in the way already mentioned.
[EXTRACTION OF] RIVER CRABS' WATER: it is useful against extreme thinness and consumption; it is applied on phlegmons and on hot and malignant sores.

Section on the Extraction of Purgative Waters

DANEWORT WATER: loosens the abdomen, is useful against dropsy and recurrent fever; it is distilled from the leaves of danewort (*ḫāmā aqtā*) and from its flowers and roots in a *bain-marie* after grinding. Baneberry (*aqṭā*) water is extracted in the same way.
PEACH WATER: it is purgative and emetic, it extracts phlegm, yellow bile, and the watery component; it is prepared like rosewater.
PREPARATION OF A COMPOUND PURGATIVE WATER: peach flowers have to be taken with scammony, and turpeth in equal parts, everything must be soaked in hot water, distilled in a *bain-marie*, and stored away for the moment of need.
PREPARATION OF THE PURGATIVE WATER OF BLACK HELLEBORE: fresh roots of black hellebore must be taken, ground, soaked in wine sublimate (*ṣā'id al-šurāb*) for a consistent number of days; then it has to be incidentally distilled with a retort (*mā'il al-ruqba*) or an *iflaṭūnī*; then, what comes out must be distilled with an alembic until the *'araq*[28] comes out while the hellebore remains in the lower part of the gourd. This water must be taken, made into a julep with sugar and given to drink.
PREPARATION OF A COMPOUND PURGATIVE WATER: cinnamon wood, turpeth, aloe, colocynth grease must be taken – one ounce of each – along with clover and saffron – one third of a dirham of each – one dirham of musk, one *raṭl* of sweet violet julep, two *raṭl* of wine sublimate; everything must ferment in a place for twelve days; then it must be drawn with a pulley a number of times until it has become pure and can be stored away. The right dosage to drink is between three and four dirhams.
PREPARATION OF ANOTHER WATER: some aloe, anise, rosemary, celery, *dūqū* (?), jalap plant, *maǧūqān* (?), *ġūtāġniyā* (?), and cinnamon must be taken – in proportion

28 *'Araq* probably refers here to a transparent and alcoholic substance produced in the course of distillation.

to the nature and conforming with the temperament – the colour must be extracted with wine sublimate and given to drink against the cold diseases that are specifically related to 'drum-like' and 'rich' dropsy, and this will be very useful.

PREPARATION OF MINERAL TURPETH: half *raṭl* of mercury washed with wine must be taken, one *raṭl* of vitriol spirit, everything must be put in the gourd and distilled; the substance must be distilled three times and what remains in the lower part of the gourd is a white powder. This must be taken, washed with *ʿaraq* until the acid (lit., 'sharp') component has left it, then it has to be dried. It is useful against the chronic diseases specifically related to the pustules of the Franks.

Section on Emetic Water

Bryony roots must be taken with cyclamen – one ounce of each – six ounces of white hellebore, everything ground together, immersed in two fingers of *ʿaraq*, and it must be left in a warm place for five days. Then it must be distilled in a *bain-marie* and, if what has been distilled is mixed again with the distillation residue, this will be stronger; eat a spoonful of it.

Section on Distilled Waters That Strengthen the Nature

SUGAR WATER: strengthens the weak parts, unblocks obstructions of the bladder and the liver, calms the pains; four *raṭl* of sugar must be taken, with one *raṭl* of Byzantine zodoary, one once of Chinese zodoary, one *raṭl* of mercury [plant], everything covered with wine and left to macerate in a warm place or in the 'horse belly' for six days, then it must be distilled in a *bain-marie* and stored.

SUGAR WATER AGAIN: strengthens the eyesight and stops the flowing of tears; half *raṭl* of sugar must be taken, half *raṭl* of fruit, everything has to be kept in a bath for eighty days, then distilled and stored.

HONEY WATER: it is useful against malignant ulcers, makes glaucoma cease, strengthens the eyesight, prevents hair loss, prevents catarrh and chronic cough, strengthens the body, prevents decay, and repels poisons. The distillation of honey is difficult: in fact, it is necessary to mix it with sand and distill it in a *bain-marie*; the gourd must be curved at one quarter [of its height]. At the first distillation, [the distillate] will be white, then, after this, it will be yellow, then red; the white and the yellow ones are for internal application; as for the red one, it has a burning effect and ought to be used for external application.

CAPON WATER:[29] increases the strength and is invigorating, is useful against the loss of strength and the weakness that comes with diseases; a capon must be taken, cooked with 'bull tongue', or the 'bull tongue' must be cooked with it until it breaks in pieces and then the broth must be cleared from it; then it must be mixed with sweet violet

29 Castrated cock.

jam and jam of 'bull tongue' flowers – one and a half ounces of each – one ounce of *kalbaškar* (?), six [ounces] of gold leaves, distilled in a *bain-marie* and administered in the morning with a spoon.

CAPON WATER: helps with fevers; a capon must be taken, its abdomen must be properly cleaned even from fat, then cut into small pieces, washed with lettuce water or nenuphar water, mixed with sweet violet jam and nenuphar jam – one ounce of each – and three ounces of 'bull tongue' jam, seeds of white poppy and lettuce seeds – one dirham of each – one dirham and a half of pearl powder, two *raṭl* of musk apple juice, everything mixed together, distilled in a *bain-marie* and given to drink.

[*Compound water*]

BEZOAR WATER: it has affect against poisons, plague, epidemic fevers, and malignant fevers, strengthens the brain and all the senses; some greater celandine must be taken as it is with its roots and leaves, along with one handful of rue, two handfuls of scordion, one handful each of dittany and holy thistle, two dirhams of long-leaved violet willow (*zanabād*, or saffron), rose, angelica and clover, two dirhams and a half of citron peel, lemon peel, and cinnamon, everything spread and soaked in three *raṭl* of wine sublimate [mixed with] half *raṭl* of vinegar and half *raṭl* of good wine, everything distilled in a *bain-marie*. Then, three ounces of theriac must be diluted in the distillate, with one and a half ounces of Mithridatum, one third of a dirham of camphor, one dirham each of viper compress, mace, and a cold paste of precious stones, one dirham and a half of cinnamon, clover, deer horn; everything distilled in a *bain-marie* and stored away for the moment of need: and this is one of the wonders of the field of antidotes.

[*Water called 'elixir of life' – Water called* akwātiryākis ('aqua theriacis') *Another theriac water – Preparation of 'sweating water' – Preparation of tartar water – Preparation of stroke water – Preparation of water useful against epilepsy, paralysis and cold diseases of the brain – Preparation of water called 'balsam of the foetus'*]

PREPARATION OF SWALLOW WATER: strengthens the brain, is useful against all the warm and cold diseases that specifically affect it, this remedy has been tried out against epilepsy, nightmares (*kābūs*), stroke, paralysis, terror, convulsions, and there is nothing more useful against the constriction of the uterus; forty swallow chicks must be taken, slaughtered and cleaned, with one and a half ounces of castoreum, two ounces of peony with its roots and wood of limy oak, one ounce of white dittany, a handful of rue leaves, linden flowers, baneberry flowers, mountain rosemary and small mullein, three dirhams of cinnamon, two dirhams of cubeb pepper and mace with their stem; everything spread and soaked in eight *raṭl* of wine and two *raṭl* of scilla vinegar for three days in a warm place, then distilled in a *bain-marie* or over hot ashes.

PREPARATION OF SMALL SWALLOW WATER: this has many applications now in Europe (lit., 'the country of the Franks') for different kinds of disease of the brain and the nerves; forty swallow chicks, slaughtered and cleaned, must be taken, with two handfuls of fresh rue, one ounce of castoreum, macerated in water as explained for the previous [swallow] water, distilled in a *bain-marie* or over hot ashes.

[*Preparation of cheering water – Preparation of precious stones water – Preparation of water called 'balsam for the stomach' – Preparation of water for sleep – Another water for sleep – Preparation of water that strengthens the eyesight and protects the health of the eye – Preparation of another water that strengthens the eyesight and removes the glaucoma from the eye – Preparation of another water that is useful against cataracts and is useful against the water that drips from the eye – Preparation of water that is useful to those affected by consumption – Preparation of water that is useful to those affected by consumption and extreme thinness*]

PREPARATION OF TARTAR WATER ASCRIBED TO PARACELSUS: this strengthens the stomach and liver, unblocks obstructions; one *raṭl* of cinnamon must be taken, soaked in four *raṭl* of rosewater, mixed with one and a half ounces of dissolved tartar (potassium bitartrate), left for one day and one night, then distilled in a *bain-marie*.

References

Berthelot, Marcellin, *La chimie au moyen âge* (Paris: Imprimerie Nationale, 1893).

Forster, Regula, *Wissensvermittlung im Gespräch. Eine Studie zu klassisch-arabischen Dialogen* (Leiden: Brill, 2017).

Ghāda al Hijjāwī al-Qaddūnī, *Book of Gifts and Rarities (Kitāb al-Hadāyā wa al-Tuḥaf). Selections Compiled in the Fifteenth Century from an Eleventh-Century Manuscript on Gifts and Treasure* (Cambridge, MA: Harvard University Press, 1996).

Holmyard, Eric J., "An Alchemical Tract Ascribed to Mary the Copt," *Archeion*, 8 (1927): 161–167.

Ibn Sallūm, *Ġāyat al-itqān fī tadbīr badan al-insān*, taḥqīq wa-dirāsa Muḥammad Yāsar ibn Muḥammad Ğamīl Zakkūr (Beirut: Dar al-Kotob al-Ilmiyah, 1971).

Meyer, Gustav, "Türkische Studien. I. Die griechischen und romanischen Bestandtheile im Wortschatze des Osmanisch-Türkischen", in *Sitzungsberichte der philosophisch-historischen Classe der Kaiserlichen Akademie der Wissenschaften*, vol. 128 (Wien: In Commission bei F. Tempsky, 1892).

Patai, Raphael, *The Jewish Alchemists* (Princeton, NJ: Princeton University Press, 1995).

Principe, Lawrence M., *The Secrets of Alchemy* (Chicago, IL/London: University of Chicago Press, 2013).

Rashed, Roshdi, ed., *Encyclopedia of the History of Arabic Science* 3 (London/New York: Routledge, 1996).

Relandi, Hadriani, *De Religione Mohammedica Libri Duo* (Trajecti ad Rhenum: Libraria Guglielmi Broedelet, 1717).

Sarı, Nil, and M. Bedizel Zülfikar, "The Paracelsusian Influence on Ottoman Medicine in the Seventeenth and Eighteenth Centuries", in Ekmeleddin İhsanoğlu, ed., *Transfer of Modern Science & Technology to the Muslim World. Proceeding of the International Symposium on "Modern Sciences and the Muslim World" Science and Technology Transfer from the West to the Muslim World From the Renaissance to the Beginning of the XXth Century (Istanbul, 2–4 September 1987)* (Istanbul: Research Centre for Islamic History, Art and Culture IRCICA, 1992), pp. 157–179.

Savage-Smith, Emilie, "Drug Therapy of Eye Diseases in Seventeenth-Century Islamic Medicine," *Pharmacy in History*, 29 (1987): 3–28.

Sezgin, Fuat, *Geschichte des arabischen Schrifttums, Band IV. Alchimie – Chemie – Botanik – Agrikultur bis ca. 430 H* (Leiden: Brill, 1971).

Ullmann, Manfred, *Die Natur- und Geheimwissenschaften im Islam* (Leiden: Brill, 1972).

Ullmann, Manfred, *Medizin im Islam* (Leiden: Brill, 1970).

Von Lippmann, Edmund, "Zur Geschichte des Wasserbades", in Paul Diergart, ed., Beiträge aus der Geschichte der Chemie: dem Gedächtnis von Georg W. A. Kahlbaum (Leipzig: Deuticke, 1909), pp. 143–157.

Von Lippmann, Edmund, *Enstehung und Ausbreitung der Alchemie* (Berlin: Julius Springer, 1919), vol. I.

Watt, Montgomery, *Muhammad at Medina* (Oxford: Clarendon Press, 1956).

CHAPTER 2

Maria's *Practica* in Early Modern Alchemy

Matteo Martelli

1 Introduction

Maria features prominently in the history of alchemy, from Graeco-Roman antiquity to the early modern period. Called by many names throughout the centuries – the divine Maria (Gr. ἡ θεῖα Μαρία), Maria the Copt (Ar. *Māriya al-Qubṭiya*), or Maria the prophetess (Lat. *Maria prophetissa*) to name but a few – she was credited with different writings by Graeco-Egyptian, Byzantine, Arabic, and Latin alchemical authors, who also attributed various technical innovations to her. Zosimus of Panopolis (third to fourth centuries) counted the woman among the Jews who openly wrote about the alchemical art. He cites a *Treatise on Furnaces* (καμινογραφία) ascribed to Maria,[1] which Zosimus extensively used as a reference work for the description of different alchemical devices: in particular, an alembic called *tribikos* (τρίβικος), namely a still with three reception vases (*bikoi*),[2] and the *kērotakis* (κηροτακίς), an instrument where metallic leaves were exposed to the vapors of dyeing substances.[3] Byzantine manuscripts preserve illustrations of these pieces of equipment and the drawing of a *kērotakis* was interpreted by the French chemist Marcelin Berthelot as an early reference to the so-called *bain-marie*, even though this technical term never occurs in Byzantine alchemical treatises.[4]

1 Zosimus, *First Book of the Final Account*, § 2 in Andrè Festugière, *La révélation d'Hermès Trismégiste, vol. 1: L'astrologie et les sciences occultes* (Paris: Les Belles Lettres, 1944), p. 365. The same book – simply referred to as *On Furnaces* (Περὶ καμίνων) and *On Furnaces and Devices* (Περὶ καμίνων καὶ ὀργάνων) – is mentioned in the Zosimus' *On the Letter Omega*, §§ 1 and 17: see Michèle Mertens, *Zosime de Panopolis, Mémoires authentiques* (Paris: Les Belles Lettres, 1995), pp. 1 (with related note 12 on pp. 63–65) and 8.

2 See Zosimus, *On the Tribikos and the Tube*, in Mertens, *Zosime de Panopolis*, pp. 14–15.

3 See Zosimus, *On Devices and Furnaces*, §§ 3 and 6 in Mertens, *Zosime de Panopolis*, pp. 24–25 (see also pp. CXXX–CLII).

4 See Marcelin Berthelot, Charles-Émile Ruelle, *Collection des anciens alchimistes grecs*, 3 vols. (Paris: G. Steinheil, 1887–1888), vol. 1, p. 146.

 | DOI:10.1163/9789004512610_004

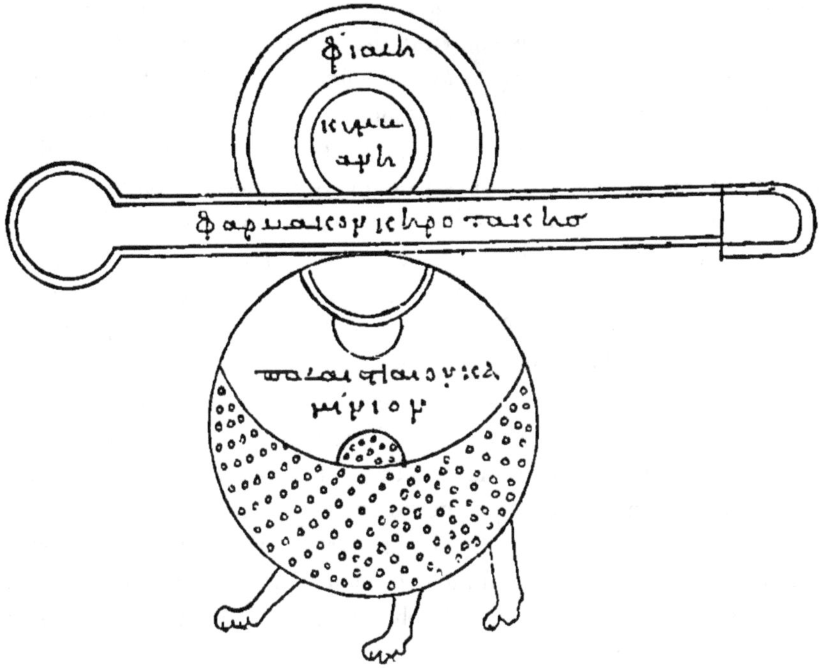

FIGURE 2.1 MS *Marcianus* gr. 299, fol. 195v as reproduced and interpreted by Marcelin Berthelot, Charles-Émile Ruelle, *Collection des anciens alchimistes grecs* (Paris, 1887–1888), vol. 1, p. 146

This paper will focus on Maria the alchemist and her writings in the early modern period. A path through which sixteenth- and seventeenth-century scholars and practitioners came to know this figure is certainly the Arabo-Latin medieval tradition. Both Maria and the famous heating method commonly associated with her name – namely the *balneum Mariae* or, according to Ullmann, *ḥammām Māriya* in Arabic[5] – are already prominent in medieval alchemical texts, which circulated, often in printed editions, during the early modern period. Arabo-Latin treatises such as the *Turba philosophorum* or the *Responsiones Aros philosophi ad Nephes regem* often refers to the authority of Maria,[6] and the Latin translation of the Arabic work *Letter of Maria,*

5 Manfred Ullmann, *Die Natur- und Geheimwissenschaften im Islam* (Leiden: Brill, 1972), p. 183. On this subject, see Lucia Raggetti's contribution in this volume.

6 See the *indexes* at the end of the most recent editions: Grégoire Lacaze, *Turba Philosophorum. Congrès pythagoricien sur l'art d'Hermès* (Leiden: Brill: 2018); Sylvain Matton, *Responsiones Aros philosophi ad Nephes regem de philosophia malis et improbis occulta et sapientibus manifesta. Avec un fac-similé du manuscrit arabe Chester Beatty 4121* (Paris: SÉHA, 2017).

Daughter of [the Queen of] Sheba, the Copt, to Aras (*Risāla Māriya bint Sāba al-malik al-qibṭī ilā Aras*)[7] was known under the title of *Practica Mariae prophetissae in artem alchemicam* (*Maria the Prophetess' Practice in the Alchemical Art*). This text was often quoted and commented on by early modern alchemists, who tried to promote their own practices and procedural methods by interpreting the ancient words of a woman who ranked among the founders of the alchemical art and was often identified with the sister of the patriarch Moses. Such exegetical attempts can be recognized, for instance, in the works of authors such as George Ripley, Andreas Ortelius, and Isaac Newton, which will be discussed in the following pages. Their interpretations of Maria's *Practica* exemplify a both antiquarian and experimental attitude, which has been meaningfully described as "practical exegesis" in a seminal article by Jennifer Rampling.[8] Moreover, along with her alchemical instructions, Maria's identity and the authenticity of her work were scrutinized by humanists, physicians, and alchemists, who, at some point, started uncovering the relics of Greek alchemical literature scattered in Byzantine manuscripts. As we shall see, different streams of tradition were compared and combined, and this cultural process fostered a complex and polyphonic discussion on Maria and her alchemical achievements.

2 Interpreting Maria's Instructions: From Isaac Newton Back to the Arabo-Latin Tradition

One might take Isaac Newton's massive writings on alchemy as a vantage point from which to survey the long-lasting historical process leading to the assimilation of the rich Arabo-Latin alchemical tradition in the early modern period.[9] Newton ranged Maria the prophetess among the earliest alchemical writers, at least according to various lists of ancient practitioners that he compiled in

7 *Risāla Māriya bint Sāba al-malik al-qibṭī ilā Aras*, see Ullmann, *Die Natur- und Geheimwissenschaften*, p. 181. The title *Risāla Māriya ilā Aras wa-su'āluhū wa-ǧawābuhā lahū*, "Letter of Mary to Aras, his question and her answer to him", is recorded in Fuat Sezgin, *Geschichte des arabischen Schrifttums, Band IV. Alchimie – Chemie – Botanik – Agrikultur bis ca. 430 H.* (Leiden: Brill, 1971), p. 72.

8 Jennifer M. Rampling, "Transmuting Sericon: Alchemy as 'Practical Exegesis' in Early Modern England," *Osiris*, 29 (2014): 19–34.

9 On Isaac Newton's alchemy, see William R. Newman, *Newton the Alchemist. Science, Enigma, and the Quest for Nature's "Secret Fire"* (Princeton and Oxford: Princeton University Press, 2019).

Keynes Ms. 13 (King's College, Cambridge): in chronological order, Maria the prophetess (*Maria prophetissa*) is mentioned after Hermes in the first list of the manuscript (fol. 1*r*) as well as in two following lists devoted to "the most ancient (alchemical) authors".[10] Newton even copied a Latin text attributed to Maria in his own hand. Indeed, Keynes Ms. 45 (Cambridge, King's College Library) transmits the *Practica Mariae prophetissae*,[11] a treatise that Newton could read in various printed copies kept in his library.

The first lines of Maria's *Practica* set the narrative framework of the entire text, namely a dialogue between Maria and the philosopher Aros:[12] Aros visited the woman, explicitly identified with Moses' sister, after hearing that she was able to whiten the "stone" (*lapis*, presumably the philosophers' stone) in only one day. Maria did confirm this reputation and replied that Hermes could even do the same just in one hour. A set of instructions follows in the dialogue. First, Maria lists the key ingredients: alum from Spain, white gum, and red gum, that is, sulphur, which is called *kibrik* in the text.[13] The white and the red gums are to be joint with a true matrimony (Keynes Ms. 45, fol. 1*r*: *matrimonifica gummi scilicet album cum gummi scilicet rubeo, vero matrimonio*) to make them like running water (*fac illa sicut aquam currentem*) and vitrify this water with the two *zubech*, a term that Newton transliterated in Hebrew script and translated either as "gum" or as "tincture".[14] One may wonder whether he agreed with Michael Maier (1568–1622), who believed that the original version of Maria's *Practica* had been written in Hebrew.[15]

Maria stresses how practitioners were required to master the art of fire perfectly, since the success of the described procedures very much depended on

---

10 See Keynes Ms. 15, fol. 1*v*: *Authores vetustissimi*; fol. 2*r*: *Authores antiquissimi*. Both lists record Hermes' *Emerald Table*, Hermes' *Seven chapters*, and *Maria Prophetissa ad Aros*.

11 For a full transcription of the manuscript (both a diplomatic and a normalized transcription), see *The Chymistry of Isaac Newton* website: http://webapp1.dlib.indiana.edu/newton/mss/norm/ALCH00034.

12 On this figure, who features in various Arabo-Latin alchemical dialogues, see Ullmann, *Die Natur- und Geheimwissenschaften*, pp. 190–191; Sezgin, *Geschichte des arabischen Schrifttums*, pp. 68–69; Matton, *Responsiones Aros philosophi*, pp. V–XIV.

13 That is, the Arabic term *kibrīt*, "sulphur". See Fabrian Käs, *Die Mineralien in der arabischen Pharmakognosie*, 2 vols. (Wiesbaden, Harrassowitz, 2010), vol. 2, pp. 917–924.

14 See Keynes Ms. 35, fol. 1*r* and note 1 in *The Chymistry of Isaac Newton* website. One must note that the Arabic *zi'baq* refers to "mercury" (see Käs, *Die Mineralien*, vol. 2, pp. 692–696), as Newton himself recorded in the *Index chemicus* (Keynes Ms. 30/1, fol. 94*r*, s.v. *zibach*).

15 On Maier, see below, pp. 59–60. We must note that Maier's alchemical writings range among the favourite sources of Newton: see Meagan S. Allen, "Revisiting Isaac Newton's *Index Chemicus*: A Response to Richard S. Westfall," *Ambix*, 66 (2019): 72–81, pp. 76–77.

the use of the right heating source. The fire was to be like "the measure of the heat of the sun in the month of June or July" (fol. 1*v*: *mensura caliditatis Solis in mense Junii vel Julii*).[16] Such a control of the temperature was mandatory to manipulate the enigmatic ingredient that is introduced in the middle of the dialogue, namely a white herb which grows upon the small mountains (*herbam albam ... crescentem super monticulis*). After insisting on this substance, Maria goes back to the already mentioned *kibrik* and *zubech*, which she identifies with two smokes of different colors. They are the roots of the art, the dyeing tincture, while the solid body that receives this tincture is referred to as "the heart of Saturn" (*cor Saturni*). The whole operation, Maria concludes, requires a particular type of vessel, that is called "the vessel of Hermes" (Keynes Ms. 45, fol. 2*r*):

> *Istud est vas Hermetis, quod Sophi occultaverunt, et non est vas ignorantium* [*id est vulgi*], *sed est mensura ignis tuis.* [...]
>
> *Omnia docent Philosophi praeter vas Hermetis, quia illud est divinum, et de sapientia domini gentibus occultatum. Et illi qui illud ignorant, nesciunt regimen veritatis, propter vasis Hermetis ignorantiam.*
>
> This is the vessel of Hermes, which was hidden by the sages;[17] it is not the vessel of unlearned people (that is, of common people), but the measure of your fire. [...]
>
> The sages teach all these things rather than the vessel of Hermes, because this is divine, and, by the wisdom of the Lord, hidden from the people. And those who disregard it do not know the regimen of truth, because of their ignorance of the vessel of Hermes.[18]

After copying the *explicit* of this text, Newton adds a few references to the printed books that included Maria's work. He writes (Keynes MS. 45, fol. 2*v*): "Extat hoc opus in Theat. Chem. Vol. 5 & Vol. 6 p. 479 et Art. Aurif. vol. 1 et in Bibliotheca Gallica".

The reference to the *Auriferae artis* is clear. In fact, the first printed edition of Maria's *Practica* appeared in an important anthology of alchemical writings,

16 English translation by Raphael Patai, *The Jewish Alchemists. A History and Source Book* (Princeton: Princeton University Press, 1994), p. 72.

17 Both in the *Artis auriferae* and in the *Theatrum chemicum* (on these printed editions, see below), the Latin text reads: *Istud est vas Hermetis, quod Stoici occultaverunt, et non est vas nigromantium etc.* On this mention of the Stoics, see Sylvain Matton, "Alchimie et Stoicisme: à propos de récentes recherches," *Chrysopoeia*, 5 (1992–1996): 5–144, pp. 18–20.

18 English translation (slightly modified) by Patai, *The Jewish Alchemists*, pp. 73–74.

the *Auriferae artis, quam Chemiam vocant, antiquissimi authores* (*The earliest authors of the aurific art that is called* chemia) printed by Pietro Perna in Basel in 1572. In 1593 the anthology was reprinted by Konrad Waldkirch, and a third edition in three volumes appeared in 1610. Newton had a copy of this last edition in his library.[19]

A few years afterwards, Maria's text was printed again in the fifth volume of the *Theatrum chemicum* (1622), in the long section entitled *Allegories of the wises about the book* (*of the*) *Turba: Twenty-nine sections.*[20] Section 27, in fact, includes the full text of Maria's *Practica*, which is not introduced by any particular title.[21] Moreover, some passages taken from this dialogue were also printed in the sixth volume of the *Theatrum chemicum* (1661) at p. 479, as Newton recorded in the above-mentioned note in Keynes Ms. 45.[22]

We cannot exclude that Newton was interested in these excerpts because they include dense exegetical notes by Andreas Ortelius (1600–1657), who insisted on the main technical parts of Maria's dialogue.[23] Maria's description of the two smokes as well as of *kibrik* and *zubech* were interpreted by Ortelius as referring to two mercuries, namely silver's and gold's mercury, which he identified with the mercury of the moon and the mercury of the sun. Then Ortelius provided a detailed description of their treatments, which he explained by referring to Paracelsus. After being distilled and liquified, the mercuries are to be solidified with the addition of lime (*calx*), which would produce, after 12 days, coagulated "mercury of Saturn". All these processes must be carried on by means of the sun's heat, to which the second part of Ortelius' commentary is devoted. Maria's instruction of performing distillations in June and July were justified in astrological terms, since the operations had to be done when the sun is in Leo. Moreover, after commenting on the distinction between terrestrial

19 See item number H90 in John Harrison, *The Library of Isaac Newton* (Cambridge: CUP, 1978). In the 1610 edition, Maria's *Practica* is to be found in vol. 1, pp. 205–208. For an up-to-date description of Newton's library, see the *Newton Project* website (http://www.newtonproject.ox.ac.uk/his-library/books-in-newtons-library).

20 *Theatrum chemicum*, vol. 5 (Straßburg, 1622), pp. 1–57: *Allegoriae sapentium supra librum Turbae: XXIX distinctiones*. On these allegories, see Julius Ruska, *Turba philosophorum. Ein Beitrag zur Geschichte der Alchemie* (Berlin: Julius Springer, 1931), pp. 328–333.

21 *Theatrum chemicum*, vol. 5, pp. 94–96.

22 Both the fifth and the sixth volumes certainly belonged to Newton's library: see item H1608 in Harrison, *The Library*.

23 *Theatrum chemicum*, vol. 6 (Straßburg, 1661) pp. 480–486. A German and Latin mixed version of this commentary was already printed in *Epilogus et recapitulatio in Michaelis Sendivogi Poloni Novum Lumen Chymicum, opera et studio Ἀνδρός Ὄρθος Ἥλιος* (Erfurt, 1624), pp. 130–149. This volume appeared as a separated appendix to the 1624 Erfurt edition of Sendivogius' alchemical writings, on which Ortelius wrote a commentary.

(or inferior) and celestial (or superior) sun, Ortelius urged using burning mirrors, which could intensify the sun beams, irradiate the treated substances, and ignite the hot ashes used to heat a proper alchemical apparatus. According to the same method, Ortelius remarked, the alchemist Alexander von Suchten (*c.*1520–1575) prepared his regulus of antimony.[24] Indeed, calcination of antimony by the sun's heat, which was intensified through burning-glasses, became a popular experiment in seventeenth-century chemistry, since the metal noticeably increases its weight after being reduced into ashes.[25] "Solar calcination" is recorded in various textbooks of chemistry, such as Lemery's *Cours de chymie* and Le Fèvre's *Traicté de chymie*, which also includes a picture to visualize the procedure.[26]

Maria's *Practica* and its commentaries did not only circulate in Latin collections. A French translation of the dialogue was printed in the second volume of Nicolas Salomon's *Bibliothèque des philosophes* [*chimiques*], which in all likelihood corresponds to the *Bibliotheca Gallica* quoted by Newton in the final note of Keynes Ms. 45.[27] As we shall see, Salomon also added textual notes to his own translation of Maria's work and discussed Maria's chronology in his introduction (*Préface*) to the second volume of the anthology.[28] However, neither in the introduction nor in the notes to the Latin text he commented on the technical instructions given in the treatise. He only considered the *bain-marie* in the list of alchemical terms that closes the first volume of the *Bibliothèque*.[29]

24 On Suchten's alchemy of antimony, see William R. Newman and Lawrence M. Principe, *Alchemy Tried in the Fire: Starkey, Boyle, and the Fate of Helmontian Chymistry* (Chicago and London: The University of Chicago Press, 2002), pp. 50–56.

25 See Henry Guelarc, *Lavoisier. The Background and Origin of Hist First Experiments on Combustion in 1772* (Ithaca, New York: Cornell University Press, 1961), pp. 114–120.

26 See Nicaise Le Fèvre, *Traicté de la chymie*, vol. 2 (Paris, 1660), pp. 899–902; Nicolas Lemery, *Cours de chymie* (Paris, 1687^{6}), p. 252.

27 According to Harrison's catalogue, Newton only possessed the first volume of this work (see item H221), whose full title reads: *Bibliothèque des philosophes* [*chymiques*], *ou recueil des œuvres des auteurs les plus approuvez qui ont écrit sur la pierre philosophale. Tome premier.* (…) *Par le sieur S.D.E.M.* (Paris, 1672). The French translation of Maria's *Practica*, however, was included in the second volume published in 1678 (pp. 82–87). Only in Jean Maugin de Richebourg's expanded edition of the *Bibliothèque* (1740), the treatise was moved to the first volume. On the French physician Nicolas Salomon, see Jean-Marc Mandosio, "Nicolas Salomon et sa *Bibliothèque des philosophes* [*chimiques*] (1672–1678)", *Chrysopoeia*, 7 (2000–2003): 343–378.

28 See Salomon, *Bibliothèque des philosophes* [*chymiques*], vol. 2 (Paris, 1678), preface (2nd part), section entitled *Dialogue de Marie et d'Aros sur le magistere de Hermès* (page numbers are not indicated in the preface).

29 See the last twenty pages (that are not numbered) in the first volume of the *Bibliothèque des philosophes* [*chymiques*].

FIGURE 2.2 Solar calcination of antimony, Le Fèvre, *Traicté de la chymie* (Paris, 1660)

Here Salomon provided a short explanation of the *bain marin* (lit. "sea bath"), a double boiler in which the outer vessel contained hot water used to heat the inner container. Particularly suitable for distillation (in which case, the inner container would be an alembic), it was also referred to as *le vicaire du ventre du cheval*, since horse's manure could be used to fill the outer vessel. Salomon further specified that this device was also called *Bain Marie*, from the name of Maria the prophetess, Moses' sister. Without questioning this identification – which he explicitly rejected in his *Préface*[30] – Salomon simply added that unlearned apothecaries misread this name and wrote *bain marin* instead.

The term *balneum Mariae* attracted the attention of Newton as well. The expression is recorded in his *Index chemicus*, a rich handwritten vocabulary of alchemical terms, which also functions as a subject-index, since each entry

30 See below, pp. 58–59.

is complemented by a series of references to alchemical primary sources:[31] almost 2.000 references taken from at least 1000 authors have been detected by analyzing the 46 longest entries of the *Index*.[32] Maria's *Practica* must certainly be counted among these sources and at least 5 out of the 11 entries which mention Maria's name comes from the Arabo-Latin dialogue. Newton recorded, for instance, "white and red gum", "mystical vessel that is the measure of the fire", "the herb under the mountains", and he explained the expression "Triptolemus' snakes" as a reference to the vessel of Hermes mentioned at the end of Maria's *Practica*.[33] It comes as no surprise that one of the entries is devoted to the *Balneum Mariae*, which Newton presented as a synonym of *Balneum Solis et Lunae* ("Sun's and Moon's bath"). The following four references are listed for the term (Keynes MS. 30/1, fol. 15*r*): "Balneum Mariae idem, Artef. p. 3; 12. Maier Embl. p 100, 101. Instruct patr. ad Fil. p. 191. Hydropyrogr. p. 1".

Three quoted sources correspond to early modern alchemical treatises, namely Maier's *Symbola aureae mensae* and two anonymous writings entitled respectively *Hydropyrographum hermeticum* (*Hermetic Fire-Water Text*) and *Instructio patris ad filium de arbore solari* (*The Father's Teaching to his Son about the Golden Tree*).[34] The last work, in a chapter devoted to the cooking of mercury and of the seed of gold, lists "the bain-marie, where an egg can be cooked" as an example of very light fire.[35] Along with these treatises, Newton also mentioned the medieval alchemist Artephius, abbreviated as "Artef. p. 3". Despite the variety of works attributed to this enigmatic figure,[36] we can safely identify Newton's reference with Artephius' *De arte occulta et lapide philosophorum liber secretus* (*Secret Book on the Occult Art and the Philosophers' Stone*). This Latin text was first printed in Amsterdam in 1678, while French and

31 The digital edition of this text, from which the following quotations are taken, is available at *The Chymistry of Isaac Newton* website: http://webapp1.dlib.indiana.edu/newton/mss/norm/ALCH00200.

32 Richard S. Westfall, "Isaac Newton's *Index Chemicus*," *Ambix*, 1975, 22: 174–185, p. 178.

33 See Keynes Ms. 30/1, fol. 31*r*: *Dracones Triptolemi (...) suntque illud vas quod Maria dicit non esse ignorantium sed regimen ignis tui.* See also fol. 41*v*: *Gummi album et rubrum*, for which *Maria prophetissa* is mentioned as a source, with no reference to a specific text; fol. 42*v*: *Herba super monticulis, aurum vivum. (...) Maria*; fol. 88*r*: *Vas mysticum, mensura ignis tui, Maria.*

34 On Maier's text, see below. On the other two sources, see Newman, *Newton the Alchemist*, pp. 206–207 and 292–293.

35 The anonymous *Instructio* was printed in *Theatrum Chemicum*, vol. 6, pp. 163–194; see p. 191: *Balneum Mariae, in quo ova coqui possunt.*

36 See Sébastien Moureau, "A Note on Artephius," published as an appendix in "*Elixir atque Fermentum*: New Investigations about the Link between Pseudo-Avicenna's alchemical 'De anima' and Roger Bacon: Alchemical and Medical Doctrines," *Traditio*, 68 (2013): 277–325, pp. 324–325.

English versions already circulated in earlier printed books.[37] Newton transcribed two full copies of this text, both in English (Keynes Ms. 14) and in Latin. The last one is transmitted in a small booklet which has been bound with other bundles of alchemical extracts and now constitutes Ms. Var. 259 (Jerusalem, National Library of Israel).[38] The second booklet, which counts 8 folia, contains Artephius' Latin text (fol. 2.1*r*–2.8*v*), which mentions the *balneum Mariae* as the name of a dyeing water of sulphur.[39] Moreover, the third booklet in the same manuscript (fol. 3.1*r*–3.4*v*) transmits the English translation of *Le livre des figures hiéroglyphiques* by Nicolas Flamel.[40] A section of chapter 2 lists the term *balneum Mariae* among the names given to a triple vessel or "vessel of earth", which Newton tried to depict in a little drawing at the margin of the manuscript (fol. 3.1*v*). This drawing slightly differs from the picture available in the printed edition of the text and seems to better capture some details of how the vessel is described in the text:

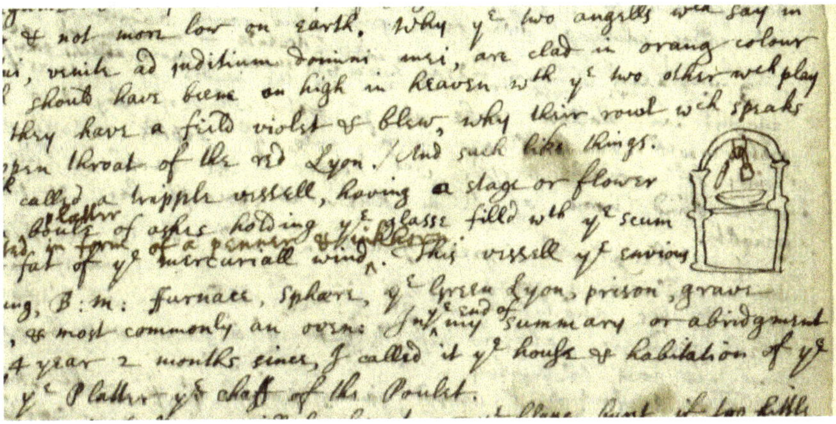

FIGURE 2.3 Triple vessel in Newton's handwritten copy of Flamel's Exposition of the *Hierogyphical Figures* (Ms. Var. 259), fol. 3.1*v*. *The Chymistry of Isaac Newton*: http://webapp1.dlib.indiana.edu/newton/mss/norm/ALCH00200

37 See John Ferguson, *Bibliotheca Chemica: A Catalogue of the Alchemical, Chemical and Pharmaceutical Books in the Collection of the Late James Young of Kelly and Durris* (Glasgow: James MacLehose and Sons, 1906), vol. 1, p. 51.

38 See Isaac Newton, "MS. Var. 259", in *The Chymistry of Isaac Newton.* Ed. 2013. Retrieved October 31, 2020 from: http://purl.dlib.indiana.edu/iudl/newton/ALCH00063.

39 Ms. Var. 256, fol. 2.3*r*.

40 An English translation of the treatise by a certain Eirenaeus Orandus was published in London, 1624: *Nicholas Flamel, His Exposition of the Hierogyphical Figures* (…) *faithfully and religiously done into English out of the French Copy, by Eirenaeus Orandus, quiest, Vera veris enodanus.*

> (Ms. Var. 259, fol. 3.1*v*) This Vessell of earth is called a tripple vessell, having a stage or flower in the midst on which is a boule < or > platter of ashes holding the glasse filld with the scum of the redd sea & the fat of the mercuriall wind which thou seest painted in form of a penner & inkhorn. This vessel the envious call an Athanor, sive, dung, Balneo Mariae Furnace, sphaere, the Green Lyon etc.[41]

The association between "dung" and the *Balneum mariae* appeared in various late medieval and Renaissance alchemical writings. For instance, within the vast *corpus* of alchemical treatises ascribed to Arnaldus de Villa Nova, the *Cathena aurea* (*Golden chain*) seems to equate "horse manure" and the *bain-marie* as sources of moist heat. The treatise is an early fifteenth-century alchemical vision, which re-uses the biblical language of prophecy to explain the central role played by mercury in the making of the philosophers' stone.[42] The author saw an old wise man holding a book with seven seals, each one revealing a secret step in the treatment of mercury. Even though the work already circulated in manuscripts under the title of *Visio Marie* (*Maria's vision*) and was reshaped into a French poem entitled *Le Sept Visions de Marie la Prophetesse*,[43] the only explicit reference to Maria is found in the description of the first seal:

> *vidi in prima parte cathenae aureae, id est primae operationis, sigillum libris argenteis scriptum, id est Mercurium dicentem: "in calido et humido pone me et ex illis pasce me, quoniam debilis sum". Super hoc interrogatus senex, quid haec scriptura significat, respondit istam scripturam clavem et primam operationem lapidis esse, quae dicitur Philosophorum putrefactio, sive digestio totius lapidis philosophi, in primam materiam soluti per suum tempus, puta per 40. dies, in igne calido et humido, in fimo equi, vel in balneo Mariae philosophico.*[44]

41 Digital edition in *The Chymistry of Isaac Newton* website (see above, n. 38). Newton's version is shorter that the description provided in the 1624 printed book of Flamel's *Exposition of the Hierogyphical Figures* (p. 19).

42 Antoine Calvet, *Les œuvres alchimiques attribuées à Arnauld de Villeneuve. Grand œuvre, médecine et prophétie au Moyen-Âge* (Paris – Milan: S.É.H.A – Archè, 2011), pp. 250–252.

43 See Calvet, *Les œuvres alchimiques attribuées à Arnauld de Villeneuve*, p. 250; Didier Kahn, "Les sept visions de Marie la prophétesse sur l'œuvre de la pierre des philosophes," *Chrysopoeia*, 2 (1998): 369–384.

44 Calvet, *Les œuvres alchimiques attribuées à Arnauld de* Villeneuve, p. 550.

> I saw in the first part of the golden chain, that is, of the first operation, the seal written with silver letters, that is, mercury saying: "put me in a moist heat and let me graze in this condition, since I am weak." The old man, being asked on the meaning of this inscription, answered that this inscription is the key and the first treatment of the stone, which is given the name of "philosophers' putrefaction" or "digestion of the all philosophers' stone", dissolved in the first matter for a given time, let's say 40 days, with a hot and humid fire in horse manure or in Maria's philosophical bath.

Another reference to the *balneum Mariae* occurs in the seminal treatise usually known as the *Rosarium philosophorum* (end of the fifteenth century). The work, which counts Arnaldus de Villa Nova's homonymous writing among its sources,[45] was first printed in the second volume of the alchemical anthology known as *De alchimia opuscola complura veterum philosophorum* (Frankfurt, 1550).[46] The *Rosarium* is a doxographical work, where the opinions of various alchemists are reported on different matters. In a section ascribed to the Arabo-Latin author Geber, the *bain-marie* (*balneum Mariae*) is used to coagulate mercury, which is putrefied in "a moist heat, that is horse manure" (*in calido et humido, id est in fimo equino*).[47] The instruction is similar to what we already read in the *Cathena aurea*, even though the *balneum Mariae* is here linked to coagulation. In fact, the author explains that only horse manure produces a fire suitable for putrefaction, whereas the fire of the *bain-marie* (*ignis balnei Marie*) is too weak for this operation.[48] Conversely, this 'weakness', so to speak, was praised in Giovanni Aurelio Augurello's allegorical poem on the making of gold (1515), which mentions the *balneum Mariae* as a source of gentle and moist heat, suitable for treating primary substances, such as sulphur and mercury, and making them germinate like seeds.[49]

45 Calvet, *Arnauld de Villeneuve*, pp. 130–1.

46 The first volume includes, among others, alchemical works by Ps.-Avicenna, Ps.-Lull and Ps.-Aristotle. The first edition of this collective work, published by the tireless printer of alchemical anthologies Johannes Petreius (1497–1550) in Nuremberg, 1541, had a different choice of treatises (Ps.-Geber, Roger Bacon. Khalid ibn Yazid, Hermes).

47 Joachim Telle, *Rosarium philosophorum. Ein alchemisches Florilegium des Spätmittelalters*, 2 vols. (Berlin: Akademie Verlag, 1992), vol. 1, p. 23.

48 Telle, *Rosarium*, vol. 1, p. 26.

49 Augurello, *Chrysopoeia*, book 2, vv. 401–408; critical edition in Matteo Soranzo, *Giovanni Aurelio Augurello (1441–1524) and Renaissance Alchemy* (Leiden: Brill, 2020), pp. 222–223.

Along with the heating method associated with her name, various instructions taken from Maria's *Practica* are mentioned and discussed in fifteenth-century alchemical literature. For instance, the *Rosarium* includes a verbatim quotation on the use of Arabic gum; in a section of the large compendium usually known as *Bosome Book*, George Ripley (*c.*1415–1490) interpreted the dissolution of the "earth of Saturn" in two *Zaibeths* as a reference to the dissolution of *sericon* in mineral and vegetal waters.[50]

3 Maria the Prophetess and the Byzantine Alchemical Tradition

In most passages discussed so far, Maria is given the rank of prophetess (*prophetissa* in Latin) and identified with the sister of Moses. The above-mentioned *Rosarium philosophorum*, for instance, introduces Maria as an *auctoritas* in the list of alchemists (*artifices*) that opens the treatise: *Adam, Noe, Idrid* (…) *Moyses et soror eius Maria* ("Moses and his sister Maria") etc.[51] Maria appears after Moses and is presented as his sister, in accordance with the information provided by Maria's *Practica*. As already seen, the influence of this Arabo-Latin text certainly extended into the sixteenth century. The name of Maria is quoted in a long list of alchemical authors compiled by the Italian alchemist Giovanni Battista Nazari in his *Il metamorfosi metallico et humano* (*Metallic and Human Metamorphosis*), a series of alchemical dreams printed in Brescia in 1564. In the second dream, Nazari saw a complex building with many niches, each containing a statue to which a book was tied with a silver chain. Nazari recorded all the titles he could read – mainly titles of medieval alchemical writings – among which one finds: *Artus rex, Conversatio philosophorum, Phoenix ad Martium regem Arragonensem, Mariae prophetissae,* etc.[52]

50 Jennifer M. Rampling, "The Catalogue of the Ripley Corpus: Alchemical Writings Attributed to George Ripley," *Ambix*, 57 (2010): 125–201, pp. 141–144. On the metallic substance called *sericon* and its role in the production of the vegetable stone, see Rampling, "Transmuting Sericon". Another refence to Maria's text can also be found in the *Accurtations of Raymond*: see Jennifer M. Rampling, "Establishing the Canon: George Ripley and his Alchemical Sources," *Ambix*, 55 (2008): 189–208, pp. 201–204.

51 Telle, *Rosarium*, vol. 1, p. 2.

52 *Il metamorfosi metallico e humano di Gio. Bat. Naz. Bre.* (Brescia, 1564), p. 26, 2nd column. Didier Kahn has recently discovered a similar list, rearranged in alphabetical order, in Ms. Ital. 929 of the Bibliothèque Nationale de France (sixteenth century), as part of the anonymous treatise *Trilogio della trasmutatione de' metalli, tra il Filosofo, il Theorico et il Pratico* (*c.*1564). At fol. 162v (column a), under the letter M, the following titles are recorded: *Morieni tractatus, Miscendes philosophus, Merlini allegoria, Marie prophetisse, Magnesie liber*, etc. See Didier Kahn, "Aristotélisme et alchimie dans l'anonyme *Trilogio*

In 1572, an expanded edition of Nazari's treatise was published in Brescia under the title *Della trasmutatione metallica sogni tre* (*Three Dreams on the Metallic Transmutation*). In this version, Nazari re-arranged alphabetically the list of alchemical titles that he saw in his second dream.[53] In this new list, now part of Nazari's third dream, the name of *Maria prophetissa* is no longer readable under the letter M; one finds instead *Mariae Iudeae epistola, vel dialogus et pratica* (*Maria the Jewess' Letter, or Dialogue and Practice*), which is followed by other titles, such as *Michaelis Pselli de conficiendo auro* (*Michael Psellos, On the Making of Gold*).[54] The mention of the Byzantine scholar Michael Psellos (eleventh century), who wrote a short alchemical treatise on the making of gold, might point to a certain familiarity of Nazari with the manuscript collections of Graeco-Egyptian and Byzantine writings.

The new title *Mariae Iudea epistola* makes explicit reference to the 'Jewishness' of Maria, which was already pointed out by Hermolao Barbaro (1454–1493), one of the first Renaissance scholars who certainly had a first-hand knowledge of the Greek alchemical literature only available in Byzantine manuscripts.[55] In his commentary on Dioscorides' *De materia medica* (published posthumous in 1516), Barbaro referred to many Graeco-Egyptian and early-Byzantine authors of alchemical treatises (Ostanes, Zosimus, Democritus, etc.) and mentioned three women who distinguished themselves as experts in the field:[56]

> *Iam feminae tres, Cleopatra, Virgo Taphnutia, quae minus probatur, et Iudeae gentis Maria, quam plurimi secuntur.*
>
> There are also three women, Cleopatra, the virgin Taphnutia, who is less esteemed, and Maria the Jewess, who had many followers.

---

della trasmutatione de' metalli", in *Alchimie et philosophie à la Renaissance*, edited by Jean-Claude Margolin and Sylvain Matton (Paris: Vrin, 1993), pp. 223–252 (249).

53 See, for instance, Frank Greiner, "L'initiation alchimique de Giovan Battista Nazari," *Bulletin de l'Association d'étude sur l'humanisme, la réforme et la renaissance*, 41 (1995): 9–35.

54 *Della tramutazione metallica sogni tre, di Gio. Battista Nazari Bresciano* (Brescia, 1599), p. 141.

55 On the reception of the Greek alchemical corpus in the Renaissance, see Sylvain Matton, "L'influence de l'humanisme sur la tradition alchimique," *Micrologus*, 3 (1995): 279–345, in particular, pp. 309–341.

56 *In Dioscoridem corollarii libri quinque* (Venice, 1516), Book V, DCCCCLIX, *s.v.* sulphur, fol. 100v. See Matton, "L'influence de l'humanisme", p. 323.

This threefold list was then reproduced by Agricola in the preface to his *De re metallica* and by Robertus Vallensis in his booklet *De pulveris* (...) *materia et compositione* (*On the Matter ... and Composition of the Powder*).[57] This last work was printed after Vallensis' *De antiquitate et veritate artis chemicae* (*On the Antiquity and Truth of the Alchemical art*), which was first published in Paris in 1561: here, Maria is listed among the Arabic authors who wrote on alchemy and no information seems to have been extracted from the three Byzantine alchemical manuscripts kept in Paris, which Vallensis did mention as witnesses of Zosimus' and Blemmides' alchemical works.[58]

As we have already seen, in his writings Zosimus of Panopolis counted Maria among the Jews who wrote on alchemical instruments.[59] The woman features in other Greek treatises as well. For instance, in his *Ninth Lecture on the Making of Gold*, the Byzantine alchemist Stephanus of Alexandria (seventh century CE) mentions Maria along with "the Jewish writings" (Ἰουδαικαὶ γραφαί) on alchemy.[60] The whole *corpus* of the Greek alchemical treatises transmitted in Byzantine manuscripts was never printed in the Renaissance, either in its original language or in Latin translation. The Calabrian scholar Domenico Pizzimenti published a partial Latin translation in Padua (1573), which only contained the alchemical writings by Pseudo-Democritus, Sinesius, Pelagius, and Stephanus of Alexandria along with Psellus' *Letter on the Making of Gold*, which was listed by Nazari among the alchemical titles of his third dream.[61] Even though important sources on Maria, such as Zosimus' writings, were not translated by Pizzimenti, his translation did include the above-mentioned *Ninth Lecture* of Stephanus, where Maria is quoted along with the Jewish writings (*Ioudaicas scripturas* in Pizzimenti's rendering).[62]

57 Georg Agricola, *De re metallica* (Basil, 1556), fol. a3r; *De veritate et antiquitate artis chemicae et pulveris sive medicinae philosophorum vel auri potabilis materia et compositione* (...) *testimonia et theoremata, ex variis auctoribus* (...) *per Robertum Vallensem selecta* (Leiden: apud F. Raphelengium, 1563), p. 41. See Matton, "L'influence de l'humanisme", pp. 327 and 336–337.

58 *De veritate et antiquitate artis chemicae* (Leiden, 1563), p. 27. See also Barbara Obrist, "Views on History in Medieval Alchemical Writings," *Ambix*, 56 (2009): 226–238, pp. 227–228.

59 See above, p. 40.

60 See Maria K. Papathanassiou, *Stephanos von Alexandria und sein alchemistisches Werk* (Athens: Cosmosware, 2017), p. 216.

61 *Democritus abderita de arte magna sive de rebus naturalibus. Nec non Synesii, & Pelagii, & Stephani Alexandrini, & Michaelis Pselli in eundem commentaria. Domenico Pizimentio Vibonensi interprete* (Padua, 1573). We must note that Pizzimenti already published his Latin translation of Psellus' *Letter* in 1572.

62 *Democritus abderita de arte magna*, p. 59.

Greek alchemical literature, either in translation or through manuscripts, provided new pieces of information, which could complement and, sometimes, rectify what had been already extracted from the Arabo-Latin tradition. In his philological enquiry into the *Corpus Hermeticum*, Isaac Casaubon (1559–1614) also mentioned Hermes' and Maria's chemical works (*Hermetis et Mariae Chymica*) among the writings that were believed to disclose the earliest Egyptian wisdom, which even predated Moses' teaching.[63] He saw these writings in a Byzantine alchemical manuscript in Paris, where he also read Maria's name: *ibidem citantur Μαριὰμ καὶ Ἰουδαικαὶ γραφαὶ, Mariam et scripta Iudaica*, "in the same (manuscript) one finds the quotation *Mariàm kaì Ioudaikaì graphaì*, i.e. Mariam and the Jewish writings."[64] The quotation is taken from the above-mentioned passage of Stephanos of Alexandria, who is not explicitly named by Causabon. Being quite skeptical about this mention, he only added a terse linguistic remark: Mariam is either Moses' sister or Jesus' mother, who is the only woman bearing this name in the New Testament; all the other women (with a similar name) are called Maria.[65] Then, Casaubon focused his investigation on the *Corpus Hermeticum*, i.e., the most important collection of Hermetic writings at his time. Indeed, his exposure of this collection as a late forgery could be easily extended to the galaxy of pseudo-ancient texts that orbited around it.[66]

A similar conclusion on the dating of Greek alchemical texts was reached by Joseph Justus Scaliger (1540–1609) in his philological work on Manilius' *Astronomicon*: Greeks and Romans came to know alchemy in a late period, since the earliest reference to this art only occurs in Firmicus Maternus' astrological work.[67] While no mention of Maria occurs in this discussion, she is referred to in Scaliger's major work *Thesaurus Temporum*, where he tried to reconstruct the original Greek text of Eusebius' *Chronology*, only known in

63 *Isaaci Casauboni De rebus sacris et ecclesiasticis exercitationes XVI* (London, 1614), p. 52. See Matton, "L'influence de l'humanisme", p. 327. On Casaubon's investigation, see Anthony Grafton, "Protestant versus Prophet: Isaac Casaubon on Hermes Trismegistus," *Journal of the Warburg and Courtauld Institues*, 46 (1983): 78–93.

64 *Isaaci Casauboni De rebus sacris*, p. 56.

65 *Isaaci Casauboni De rebus sacris*, p. 56: *Mariam autem est vel soror Mosis, vel gloriosissima Virgo Deipara. Nam in Novo Testamento sola sic dicitur; quaevis reliquarum Μαρία.*

66 See Claudio Moreschini, *Hermes Christianus. The Intermingling of Hermetic Piety and Christian Thought* (Turnhout: Brepols, 2011), p. 278.

67 See Scaliger's commentary on Manilius' book 4.248 in *M. Manili Astronomicon libri quique, Iosephus Scaliger Iul. Caes. F. Recensuit, ac pristino ordine suo restituit. Eiusdem Ios. Scaligeri Commentarius in eosdem libros et Castigationum explicationes* (Paris, 1579), section *Commentarius et castigationes in lib. IIII. Manili*, pp. 218–219.

Latin translation. Among the many quotations taken from Syncellus, Scaliger also included the passage where "Maria, a Hebrew woman of learning" (Μαρία τις Ἑβραία σοφή) and Democritus appear as students of the magus Ostanes in the temple of Memphis.[68] However, in his brief exegetical notes, Scaliger tentatively attributed this passage to the late-antique historian Panodorus of Alexandria, in lines with what he already argued about alchemy's chronology. As for Maria, he openly admitted to ignore any other work on this woman.[69]

Scaliger's and Casaubon's studies were used by seventeenth-century physicians, alchemists and philologists to support different, in some cases even contrasting, views. Pierre-Daniel Huet (1630–1721) criticized Scaliger's position and used Syncellus' passages to prove the antiquity of alchemy.[70] Pietro Servio, who taught medicine in Rome (where he died in 1648), paradoxically mentioned Casaubon' writings to prove that alchemy predated Hippocrates and Galen, who would have even demonstrated some knowledge of this art in their writings. In his exegesis of Estêvão Rodrigues de Castro's essay devoted to the medical uses of whey, Servio maintained that alchemy was already practiced by Moses and uncritically reported how Casaubon saw a work by Maria, Moses' sister, at the *Regia Bibliotheca* in Paris.[71] Conversely, this information was questioned in Jean Bernier's *Essais de medecine* (1689). After visiting the library to check the actual content of the manuscript, he reported to have only found a short excerpt on the philosopher's stone attributed to the wisest Maria.[72] Since her name occurred along with many alchemists who worked

---

68 Scaliger's *Thesaurus temporum* (Leiden, 1601) includes various works with separated foliation; see *Τῶν χρονικῶν κανόνων παντοδαπῆς ἱστορίας Εὐσηβίου τοῦ Παμφίλου τὰ σωζόμενα*, p. 43. The Greek passage of Syncellus' *Chronography* is edited in Alden A. Mosshammer, *Georgii Syncelli Ecloga chronographica* (Leipzig: Teubner, 1984), p. 297; English translation in Willian Adler and Paul Tuffin, *The Chronography of George Synkellos. A Byzantine Chronicle of Universal History from the Creation* (Oxford: OUP, 2002), p. 298.

69 Scaliger, *Thesaurus temporum*; section *Iosephi Scaligeri Iulii Caesaris filii, in Graeca παντοδαπῆς ἱστορίας Eusebii, quae supersunt, notae*, p. 258.

70 See Sylvain Matton, "Pierre-Daniel Huet et l'alchimie," *Chrysopoeia*, 7 (2000–2003): 379–394.

71 See Persius Trevius (i.e. Servio's pseudonym), *Ad librum de sero lactis Stephani Rooderici Castrensis Lusitani declamationes* (Rome, 1634), chap. XVII (*Hermeticae artis antiquitas: An Hippocrati et Galeno cognita, quomodo cognita*), pp. 82–83. The treatise was written in response to *D. Stefani Roderici Castrensis* (…) *de sero lactis* (Florence, 1631), where De Castro denied that distillation was known to Galen and Hippocrates, being it only practiced by Arabic alchemists and physicians: see chap. XI (*Destillatio an antiquis fuerit cognita*), pp. 66–70.

72 See Jean Bernier, *Essais de medecine* (Paris: Simon Langronne, 1689), pp. 223–224 (text quoted in Matton, "L'influence de l'humanisme", p. 327). Bernier might have seen the

in the first centuries CE, Maria cannot be identified with Moses's sister and must have lived after Cleopatra, who is also mentioned in the Paris alchemical collection.

A similar interpretation of Casaubon's reference to Maria was proposed by the alchemist Daniel Sennert (1572–1637), who also dismissed the alchemical expertise often attributed to Moses:[73] he tentatively referred the mention of Mariam and the Jewish writings in the Paris manuscript to the treatise *Practica Mariae prophetissae sororis Mosis*, which Sennert knew as transmitted along with the *Turba philosophorum*. It therefore belonged to the Arabo-Latin tradition and was to be counted among the late works attributed to ancient figures.[74] Likewise, in his *De Hermetica medicina* (*On Hermetic medicine*), Hermann Corning (1606–1681) relied on Causabon' condemnation of the *Hermetica* as a forgery to debunk the alleged antiquity of alchemy and iatrochemistry.[75] Accordingly, he maintained that a certain Maria the Jewess (*Maria quaedam Iudaica*) mentioned by Greek sources must date to a late period and substantiated his conclusion by quoting the passage from Syncellus' *Chronography* edited by Scaliger.[76] Even Olaus Borrichius – who defended the antiquity of Hermes and his alchemical teaching against Conring's attack[77] – hesitated to acknowledge the authenticity of Maria's excerpt transmitted in the Paris manuscript. In his work *On the Origin and Development of Chemistry*, he listed various women who knew alchemy, such as the Egyptian goddess Isis, Cleopatra in Egypt, and "a certain Maria in Palestine"; but he immediately specified in

short collection of alchemical sayings Περὶ τοῦ λίθου τῆς φιλοσοφίας, *On the Philosopher's Stone*, in MS *Parisinus* gr.2327 (foll. 136*v*–138*v*), where a marginal note (fol. 136*v*) adds the title Λόγος τῆς σοφοτάτης (*sic*) Μαρίας περὶ τοῦ λίθου τῆς φιλοσοφίας, *Discourse by the Wisest Maria On the Philosopher's Stone*. Greek text edited in Berthelot-Ruelle, *Collection des anciens alchimistes grecs*, vol. 2, pp. 198–201.

73 On this point, see below, p. 58.

74 Daniel Sennert, *De chymicorum cum Aristotelicis et Galenicis consensu ac dissensu, Liber I* (Wittenberg, 1619) pp. 47–48.

75 See Grafton, "Protestant versus Prophet", pp. 88–89; Antonio Clericuzio, "'The White Beard of Chemistry'. Alchemy, Paracelsianism and the *Prisca Sapientia*," *Bruniana & Campanelliana*, 14 (2008): 107–116.

76 *Hermanni Conringii De Hermetica medicina libri duo* (Helmstadt: H. Muller, 1669), pp. 29–31.

77 See Robert Halleux, "La controverse sur les origines de la chimie, de Paracelse à Borrichius," in Jean-Claude Margolin, ed., *Acta Conventus Neo-Latini Turonensis*, (Paris: Vrin, 1980), vol. 2, pp. 807–819.

brackets: "I was shown a work by her (i.e. Maria) in Paris, even though it might be spurious".[78]

Women's contribution to alchemy is also stressed by Nicolas Salomon in the above-mentioned introduction to his *Bibliothèque des philosophes* [*chimiques*]. In his opinion, the figure of Maria exemplifies how women too mastered the alchemical art, even though Salomon seems to ignore the names of other female alchemists, such as Cleopatra and Taphnoutia. Indeed, he considered Maria as the only woman who stood out among ancient 'philosophers', i.e., alchemists. For this reason – he continues – the uniqueness of the text *Practica Mariae prophetissae* testifies to the difficulties that women could encounter when practicing alchemy. They certainly have all the intellectual capacities to understand the art, but they would lack the patience necessary to carry out its extremely long and time-consuming procedures.

Then, Salomon acknowledged the antiquity of Maria's treatise. However, he questioned her identification with Moses' sister as well as the alchemical expertise that many attributed to the patriarch. Indeed, Salomon disagreed with those who interpreted the biblical episode of the golden calf (*Exodus* 32:20) as a demonstration of Moses' alchemical mastery.[79] By referring to Daniel Sennert's interpretation of the biblical passage,[80] Salomon argued that Moses simply melted the golden calf and neither calcined nor reduced it into a kind of potable gold by means of an alchemical acid – a substance that was only discovered around the year one thousand, Salomon pointed out. Maria's work, he concluded, represented the Latin translation of an Arabic alchemical text, whose author might be earlier than the monk Morienus, who was believed to have taught alchemy to the Omayyad prince Ḫālid ibn Yazīd. Maria might be identified, Salomon tentatively proposed, with a certain Madera mentioned in a fifteenth-century alchemical book attributed to Bernard Trevisan, as we shall see at the end of this paper.[81]

---

78 Olaus Borrichius, *De ortu et progressu chemiae dissertatio* (Copenhagen, 1668), p. 51: … *in Palaestina Maria quaedam cuius scriptum Parisiis mihi ostensum, quamquam forsan supposititium.*

79 See Patai, *The Jewish Alchemists*, pp. 30–40; Didier Kahn, "Généalogie de l'alchimie et interprétation alchimique de la Bible au XVIe siècle: *Qui fuerint primi inventores hujus artis*," *Archives d'histoire doctrinale et littéraire du Moyen Âge*, 84 (2017): 313–347.

80 See Sennert, *De chymicorum cum Aristotelicis et Galenicis consensu ac dissensu*, pp. 46–47, where Sennert refers to the interpretation of the golden calf proposed by the Franciscan biblical exegetes Nicholas of Lyra (1270–1349). This section belongs to chap. 3 (*De inventoribus et cultoribus chymiae*), where Sennert discusses figures and episodes of ancient alchemy, such as Adam, Hermes, Democritus, and the golden fleece.

81 On the fictional figure of Bernard Trevisan and the treatise attributed to him, see Didier Kahn, "Recherches sur le livre attribué au prétendu Bernard le Trévisan (fin du XVe

A certain Madera or Medera had been already mentioned by the German alchemist and physician Michael Maier, who held a more nuanced position on the questions of Maria's chronology and the authenticity of her work. He devoted an entire chapter of his *Symbola aureae mensae* (1617) to the figure of Maria, who could not be identified with a pure fictional character.[82] He argued that the *Dialogue of Maria and Aros* (i.e., the *Practica Mariae Prophetissae*) had been originally written in Hebrew, even though only the Latin translation was known. Most part of the chapter discusses the question whether Maria was the sister of Moses. Maier did not exclude this hypothesis, since a divine inspiration was needed to accomplish her alchemical work. In her quest for the philosophers' stone, she might have relied on Moses' alchemical expertise, which Maier never questioned, but rather confirmed on the basis of various Biblical passages. However, he admitted that there was no compelling evidence to prove the identification of Maria the alchemist with Moses' sister, so that Maria might have been a different Jewess woman. In any case, the treatise *Practica Mariae Prophetissae*, which Maier briefly summarized, testifies to her great alchemical mastery and shows how women could excel in this art. Maria, indeed, was not alone, since she must be counted – Maier concluded – among other women who succeeded in this art:[83]

> *Hermolaus Barbarus Scriptor magnae authoritatis, quamvis non Chymicus, Corollar. In Dioscorid. 1.5.c.115, narrat tres foeminas lapidem philosophicum composuisse, Cleopatram, Taphnoutiam et Iudeae gentis Mariam, cuius liber passim circumfertur. His et quarta annumeranda est, Medera, cuius meminit Bernhardus Comes, cum dicit, quod Aros docuit eam arte Nephandim fratrem suum, et Saturnus Luncabur, atque sororem eius Mederam.*

> Hermolaus Barbarus, a very authoritative writer, even without being an alchemist, in his *Corollarii in Dioscorides* (1.5.c.115), says that three women have produced the philosophers' stone, namely Cleopatra, Taphnoutia and Maria, whose book is widely known. A fourth woman must be added to these, Medera, who is mentioned by Bernard *comes*, when he says:

siècle)," in Chiara Crisciani, Agostino Paravicini Bagliani, ed., *Alchimia e medicina nel Medioevo* (Firenze: SISMEL – Edizioni del Galluzzo, 2003), pp. 265–336.

82 Michael Maer, *Symbola aureae mensae duodecim nationum (...) auctore Michaele Maiero Comite* (Frankfurt a. M: L. Jennis, 1617), p. 64.

83 *Symbola aureae mensae duodecim nationum (...) auctore Michaele Maiero Comite* (Frankfurt a. M, 1617), pp. 56–64. See Patai, *The Jewish Alchemists*, pp. 76–78.

Aros taught this art to his brother Nephandis, and Saturnus Luncabur (?), and to his sister Medera.

4 Concluding Remarks

Maier's list features names that had been extracted by Hermolaus Barbarus from Byzantine alchemical manuscripts. This list, however, is enlarged with a new name taken from a fifteenth-century French alchemical book attributed to Bernard Trevisan, which Maier quoted in the Latin translation published by Gerard Dorn in 1583.[84]

Medera represents a very enigmatic figure, who adds a further question mark in an already stratified and complex picture.[85] She seems to be linked to the philosopher Aros, to whom Maria's *Practica* is also addressed. As already seen, Nicolas Salomon tentatively proposed to identify the two female alchemists. Maier's pupil Daniel Stolz von Stolzenberg (1600–1660), in his *Hortolus hermeticus* (1627), devoted two separated emblems and poems to *Medera, foemina alchimystica* ("Medera, a female alchemist") and to *Maria Hebraea Moysis Soror* ("Maria the Jewess, Moses' sister"). Interestingly, the first verses of Medera' poem mention "the vessels of Hermes" (*Hermetis vascula*), which are also referred to in the saying that encircles Medera's emblem: *qui nescit regimen Veritatis, ignorat vas Hermetis*, "those who disregard the regimen of truth do not know the vessel of Hermes".[86] This sentence clearly rephrases a quote from Maria's *Practica*[87] and seems to tighten the relationship between the two women.

---

84 *Trevisanus De chimico miracuolo, quod lapidem philosophiae appellant* (…) *per Gerardum Dorneum* (Basel, 1593), p. 5. This translation has been also reprinted in the *Theatrum Chemicum*, vol. 1 ((Straßburg, 1602), pp. 773–803. An earlier Latin translation was published by Guglielmo Gratarolo in 1567 (*Comitis antiqui Trevirensis philosophi expertissimi, περὶ χημείας opus historicum et dogmaticum*), but this version does not mention the name of Medera (p. 10). According to the recent edition by Didier Kahn, in the French original text of Trevisan's book we read the name 'Madera': see Kahn, "Recherches sur le livre attribué au prétendu Bernard le Trévisan," p. 329.

85 As recently pointed out by Kahn, one of the sources of Trevisan's alchemical book was a gloss to the fourteenth-century treatise *Textus alchimie*. This note (*Qui fuerint primi inventores hujus artis*), however, does not report the name of Medera/Madera. See Kahn, "Généalogie de l'alchimie," p. 346.

86 *Hortulus hermeticus flosculis philosophorum cupro incises confomatus et brevissimis versiculis explicatus* (…) *Autore Daniele Stolcio de Stolcenberg* (Frankfurt, 1637), pp. 8–13 The poem reads (p. 8): *Medera foemina alchymistica. Quisquis Alithiae non novit Regna beata, nescit is Hermetis vascula mira Senis.*

87 See above, p. 44.

FIGURE 2.4 Daniel Stolz von Stolzenberg, *Hortolus Hermeticus* (Frankfurt, 1637), p. 9: Emblems dedicated to Hermes, Adfar, Cleopatra and Medera

Maier's, Solomon's, and Stolz's remarks exemplify how early modern alchemists combined information taken from Greek, Arabo-Latin, and more recent sources, in various attempts to reconstruct the origins of their discipline. When considered through the twists and turns of this expansive textual tradition, Maria featured as a key-figure in different early modern 'histories' of alchemy and, at the same time, her *Practica* kept many practitioners busy at least until

Newton, who still continued to copy and interpret the text. If Maria's name has remained firmly attached to the heating method that is still called *bain-marie*, her identity and chronology were often reshaped across the centuries. As Moses' sister, she was often evoked in various discussions on the alchemical exegesis of the Bible and the role that patriarchs played in the discovery and development of the art. At the same time, philologists scrutinized the scattered excerpts and information included in Byzantine sources, while physicians, iatro-chemists and alchemists often relied on this work to argue either against or in favor of the remote origins of alchemy. This historical and textual inquiries allowed early modern scholars to redesign the chronological and geographical boundaries of the alchemical art and to root their own theories and practices into this reconstructed past, namely that past which produced the 'ancient' alchemical literature that they never stopped to reinterpret and recontextualize.

Acknowledgements

This paper is published in the framework of the ERC project *AlchemEast*, funded by the European Research Council (ERC) under the European Union's Horizon 2020 research programme (G.A. 724914) and of the related FARE project *The Western AlchemEast*.

References

Adler, William, and Paul Tuffin, *The Chronography of George Synkellos. A Byzantine Chronicle of Universal History from the Creation* (Oxford: Oxford University Press, 2002).

Agricola, Georg, *De re metallica* (Basil, 1556).

Allen, Meagan S., "Revisiting Isaac Newton's *Index Chemicus*: A Response to Richard S. Westfall," *Ambix*, 66 (2019): 72–81.

Bernier, Jean, *Essais de medecine* (Paris: Simon Langronne, 1689).

Berthelot, Marcelin, Charles-Émile Ruelle, *Collection des anciens alchimistes grecs*, 3 vols. (Paris: G. Steinheil, 1887–1888).

Borrichius, Olaus, *De ortu et progressu chemiae dissertatio* (Copenhagen, 1668).

Calvet, Antoine, *Les œuvres alchimiques attribuées à Arnauld de Villeneuve. Grand œuvre, médecine et prophétie au Moyen-Âge* (Paris – Milan: S.É.H.A – Archè, 2011).

Casaubon, Isaac, *Isaaci Casauboni De rebus sacris et ecclesiasticis exercitationes XVI* (London, 1614).

Clericuzio, Antonio, "'*The White Beard of Chemistry*'*. Alchemy, Paracelsianism and the Prisca Sapientia*," *Bruniana & Campanelliana*, 14 (2008): 107–116.

Corning, Hermann, *Hermanni Conringii De Hermetica medicina libri duo* (Helmstadt: H. Muller, 1669).

De Castro, Stefano, *D. Stefani Roderici Castrensis* (…) *de sero lactis* (Florence, 1631).

Ferguson, John, *Bibliotheca Chemica: A Catalogue of the Alchemical, Chemical and Pharmaceutical Books in the Collection of the Late James Young of Kelly and Durris* (Glasgow: James MacLehose and Sons, 1906).

Festugière, Andrè, *La révélation d'Hermès Trismégiste, vol. 1: L'astrologie et les sciences occultes* (Paris: Les Belles Lettres, 1944).

Festugière, André, *La révélation d'Hermès Trismégiste*, vol. 1. *L'astrologie et les sciences occultes* (Paris: Les Belles Lettres, 1950).

Grafton, Anthony, "Protestant versus Prophet: Isaac Casaubon on Hermes Trismegistus," *Journal of the Warburg and Courtauld Institues*, 46 (1983): 78–93.

Greiner, Frank, "L'initiation alchimique de Giovan Battista Nazari," *Bulletin de l'Association d'étude sur l'humanisme, la réforme et la renaissance*, 41 (1995): 9–35.

Guelarc, Henry, *Lavoisier. The Background and Origin of Hist First Experiments on Combustion in 1772* (Ithaca, New York: Cornell University Press, 1961).

Halleux, Robert, "La controverse sur les origines de la chimie, de Paracelse à Borrichius," in *Acta Conventus Neo-Latini Turonensis*, edited by Jean-Claude Margolin (Paris: Vrin, 1980).

Harrison, John, *The Library of Isaac Newton* (Cambridge: Cambridge University Press, 1978).

Hortulus hermeticus flosculis philosophorum cupro incises confomatus et brevissimis versiculis explicatus (…) *Autore Daniele Stolcio de Stolcenberg* (Frankfurt, 1637).

Kahn, Didier, "Aristotélisme et alchimie dans l'anonyme *Trilogio della trasmutatione de' metulli*", in Jean-Claude Margolin and Sylvain Matton, ed., *Alchimie et philosophie à la Renaissance*, (Paris: Vrin, 1993), 223–252.

Kahn, Didier, "Généalogie de l'alchimie et interprétation alchimique de la Bible au XVIe siècle: *Qui fuerint primi inventores hujus artis*," *Archives d'histoire doctrinale et littéraire du Moyen Âge*, 84 (2017): 313–347.

Kahn, Didier, "Les sept visions de Marie la prophétesse sur l'œuvre de la pierre des philosophes," *Chrysopoeia*, 2 (1998): 369–384.

Kahn, Didier, "Recherches sur le livre attribué au prétendu Bernard le Trévisan (fin du XVe siècle)," in Chiara Crisciani and Agostino Paravicini Bagliani, ed., *Alchimia e medicina nel Medioevo* (Firenze: SISMEL – Edizioni del Galluzzo, 2003), pp. 265–336.

Käs, Fabrian, *Die Mineralien in der arabischen Pharmakognosie*, 2 vols. (Wiesbaden, Harrassowitz, 2010).

Lacaze, Grégoire, *Turba Philosophorum. Congrès pythagoricien sur l'art d'Hermès* (Leiden: Brill, 2018).

Le Fèvre, Nicaise, *Traicté de la chymie*, vol. 2 (Paris, 1660).

Lemery, Nicolas, *Cours de chymie* (Paris, 1687).

Maer, Michael, *Symbola aureae mensae duodecim nationum* (…) *auctore Michaele Maiero Comite* (Frankfurt a. M: L. Jennis, 1617).

Mandosio, Jean-Marc, "Nicolas Salomon et sa *Bibliothèque des philosophes* [*chimiques*] (1672–1678)", *Chrysopoeia*, 7 (2000–2003): 343–378.

Matton, Sylvain, "Alchimie et Stoicisme: à propos de récentes recherches," *Chrysopoeia*, 5 (1992–1996): 5–144.

Matton, Sylvain, "L'influence de l'humanisme sur la tradition alchimique," *Micrologus*, 3 (1995): 279–345.

Matton, Sylvain, "Pierre-Daniel Huet et l'alchimie," *Chrysopoeia*, 7 (2000–2003): 379–394.

Matton, Sylvain, *Responsiones Aros philosophi ad Nephes regem de philosophia malis et improbis occulta et sapientibus manifesta. Avec un fac-similé du manuscrit arabe Chester Beatty 4121* (Paris: SÉHA, 2017).

Mertens, Michèle, *Zosime de Panopolis, Mémoires authentiques* (Paris: Les Belles Lettres, 1995).

Moreschini, Claudio, *Hermes Christianus. The Intermingling of Hermetic Piety and Christian Thought* (Turnhout: Brepols, 2011).

Mosshammer, Alden A., *Georgii Syncelli Ecloga chronographica* (Leipzig: Teubner, 1984).

Moureau, Sébastien, "*Elixir atque Fermentum*: New Investigations about the Link between Pseudo-Avicenna's alchemical 'De anima' and Roger Bacon: Alchemical and Medical Doctrines," *Traditio*, 68 (2013): 277–325.

Nazari, Giovanni Battista, *Della tramutazione metallica sogni tre, di Gio. Battista Nazari Bresciano* (Brescia, 1599).

Newman, William R., and Lawrence M. Principe, *Alchemy Tried in the Fire: Starkey, Boyle, and the Fate of Helmontian Chymistry* (Chicago and London: The University of Chicago Press, 2002).

Newman, William R., *Newton the Alchemist. Science, Enigma, and the Quest for Nature's "Secret Fire"* (Princeton and Oxford: Princeton University Press, 2019).

Newton Project (http://www.newtonproject.ox.ac.uk/his-library/books-in-newtons-library).

Obrist, Barbara, "Views on History in Medieval Alchemical Writings," *Ambix*, 56 (2009): 226–238.

Papathanassiou, Maria K., *Stephanos von Alexandria und sein alchemistisches Werk* (Athens: Cosmosware, 2017).

Patai, Raphael, *The Jewish Alchemists. A History and Source Book* (Princeton: Princeton University Press, 1994).

Pizimenti, Domenico, *Democritus abderita de arte magna sive de rebus naturalibus. Nec non Synesii, & Pelagii, & Stephani Alexandrini, & Michaelis Pselli in eundem commentaria. Domenico Pizimentio Vibonensi interprete* (Padua, 1573).

Rampling, Jennifer M., "Establishing the Canon: George Ripley and his Alchemical Sources," *Ambix*, 55 (2008): 189–208.

Rampling, Jennifer M., "The Catalogue of the Ripley Corpus: Alchemical Writings Attributed to George Ripley," *Ambix*, 57 (2010): 125–201.

Rampling, Jennifer M., "Transmuting Sericon: Alchemy as 'Practical Exegesis' in Early Modern England," *Osiris*, 29 (2014): 19–34.

Ruska, Julius, *Turba philosophorum. Ein Beitrag zur Geschichte der Alchemie* (Berlin: Julius Springer, 1931).

Salomon, Nicolas, *Bibliothèque des philosophes* [*chymiques*] (Paris, 1678).

Scaliger, Julius, *Thesaurus temporum* (Leiden, 1601).

Sendivogius, Michael, *Epilogus et recapitulatio in Michaelis Sendivogi Poloni Novum Lumen Chymicum, opera et studio Ἀνδρός Ὄρθος Ἥλιος* (Erfurt, 1624).

Sennert, Daniel, *De chymicorum cum Aristotelicis et Galenicis consensu ac dissensu, Liber I* (Wittenberg, 1619).

Sezgin, Fuat, *Geschichte des arabischen Schrifttums, Band IV. Alchimie – Chemie – Botanik – Agrikultur bis ca. 430 H.* (Leiden: Brill, 1971).

Soranzo, Matteo, *Giovanni Aurelio Augurello (1441–1524) and Renaissance Alchemy* (Leiden: Brill, 2020).

Telle, Joachim, *Rosarium philosophorum. Ein alchemisches Florilegium des Spätmittelalters*, 2 vol. (Berlin: Akademie Verlag, 1992).

The Chymistry of Isaac Newton, http://webapp1.dlib.indiana.edu/newton/mss/norm/ALCH00034.

Trevisan, Bernard, *Trevisanus De chimico miracuolo, quod lapidem philosophiae appellant (…) per Gerardum Dorneum* (Basel, 1593).

Trevius (Servio), Persius, *Ad librum de sero lactis Stephani Rooderici Castrensis Lusitani declamationes* (Rome, 1634).

Ullmann, Manfred, *Die Natur- und Geheimwissenschaften im Islam* (Leiden: Brill, 1972).

Vallensis, Robertus, *De veritate et antiquitate artis chemicae et pulveris sive medicinae philosophorum vel auri potabilis materia et compositione (…) testimonia et theoremata, ex variis auctoribus (…) per Robertum Vallensem selecta* (Leiden: apud F. Raphelengium, 1563).

Westfall, Richard S., "Isaac Newton's *Index Chemicus*," *Ambix*, 22 (1975): 174–185.

PART 2

Domestic and Apothecary Workshops: Food and Pharmacy in the Seventeenth Century

CHAPTER 3

Cheese-making and Knowledge-making: Women's Expertise and Men's Explanation

Paolo Savoia

1 Introduction

In 1789 English agrarian reformer William Marshall published a long book titled *The rural economy of Gloucestershire.* In this book, Marshall advanced a series of measures to rationalize and maximize the production of cheese according to the new enlightened economic and scientific principles, and to take it away from the control of dairywomen who treated cheesemaking as a secret. By the middle of the seventeenth century cheese, besides being one of the cornerstones of peasant diet all over Europe, had become an important commodity on the global markets. Italian, English, French and Dutch cheeses represented an important portion of the global market of luxury goods.

Marshall wrote that because of "the imperfectness of the art […] dairywomen are not able to reach any degree of certainty, much less perfection […]. The art is evidently destitute of principles. So far from being scientific, it is altogether immechanical […]. It may be said to be at present, a knack involved in mystery."[1] Marshall was taken by the idea of measuring everything – the weight of the presses used to squeeze whey out of the curds, the length of the handle of the knives used to break the curdled mass down, the diameter of the pots used to cook milk with rennet – and his aim was to pile up as much empirical information as he could. "The dairy room is consecrated to the sex [women], and it is generally understood to require some interest, and more address, to gain full admission to its rites". Cheesemaking was "a private manufactory – a craft – a mystery – secluded from the public eye […]. Individual practice is the rule, known to women and perhaps a few neighbours."[2] Marshall described all the details of the embodied knowledge involved in the mastering of a long and variable natural process of transformation. In particular, he was fascinated by the knowledge embodied in the hands of those women who dealt with milk, since with their touch only they were always and everyday, able to

1 William Marshall, *The Rural Economy of Gloucestershire* (Gloucester, 1789), vol. 2, p. 186.
2 Ibid., vol. 2, pp. 184–85.

 | DOI:10.1163/9789004512610_005

measure the right temperature at which milk curdled. Marshall gained permission to bring a thermometer with him and took diligent notes. He wrote: "What an accuracy of judgment here appears to be displayed! Let the state of the air be what it will, we find the heat of the whey, when the curd is sufficiently coagulated, exactly 80 degrees [about 26 C]; and this, without the assistance of a thermometer, or any other artificial help. But what will not daily practice, natural good sense, and minute attention accomplish." Marshall clearly had in mind a whole set of analogies between these artisanal procedures and the protocols of observation, manipulation, and control of the chemical experiments in the European laboratories. However, he could not but conclude that touch, and the senses in general, could always lead to errors, and that "the natural judgment, in the art under consideration, may frequently prove to be right, but never can be certain."[3]

In the same period, empress Maria Theresa in Lombardy became renown for a moderate sympathy towards the reformer's tendencies of the Italian Enlightenment. Among other things, the empress became sensible to questions of political economy and agrarian reforms, and encouraged scholars operating in those fields. In 1784, the *Società patriottica* in Milan invited scholars to compete for a prize for the best essay on how to improve cheesemaking, to be awarded to those who could explain in the most analytical way all the phases of the making of such an economically important product. The prize was awarded to Carlo Castelli, whose account remained in manuscript form. The essay by another participant was immediately published: Girolamo Ottolini's *Prodromo intorno alla maniera di migliorare la fabbrica dei formaggi*. Despite it being much shorter dissertation than Marshall's book, Ottolini's work was marked by the same fascinations and concerns. In the preface the author stated that cheesemaking had to be pulled away from the hands of "mechanical untrained persons" who worked with no rule, with no system, and with no order because no one until then had thought about writing the process down – a gross exaggeration, as we will see. Ottolini visited the farmhouses where the "cheesemakers" worked, and he became aware that those cheese-makers – both men and women – needed precise instructions, as they still measured the temperature of curdled milk with their own hands. A thermometer had to be installed in those workspaces to exactly measure the temperature of both the curds and the environment, since "sensation alone" easily brought about "errors and confusion."[4]

3 Ibid., vol. 1, pp. 297–99.

4 Gerolamo Ottolini, *Prodromo dell'abate Gerolamo Ottolini, intorno alla maniera di migliorare la fabbrica de' formaggi* (Milano: Fratelli Pirola, 1785), p. 33.

This was a pan-European trend in the late eighteenth century, and examples could be multiplied.

2 Observing and Learning

Early modern cheesemaking is accompanied by a rather constant scientific interest to understand such a peculiar natural process of transformation of natural matter by collecting information from those who mastered it in their everyday life. It turns out that in early modern Europe the art of transforming milk into cheese and butter was almost exclusively the domain of women makers. By showing connections between ideas about transformation of natural matter, gender, and techniques of manipulation of nature, I argue that these observations and artisanal practices contributed to changing the way knowledge was made and nature was known by physicians, natural philosophers, and natural historians around the time of the "scientific revolution." This contribution was a process of transformation of expertise into explanation.

Many intertwined threads run through the history of the practices of cheesemaking if looked at in connection with the history of early modern European knowledge – from early chemical analyses to medical dietetics, from skilled artisanship to husbandry, from cosmology to embryology, from microscopic observation to folk practices.[5] In this chapter I will only focus on the issues of the observation, translation and eventually appropriation of both artisanal practical skills embodied in women's cheese-making, and of a supposedly female intimate knowledge with milk and dairy products by learned natural historians and natural philosophers from the late fifteenth century to the seventeenth century. In fact, one of the arguments of this article is that the aims and the rhetoric of the agrarian and economic reformers of the eighteenth century – while presenting an obvious change of pace – had their roots planted in the Renaissance and in the so-called "age of the new." The same ambiguity between admiration, suspicion, and will to know – always mediated by the norms of gender roles – showed by Marshall and Ottolini can in fact be found much earlier in the writings of European naturalists, physicians, and natural philosophers.

As far back as 1942, Edgar Zilsel argued in the *American Journal of Sociology* that a vast group of "superior" artisans – including surgeons, artists, gunners,

5 I have discussed some economic aspects of cheesemaking in early modern Europe in Paolo Savoia, "Cheesemakig in the Scientific Revolution: A Seventeenth-Century Royal Society Report on Dairy Products and the History of European Knowledge," *Nuncius*, 34, 2 (2019): 427–455.

navigation experts, instrument makers, etc. – were sociologically and epistemologically at the core of the innovations that came to characterize the "scientific revolution." Zilsel believed that these artisans' sensitivity towards manual experimentalism and empirical observation eventually taught the scholars a new attitude towards the natural world.[6] It had certainly not occurred to him to include home-based and farm-based untrained, even non-artisan men and women trafficking with natural, medical, and edible stuff, often without the ability to read and write, within the list of people bringing about the "scientific revolution."

Much time has passed since Zilsel's work. In more recent years the idea that Renaissance and early modern artisanal knowledge – or "epistemology" – led the way to the empiricism, the experimentalism, and the focus on observation and description of the new sciences has bloomed anew. Bodily engagement with natural matter and active pursuit of the inner workings and processes of transformation of nature have been tracked back to "invisible technicians" and artisans of all kinds and genders, in all kinds of settings – from the court to the marketplace, from the household to the alchemical workshop.[7] Elaine Leong, discussing recipe collections, has pushed forward this historiographical trend by describing what she calls "household science," bringing about a kind of knowledge gained in pursuing collaboratively everyday activities that had much in common with the protocols of manipulation, observation, recording, and transmitting information typical of seventeenth century "high" science.[8] And sometimes, as argued by Pamela Long, the "high" and the "low," the artisanal and the learned become indistinguishable in the eyes of both the

6 Edgar Zilsel, "The Sociological Roots of Science," in Id., *The Social Origins of Modern Science*, ed. by Diederick Raven, Wolfgang Krohn, and Robert S. Cohen (Dordrecht: Springer, 2003), pp. 7–21.

7 Steven Shapin, "The Invisible Technician," *American Scientist*, 1989, 77, 6: 554–563. For recent scholarship on crafts, markets and science see, as examples, Pamela Smith, *The Body of the Artisan: Art and Experience in the Scientific Revolution* (Chicago: The University of Chicago Press, 2004); Harold J. Cook, *Matters of Exchange: Commerce, Medicine, and Science in the Dutch Golden Age* (New Haven: Yale University Press, 2008); Graeme Gooday "Placing or Replacing the Laboratory in the History of Science?," *Isis*, 99 (2008): 783–95; Eva Struhal, "Who Can Read the Book of Nature? Early Modern Artists and Scientists in Dialogue," *Nuncius* 32, 3 (2017): 501–513; Ursula Kelin, Emma Spary, "Introduction: Why Materials?" in Ursula Klein, Emma Spary, ed., *Materials and Expertise in Early Modern Europe: Between Market and Laboratory* (Chicago: The University of Chicago Press, 2010), pp. 1–23. For an historiographical overview on households as sites of scientific knowledge see Donald L. Opitz, Staffan Bergwik, and Brigitte Van Tiggelen, "Domesticity and the Historiography of Science," in Donald L. Opitz, Staffan Bergwik, Brigitte Van Tiggelen, ed., *Domesticity in the Making of Modern Science* (New York: Palgrave, 2016), pp. 1–15.

8 Elaine Leong, *Recipes and Everyday Knowledge: Medicine, Science, and the Household in Early Modern England* (Chicago: The University of Chicago Press, 2018), pp. 4–10.

contemporaries and the historians.[9] So far however, studies on women experimenters have focused on elite women operating in their home spaces and codifying their knowledge in layered collections of recipes blending alchemy, medicine, pharmacology, and cookery.[10]

This chapter aims to move forward along these lines. Cheese-making practices represent the chance to both focus on one single set of techniques and ideas, and to offer a broad overview spanning different European regions. This chapter has two broad goals: 1) including non-elite, lower-class, even anonymous women in the narrative of women and science in the early modern period, which for understandable reasons has focused mostly on elite women. 2) To study how cultural notions of gender were involved not only in representations of men and women working with nature, but also in the actual gathering of empirical knowledge, in the translation of empirical knowledge, and in theory-making by male natural philosophers, natural historians, and physicians. To adopt the terminology of Marxist feminism, which will adapted to the nuances of the specific case that I am exploring in this chapter, the relationship between female expertise and male explanation can be thought of in terms of the distinction between the "social reproduction" of practical knowledge, and the "production" of theoretical knowledge.[11] In other words, the production of theories of transformation of nature relied on a social reproduction of practi cal skills happening mostly in the domestic sphere, or in any case away from the spotlight of the learned book market and of scientific societies. Finally, this social reproduction of practical knowledge, performed by women, relied on the institution of gender roles.

As historians like Emma Spary have argued, it is becoming increasingly difficult to trace a clear boundary between artisanal food production and

9 Pamela O. Long, *Artisan/Practitioners and the Rise of the New Sciences, 1400–1600* (Portland: Oregon State University Press, 2011), p. 7.

10 See for example Alisha Rankin, *Panaceia's Daughters: Noblewomen as Healers in Early Modern Germany* (Chicago: The University of Chicago Press, 2013); Leigh Ann Whaley, *Women and the Practice of Medical Care in Early Modern Europe, 1400–1800* (Palgrave Macmillan: Basingstoke, 2011); Michelle Di Meo, Sara Pennell, ed., *Reading and Writing Recipe Books 1550–1800* (Manchester: Manchester University Press, 2013); Sheila Barker, "The Contributions of Medici Women to Medicine in Granducal Tuscany and Beyond," in Alessio Assonitis and Brian Sandberg, ed., *The Medici and Their Archive: Power and Representation in Early Modern Tuscany* (Turhnout: Harvey Miller/Brepols, 2016), pp. 101–116; Edith Snook, "'The Women Know': Children's Diseases, Recipes and Women's Knowledge in Early Modern Medical Publications," *Social History of Medicine* 30, 1 (2017): 1–21; Mary E. Fissell, "Introduction: Women, Health, and Healing in Early Modern Europe," *Bullettin of the History of Medicine*, 82 (2008): 1–17; Sharon Strocchia, *Forgotten Healers: Women and the Pursuit of Health in Late Renaissance Italy* (Cambridge: Harvard University Press, 2019), pp. 1–13.

11 Tithi Bhattacharya, ed., *Social Reproduction Theory: Remapping Class, Recentering Oppression* (London: Pluto Press, 2017).

industrialized, standardized food production between the late 18th century and the early 19th century.[12] Modern food chemists and gastronomists – both claiming to operate under the umbrella of "science" – were born at the same time. Indeed, early modern attempts at gathering information on such folk or "old" technologies can be traced back from the late fifteenth onto the eighteenth centuries. Agronomists, physicians, natural historians, and natural philosophers' efforts at knowing cheese-making practices was accompanied by a certain fascination for the themes of rotting, decaying, and rebirth associated with cheese. The history of cheese-making should not be represented only as a diachronic development from artisans to factory workers, from personal skills to anonymous technologies. Rather, it could be described as a synchronic coexistence of the artisanal, the technological, and the cosmological throughout the early modern period.

After briefly describing the intellectual theories of milk transformation before and after the middle of the seventeenth century, this article examines examples of the dynamics of expertise/explanation: the late fifteenth-century book on dairy products written by Piedmontese physician Pantaleone of Confienza; the reports and experiments on cheese-making at the Royal Society in London and the Académie des Sciences in Paris in the 1660s; the 1664 booklet on people disgusted by cheese by the Dutch theologian and natural philosopher Martin Schook.

3 Cheese-Making as Transformation

Around 1500 Aristotelian biology, on the one hand, and medical dietetics (i.e. that part of Hippocratic-Galenic medicine dealing with regulating human complexions through balancing the intake of food and drink) on the other, constituted the most important intellectual traditions concerning milk and dairy products. Aristotle used cheese as an analogy, with this dairy substance becoming an explanatory tool in different parts of his work, and for different purposes. The process of coagulation of milk through the action of rennet (which came mostly from the stomachs of calves, lambs and goats) was useful for illustrating: the process of generation of a foetus *in utero* (the rennet on milk acted just as active male semen on passive menstrual blood in order to create and shape human life); the formation of human skin as a crust made of

12 Emma Spary, *Feeding France: New Sciences of Food, 1760–1815* (Cambridge: Cambridge University Press, 2014), pp. 3–9; see also Emma Spary, Anya Zilberstein, "On the Virtues of Historical Entomophagy," in Emma Spary, Anya Zilberstein, ed., *Food Matters: Critical Histories of Food and the Sciences*, *Osiris*, vol. 35 (2020), pp. 1–19.

dried flesh which captured the vapours of foetal concoction inside the body was similar to how rind on cheese is formed;[13] the processes of transformation of natural matter through the action of hot and cold elements; and finally, the process through which worms and insects could be generated spontaneously from rotten matter.[14] Aristotelian theory of material change, which was detailed in the fourth book of *Meteorology*, provided the foundation for Renaissance and early modern natural philosophers and chemists' approach to dairy products.[15] Milk itself was a homeomerous substance (namely, a substance composed of apparently different parts but was still a unified, single substance) composed of two or three parts. For Aristotle, it had two components, the watery one (whey) and the earthly one (cheese), while Galen acknowledged the existence of a third part, oily and fat, the butter component.

In the Aristotelian-galenic tradition, milk was to be understood as an exceeding white liquid generated from a double process of concoction in the female body. Food nourished the body as it was transformed into blood by the physiological process of digestion. Menstrual blood, derived from the concoction of food and its transformation into blood, was divided in two parts: the purest one went from the vagina to the breast through the veins, where a second process of "concoction" took place, in which the blood became white and assumed the complexion of milk. The "impure" part of menstrual blood was expelled. In the most basic terms, milk was white blood.

By the first decades of the seventeenth century, the conception of milk and cheese as natural matter and transformation processes began to change. Corpuscular, chemical, mechanical, and atomistic philosophies of nature began to decompose milk and cheese into micro-elements.[16] Nominally Aristotelian philosopher Gabriele Nardi wrote that the only thing singular and unified in milk was its name.[17] German "iatrochemist" Daniel Sennert wrote in 1636: "even though milk may seem one body, its whey, butter, and cheese reveal [the existence] of diverse parts mixed per minima, when they are separated. So, too, even if the blood of animals appears to be one homogeneous body,

13 Aristotle, *Generation of Animals*, tr. A.L. Peck (Cambridge: Harvard University Press, 1942), II, iv (739b, 22–23), pp. 190–193.

14 Aristotle, *History of Animals*, tr. A.L. Peck (Cambridge: Harvard University Press, 1970), v, xxxi (556b, 22–30), p. 209.

15 See Craig Martin, *Renaissance Meteorology: Pomponazzi to Descartes* (Baltimore: Johns Hopkins University Press, 2011); Aristotle, *Meteorologica*, tr. H.D.P. Lee (Cambridge: Harvard University Press), IV, ii (379b, 21–29), p. 299; see Joseph S. Fruton, *Fermentation: Vital or Chemical Process?* (Leiden: Brill, 2006), pp. 2–4.

16 Antonio Clericuzio, "Chemistry of Life: Ferments and Fermentation in 17th-century Iatrochemistry," *Medicina nei Secoli*, 15, 2 (2003): 227–245.

17 Gabriele Nardi, *Lactis physica analysis* (Florence: typis Petri Nestij, sub signo Solis, 1634), pp. 168–171.

not only are diverse parts found to exist in it which supply food to the various members of the body, but if it should be distilled, a volatile salt which was not in evidence before adheres in great quantity to the flask."[18] Before the middle of the seventeenth century, several natural philosophers believed that there was no such thing as milk as a whole, if composed, substance, but only combinations of particles.[19] Cheesemaking – which was conceived as one of the arts of separation since the times of Plato (*Soph.*, 226 b–c)[20] – ceased to be a process of separation and re-composition of milk's macro-components, but micro-molecules reacting to acid parts and re-combinating while releasing gases and sugars. This will be more fully described by French and German chemists working in their new eighteenth-century laboratories,[21] but seventeenth-century cheesemakers, as they appeared in the learned observers' records, played a fundamental role in the process. Of course, a certain degree of speculation is needed here, since these artisans left nothing in written form and left no object and almost no material traces behind themselves – in this respect, they are different from both Zilsel's superior artisans and Smith's artisan/epistemologists, and their legacy is much more difficult to capture.

4 Gendered Cheese-making

These cheesemakers were women. Parallel to the rise of new theories of milk and cheese, lay a learned interest for the practice of cheese-making which is typical of the seventeenth century. Robert Boyle stated in *The Usefulnesse of Experimental Philosophy*: "I shall not dare to think my self a true naturalist, till my skill can make my garden yield better herbs and flowers, or my orchard better fruit, or my fields better corn, or my dairy better cheese then theirs that are strangers to physiology."[22] In the spring of 1669 the Parisian Académie des

---

18 Daniel Sennert, *Hypomnemata physica* (Frankfurt: sumptibus Clementis Schleichii et consortum, 1636), pp. 113–114; on this passage see William R. Newman, *Atoms and Alchemy: Chymistry and the Experimental Origins of the Scientific Revolution* (Chicago: The University of Chicago Press, 2006), p. 165.

19 On the importance of milk and its transformations for seventeenth-century chemically inclined natural philosophy, see James Riddick Partington, *A History of Chemistry* (London: MacMillan, 1969), vol. 2, pp. 156, 246, 445.

20 Giancarlo Plazio, *La cera, il latte, l'uomo dei boschi: Mitologia e realtà sociale in una comunità prealpina* (Turin: Giappicchelli, 1979), pp. 6–7.

21 Barbara Orland, "Enlightened Milk: Reshaping a Bodily Substance into a Chemical Object," in Ursula Klein, Emma Spary, ed., *Materials and Expertise in Early Modern Europe: Between Market and Laboratory* (Chicago: The University of Chicago Press, 2010), pp. 163–197.

22 Robert Boyle, *The Usefulnesse of Natural Philosophy*, in Id., *The Works of Robert Boyle*, edited by Michael Hunter, Edward B. Davis (London: Pickering & Chatto, 1999), vol. 3, p. 295.

Sciences devoted a series of meetings to the problem of the coagulation of fluids, involving physicians, natural historians, and natural philosophers like Samuel Cottereau Duclos, Claude Perrault, and l'Abbé Mariotte.[23] The academicians desired to interview peasants and dairywomen in order to acquire the conceptual equipment capable of explaining the making of cheese out of milk. Duclos was charged of observing, reporting, and eventually replicating the "vulgar experiences" of the peasants. Some experiences – he claimed – had to be re-made at the Académie in order to "confirm or contradict the inductions made from the observation of the vulgar experiences."[24] Among the many experiments and essays to be done, Duclos listed:

1. Putting mint into milk before adding rennet in order to see if it curdles;
2. Mix spirit of wine, oil of tartar [red deposit of wine], or other similar sulphureous liquors with milk to see if they forbid curdling;
3. Adding common salt, sugar or honey to milk to see whether these substances help of forbid coagulation;
4. Separating the cream from milk to see if it curdles faster;
5. Putting curdled milk (non-skimmed) into hot water to see if some fattiness can be separated, or if this curdled milk will become viscous and hard when it cools down;
6. Doing the same thing, but with skimmed milk;
7. Mixing lacteous juices [...] with milk to see if they forbid coagulation;
8. Observing whether the milk of certain plants can curdle under the action of "acids" just like that of animals;
9. Which animal milk curdles more easily.[25]

Sometime during the 1660s, Royal Society correspondent William Jackson wrote a report to the fellows on the way in which cheese was made in Cheshire, one of the most renown regions for the production of good cheese in England. The five-page manuscript document details the work of the "dayrywomen" in all the phases of cheesemaking: from preparing rennet with the calves' stomachs to the pressing and smoothing of the curds with the hands, from pressing the cheese in a mechanical press to expel the whey to making sure that

23 *Mémoires de l'Académie royale des sciences depuis 1666 jusqu'en 1699* (Paris: par la Compagnie des libraires, 1733), Vol. 1, p. 92.

24 Archives de l'Académie des Sciences, Paris, Procès-verbaux, T5 (1669, Registre de mathématique), fol. 60r.

25 Archives de l'Académie des sciences, Paris, Procès-verbaux, T5 (1669, Registre de mathématique), fol. 60v–61r.

cats chase away rats when the cheese is seasoned in the "cheese chamber." The observer here presents the same ambivalence between admiration, envy, and dismissal attitude of the eighteenth-century reformers. In this document, dairywomen were described not only preparing rennet, but also mastering complex technological instruments such as the huge mechanical press needed to make the highly valued Cheshire cheese. These women not only had the embodied skills needed to prepare the cheese with their own hands, but they could also speculate on why some cheeses came with holes on the basis of their knowledge of animals and the effects of heat on matter. That is how they theorized about imperfections in cheese:

> I remember I have heard the huswifes in Kent impute it to the time of the Cows going to bull; and affirm that at such a time, the milke had such a working in it that was impossible to prevent cheese having eyes in it, more than at other times: which is not alltogether without a seeming reason: for tis possible when the beast is driven with such a propensity, that it must be promoted by a strong fermentation in the blood and nervous juice; which can not but have some influence on the milke at that time, sufficient to continue it in the Curds, more than at other times.[26]

Evidence of the women's skills in cheesemaking span across Europe from the Middle Ages, even if it seems that English dairywomen had a specifically recognized status as expert of milk transformations.[27] Examples from France and Italy can be found in Pantaleone of Confienza's *Summa lacticiniourum* (1477). According to Pantaleone, the cheesemakers' techniques had a prominent place in the list of the variables that could make cheese better or worse. The physician reported that he saw a woman in the French region of the Bresse who became "so famous and expert in making cheese" (*experta*) that everyone wanted it, and she sold it at a very high price. Reading Pantaleone's book together with iconographic sources one can hypothesize that cheese making (and selling) was, if not exclusively a female and domestic occupation, an activity in which women were especially involved. The *Tacuina sanitatis*[28]

---

26 I have published a commented transcription of this document in Savoia, *Cheesemaking in the Scientific Revolution*; quotation on pp. 449–450.

27 Deborah Valenze, "The Art of Women and the Business of Men: Women's Work and the Dairy Industry *c.* 1740–1840," *Past & Present*, 130, 1 (1991): 142–169.

28 *Tacuina Sanitatis* were a series of fourteenth and fifteenth century illuminated manuscripts – most of them of Milanese origin, but informed by the Arabic medical tradition – that illustrated the life of the peasant workers and the various kinds of food they produced, adding schematic but precise dietetic information on each foodstuff. By the beginning of the sixteenth century they were printed and had much success. See

FIGURE 3.1 Women squeezing whey out of the mass of curd. Royal Society, CI.P/3i/22
© The Royal Society

often present women preparing fresh cheese. The late fourteenth century fresco paintings in the Castle of Buonconsiglio in Trent, depicting the cycle of

Luisa Cogliati Aramo, *Tacuinum sanitatis* (Milano: Electa, 1973); Alixe Bovey, Tacuinum Sanitatis: *An Early Renaissance Guide to Health* (London: Sam Fogg, 2005).

FIGURE 3.2 *Tacuinum sanitatis: Recocta.* Cod. Ser. n. 2644, fol. 62r. Vienna, Österreichische Nationalbibliothek

the twelve months, show for the month of June the cycle of the transformation of milk into cheese, with female figures taking care of each step, from milking to cheesemaking through the preparation of butter.[29] Women were also aware

29 See also Piero Camporesi, "Il formaggio maledetto" in Id., *Le officine dei sensi* (Milan: Garzanti, 1985), pp. 63–65; Irma Naso, *Università e sapere medico nel Quattrocento:*

FIGURE 3.3 Women milking cows and making butter and cheese. *Cycle of the Months, June* (detail). Castle of Buonconsiglio, Trent, The Month of June (late fourteenth century)

and making use of the medical properties of dairy products: speaking of the cheese coming from the valley of Nus, Pantaleone pictured a scene of domestic medicine, the domain of female healers in the Middle Ages. He remarked that the local cheese was particularly easy to digest in fact, women (*mulieres*) gave them to ill people.[30] Remarkably, the word 'women' was used here without qualification, implying a conflation of meanings between the word "woman" and "caregiver".[31]

In Italy one could find "learned masters" (*magistri et docti*) – the physician uses here the same words usually used for both university-trained professionals and guild members – who were "very skilled and careful in considering all the variable factors in cheese-making". Moreover, as he "learned from experts" (*ut ab expertis accepi*), the hands of the cheese-maker had to be smooth and not too warm.[32] Pantaleone's accounts reveal a world of artisans of cheese that were be self-aware of their particular skills. There emerges a non-written practical knowledge proper of the cheese-makes who dealt with very complex phenomena of transformation of matter. Finally, the fact that women were expert cheese-makers must have been so self-evident that in his early seventeenth-century description of the *City of the Sun* Tommaso Campanella, while advocating for a much bigger gender equality than most of his peers, could not but acknowledge that even in the utopia "cheese making, milking, tending to the kitchen gardens [...] are assigned to women."[33]

Indeed, cheesemaking practices required a remarkable degree of tacit knowledge of matter. Modelling and salting the surface of cheese, for example, was crucial. In fact, the function of the surface of cheese was both to limit excessive evaporative moisture loss – the rottening from the inside – at the same time being porous enough to prevent excessive dehydration; and to protect the cheese from cracks and maggot infestation. While small cheeses had a salted rind which prevented excessive evaporation and the spoiling of cheese during aging, in contrast, large cheeses had much less surface area relative to their volumes. Large cheeses with high initial moisture could not dry out enough via surface evaporation before the rind formed and slowed down moisture loss. Salt was diffused less easily on large cheeses: this combination

---

Pantaleone da Confienza e le sue opere (Cuneo: Società per gli studi storici, archeologici ed artistici della provincia di Cuneo, 2000), pp. 98–99.

30 Naso, *Università e sapere medico*, p. 196.

31 Monsterrat Cabré, "Women or Healers? Household Practices and the Categories of Health Care in Late Medieval Iberia," *Bulletin of the History of Medicine*, 82, 1 (2008): 18–51.

32 Naso, *Università e sapere medico*, pp. 191–192.

33 Tommaso Campanella, *La Città del Sole: Dialogo Poetico/The City of the Sun: A Poetical Dialogue* (Berkeley: University of California Press, 1981), p. 48.

of high moisture and less saltiness could lead to fermentations and rotting. So large cheeses had to have, right after coagulation, lower initial moisture. This could be gained by squeezing most of the whey out of the curds during cheese-making, either by cooking the milk at higher temperature, or by pressing the curds with large presses. Another way of making low-moisture cheese was to salt the curds, break them down into pieces, and then to salt them again when they are pressed.[34]

The new early modern scientific habits of description, observation, and experimentation contributed to a science based on the observation of certain features of matter, of certain processes often happening on the surface of nature that were linked to women's expertise. This was also a process of translation, and appropriation, of household, farm, and vernacular knowledge into, and from, natural philosophy.

5 Disgust, Breastfeeding, and Norms of Womanhood

Besides household and farm "science," the link between cheese-making, knowledge, and gender could take yet another form, connected to gender norms of womanhood as motherhood, women's impurity and cleanliness, and women's empirical expertise as beings actually producing milk. The middle of the seventeenth century also witnessed a rise of learned philosophical and medical discourses on the benefits of mothers feeding their babies with their own milk, and a certain suspicion towards wet-nursing, although this remained a widespread practice.[35] This article does not advocate for some sort of psychohistory which explains cheese-making with a supposedly ahistorical proximity between women and feeding or caregiving; rather, its aim is to highlight another path through which men extracted knowledge from women by relying on representations of gender roles.

All of this becomes clearer if one looks at Martin Schoock's 1664 theory of the origins of "cheese phobia." Here, one can see a powerful gendered and cultural imagery linking women, milk, and cheese at play. Schoock's *De aversatione casei* is devoted to explain why so many people, including the author, feel

34 Paul Kindstedt, *Cheese and Culture: A History of Cheese and its Place in Western Civilization* (White River Junction: Chelsea Green Pub, 2012), pp. 168–170; Piero Camporesi, *The Incorruptible Flesh: Bodily Mutation and Mortification in Religion and Folklore*, tr. Tania Crofft-Murray (Cambridge: Cambridge University Press, 1988), pp. 270–281.

35 Valerie A. Fildes, *Wet Nursing: A History from Antiquity to the Present* (Oxford: Basil Blackwell, 1988); Maria Giuseppina Muzzarelli, *Nelle mani delle donne: nutrire, guarire, avvelenare dal Medioevo a oggi* (Roma-Bari: Laterza, 2013).

a sort of a phobic disgust for cheese. This apparently bizarre topic was in fact linked to vast social and cultural problems of manners for the upper classes. By the seventeenth century, manners and civility books shown a greater and greater intolerance for food aversions: it was considered more and more impolite not to eat the food that was offered during a meal, and food aversions were generally considered maters of imagination that had to be overcome by reason and training. Disgust as a bodily phenomenon had to be kept under control by men of high social status.[36] Schoock's new explanation of individual aversions to certain kinds of food indeed looks for a new physiology rather than looking back at old Galenic humoral principles. By the beginning, this work aligns to some of the most innovative principles of the new epistemology of the seventeenth century by insisting on the values of observation and the collection of cases to formulate a probable explanation.[37]

Before getting to its main issue, Schoock diligently discusses the causes of the coagulation of milk by tackling both the ancients – mostly Aristotle and Galen – and the moderns, and he opts for a moderately acceptance of the new "chemical" theories. In any case, he acknowledges that on the action of the *vis coagulandi* there is no consensus. The ancients argued, as we have seen, that when substances with the force of coagulation – namely, warm substances – are introduced into milk, cheese is brought about. But Schoock argues that experience has shown that vinegar, an acid substance, can be used in cheesemaking. He writes that vinegar and the substances reacting to it cause different parts of a substance to form mixtures. "In fact – he goes on – the most careful Chemists make us notice that an acid salt, called vitriol by the common people, which with respect to its acrimony has an attenuating or dissolving power, consequently can separate or disjoin the serous parts from the fatty parts in milk; in the same way, therefore, it can assemble the fatty and earthly parts of milk together, until cheese is made."[38] In other words, Schoock endorsed the recent theories of milk and cheese as composed of micro-components and not by different parts of one and the same substance.

But why some people hate cheese? First of all, many people mention the fact that a strong smell emanates from cheese, especially when it is rotting, which is particularly offensive to the brain. This feature of cheese hurts people whether

36 Viktoria Von Hoffmann, *From Gluttony to Enlightenment: The World of Taste in Early Modern Europe* (Urbana: University of Illinois Press, 2016), pp. 32–33. For an overview on theories of disgust see Carolyn Korsmeyer, "Delightful, Delicious, Disgusting," *The Journal of Aesthetics and Art Criticism*, 60, 3 (2002), pp. 217–225.

37 Martin Schook, *Tractatus de butyro, accessit eiusdem diatriba De aversatione casei* (Groningen: typis Johannis Colleni, 1664).

38 Ibid., pp. 244–246.

the cheese is actually putrefying or not.[39] Indeed, Bernardino Ramazzini in his work on the diseases of workers reports that unbearable stink come from the neighbourhoods where cheese is produced, such as in the city of Frankfurt.[40] However, Schoock rules out this explanation by citing an abundance of cases of people who get accustomed to foul smell, and of people who strongly dislike pleasant scents (of roses, for example). No rule can be found in the domain of smell. Another possible explanation, Schoock goes on, are the teachings of Galenic medical dietetics, which were at best sceptical, and generally negative about cheese, describing it as a dangerous foodstuff, breeding constipation, hard to digest, and exceedingly fat.[41]

But here's the real explanation. "First, if a strong female, who is giving milk to a baby, already had milk in her breasts, it would be normal to expect that such milk, according to the causes of cheesemaking [heat], would curdle." This process of curdling before conception happens because of the "preternatural heat" coming from the female breast. "Heat permeates the whole pregnant body, since the food being transformed into the matter forming chyle, which in turn serves the formation of milk, goes back to the uterus for the formation of the parts of the body [of the foetus], and generates a disproportionate amount of heat. For this reason it is sure that some milk does coagulate. So this milk that goes to the breast to nourish the baby undergoes a curdling process; this curdled milk is what most babies suckle when they begin to be fed."[42]

With a very interesting move, Schoock adds what he calls an "experiment," and is in fact an inquiry into the social and physiological habits of the women in his own family. He says he noticed that many of those who dislike (including himself) cheese have been nourished either by their mothers' or a wetnurse's milk when they were pregnant, and therefore "they began to dislike that curdled milk (*lac caseatum*)."[43] He remembers what his maternal grandmother, whose three daughters all abhorred cheese, said: that even if they knew they were pregnant, they nonetheless kept breastfeeding. Schoock's own mother "raised six children (I was the firstborn) who hate cheese, except the third one (who

39 Ibid., pp. 210–212.

40 Bernardino Ramazzini, *De morbum artificum diatriba* (Modena: typis Antonii Capponi, impressoris episcopalis, 1700), chapter XVI.

41 On the bad reputation of cheese see Piero Camporesi, "Il formaggio maledetto," pp. 47–77; Anna Maria Nada Patrone, *Il cibo del ricco e il cibo del povero. Contributo alla storia qualitativa dell'alimentazione* (Torino: Centro Studi piemontesi, 1981), pp. 349–357; Allen Grieco, "The Social Politics of Pre-Linnaean Botanical Classification," *I Tatti Studies*, 4 (1991): 131–149.

42 Schook, *De aversatione casei*, p. 266.

43 Ibid., p. 273.

became a doctor in law in 1645), because the mother had a bad abscess on her breasts and had to stop breastfeeding. […] Similarly, I know from older friends that my father, himself most averse to cheese, was sent to a wetnurse who was secretly impregnated by a servant and kept feeding him all the same."[44] In the same way, his son Isaac, who was given to a wetnurse, was milk-fed for six months; but the wetnurse had been impregnated several months before.

Royal Society fellow Kenelm Digby had made the same observations and formulated the same hypothesis. In his *Treatise on the Nature of the Body* he wrote abut the fact that so many people have a "natural antipathy" for cheese:

> A principal reason of which symptome [besides the fact that mothers had such an aversion and passed it on to their children via blood which became milk] I conceive to be, that their nurses proved with child when they gave them suck: for I have by experience found it to have been so in as many as I have made inquiry into. And it is very conformable to reason: for the nurses milk curdling on her brest upon her breeding of childe, and becomming very offensive to the childe tender stomack, (whose being sick obligeth the parents to change the Nurse, though per adventure they know nothing of the true reason that maketh their milke unnaturall) he hath a dislike of Cheese (which is strong curdled milk) ever after settled in him.[45]

Explanations like these are linked to issues of milk and female impurity. In his late sixteenth-century century bestseller on how to manage a farm, *La maison rustique*, the French physician and polymath Charles Estienne painted a picture of the world of cheesemaking as a female occupation. The household lady must coordinate her daughters and her maids and teach them "the art of governing milk" in order to make butter and cheese. Somehow anticipating Schoock, he says that mothers know that their "first" milk is not good, so the "first" milk milked by cows is not good for cheese either. Women recognize "good milk" with their senses: "its whiteness, pleasant smell, sweet taste, moderately dense texture (this can be seen by putting a drop of it on a fingernail: if it does not trickle down but stands still and does not loose its round shape then it is good)."[46] While describing the actual process of cheesemaking, Estienne

44 Ibid., p. 274.

45 Kenelm Digby, *Two treatises: in the one of which, the nature of bodies; in the other, the nature of mans soule, is looked into* (London: John Williams, 1645), pp. 416–417.

46 Charles Estienne, *La maison rustique* (Paris, 1589), p. 38v.

adds that "the chambermaids who make cheese must be clean, polite (*propres*), with their sleeves rolled up, and far away from their period."[47]

This link between women, animal milk, cheese, and the threat of pollution is confirmed by seventeenth century English cases of suspect possessed or "bewitched" women, which I took from Richard Napier's casebooks of medico-astrological consultations, made available by a team led by Lauren Kassell. For example, in 1633 one Eleanor Aylet reported that she felt strange, and strange symptoms of bewitchment appeared. Among them, she spoiled the cheese she used to make with her hands: "sinc hath bene troubled in her milke & making of cheese her milke sowrs & her cheese wth an ill sent."[48]

This is a particular case of a vast literature on how women's menses pollute food preparation, especially processes of fermentation, leavening, and coagulation.[49] Women's touch embodied knowledge which was essential to make cheese, but their "impurity" was a threat that could spoil the whole process.

6 Expertise and Explanation

Early modern sources present women's role in cheese-making as a fact, and no explanation is given why that was the case. Of course, one could say that cheese-making was part of household work and farm work traditionally entrusted to women, but they kept doing it even when, by the middle at the seventeenth-century, dairy products became a proto-industrial enterprise. Therefore, the issue might be more complex, and it might involve women's expertise not only with medicaments and recipes, but also with touching natural matter, mastering complex natural processes of transformation, and devising/utlizing technology.

47 Id.

48 MS Ashmole 211, p. 324, in Lauren Kassell, Michael Hawkins, Robert Ralley, John Young, Joanne Edge, Janet Yvonne Martin-Portugues, and Natalie Kaoukji, ed., *'Casebooks,' The casebooks of Simon Forman and Richard Napier, 1596–1634: A Digital Edition*, https://casebooks.lib.cam.ac.uk.

49 See Caroline Walker-Bynum, *Holy Feast and Holy Fast: The Religious Significance of Food to Medieval Women* (Berkeley: University of California Press, 1987), pp. 189–194 and pp. 297–302; Raffaella Sarti, *Europe at Home: Family and Material Culture, 1500–1800*, tr. Allan Cameron (Yale University Press, 2002). Menstruation, however, was not viewed as an exclusively negative bodily phenomenon in early modern Europe: see Gianna Pomata, "Menstruating Men: Similarity and Differences of the Sexes in Early Modern Medicine," in Valeria Finucci and Kevin Brownlee, ed., *Generation and Degeneration: Tropes of Reproduction in Literature and History from Antiquity to Early Modern Europe* (Durham: Duke University Press, 2001), pp. 109–152, especially pp. 141–144.

This article has been structured by the polarity between expertise and explanation, or practical skills and their scientific appropriation and translation. This polarity was reinforced by notions of gender and by the social work gender roles played in acquiring expertise, disseminating practical knowledge, and establishing new knowledge. The category of "gender" is not a disembodied or theoretical tool abstracting from the history and life of men and women, but rather can be used productively as a historiographical tool to understand processes of circulation of knowledge between women and men. Far from being only a matter of "representations," gender roles mediated, on the one hand, the observation and recording of experimental and practical knowledge; and, on the other hand, some of the processes of theory-making and induction which are typical of the scientific revolution. Women transforming milk into cheese were the object of scrutiny of men appropriating expertise and claiming explanations.

References

Aristotle, *Generation of Animals*, translated by A.L. Peck (Cambridge: Harvard University Press, 1942).

Aristotle, *History of Animals*, translated by A.L. Peck (Cambridge: Harvard University Press, 1970).

Aristotle, *Meteorologica*, translated by H.D.P. Lee (Cambridge: Harvard University Press).

Barker, Sheila, "The Contributions of Medici Women to Medicine in Granducal Tuscany and Beyond," in Alessio Assonitis and Brian Sandberg, ed., *The Medici and Their Archive: Power and Representation in Early Modern Tuscany* (Turhnout: Harvey Miller/Brepols, 2016), 101–116.

Bhattacharya, Tithi, ed., *Social Reproduction Theory: Remapping Class, Recentering Oppression* (London: Pluto Press, 2017).

Bovey, Alixe Tacuinum Sanitatis*: An Early Renaissance Guide to Health* (London: Sam Fogg, 2005).

Boyle, Robert, *The Usefulnesse of Natural Philosophy*, in Id., *The Works of Robert Boyle*, edited by Michael Hunter, Edward B. Davis (London: Pickering & Chatto, 1999), vol. 3.

Cabré, Monsterrat, "Women or Healers? Household Practices and the Categories of Health Care in Late Medieval Iberia," *Bulletin of the History of Medicine*, 82, 1 (2008): 18–51.

Campanella, Tommaso, *La Città del Sole: Dialogo Poetico/The City of the Sun: A Poetical Dialogue* (Berkeley: University of California Press, 1981).

Camporesi, Piero, *Le officine dei sensi* (Milan: Garzanti, 1985).

Camporesi, Piero, *The Incorruptible Flesh: Bodily mutation and mortification in religion and folklore*, tr. Tania Crofft-Murray (Cambridge: Cambridge University Press, 1988).

Clericuzio, Antonio, "Chemistry of Life: Ferments and Fermentation in 17th-century Iatrochemistry," *Medicina nei Secoli*, 15, 2 (2003): 227–245.

Cogliati Aramo, Luisa, *Tacuinum sanitatis* (Milano: Electa, 1973).

Cook, Harold J., *Matters of Exchange: Commerce, Medicine, and Science in the Dutch Golden Age* (New Haven: Yale University Press, 2008).

Di Meo, Michelle, Sara Pennell, ed., *Reading and Writing Recipe Books 1550–1800* (Manchester: Manchester University Press, 2013).

Digby, Kenelm, *Two treatises: in the one of which, the nature of bodies; in the other, the nature of mans soule, is looked into* (London: John Williams, 1645).

Estienne, Charles, *La maison rustique* (Paris, 1589).

Fildes, Valerie A., *Wet Nursing: A History from Antiquity to the Present* (Oxford: Basil Blackwell, 1988).

Fissell, Mary E., "Introduction: Women, Health, and Healing in Early Modern Europe," *Bullettin of the History of Medicine*, 82 (2008): 1–17.

Fruton, Joseph S., *Fermentation: Vital or Chemical Process?* (Leiden: Brill, 2006).

Gooday, Graeme, "Placing or Replacing the Laboratory in the History of Science?," *Isis*, 99 (2008): 783–95.

Grieco, Allen, "The Social Politics of Pre-Linnaean Botanical Classification," *I Tatti Studies*, 4 (1991): 131–149.

Kassell, Lauren, Michael Hawkins, Robert Ralley, John Young, Joanne Edge, Janet Yvonne Martin-Portugues, and Natalie Kaoukji, ed., *Casebooks: The Casebooks of Simon Forman and Richard Napier, 1596–1634: A Digital Edition*: https://casebooks.lib.cam.ac.uk.

Kelin, Ursula, Emma Spary, ed., *Materials and Expertise in Early Modern Europe: Between Market and Laboratory* (Chicago: The University of Chicago Press, 2010).

Kindstedt, Paul, *Cheese and Culture: A History of Cheese and its Place in Western Civilization* (White River Junction: Chelsea Green Pub, 2012).

Korsmeyer, Carolyn, "Delightful, Delicious, Disgusting," *The Journal of Aesthetics and Art Criticism*, 60, 3 (2002): 217–225.

Leong, Elaine, *Recipes and Everyday Knowledge: Medicine, Science, and the Household in Early Modern England* (Chicago: The University of Chicago Press, 2018).

Long, Pamela O., *Artisan/Practitioners and the Rise of the New Sciences, 1400–1600* (Portland: Oregon State University Press, 2011).

Marshall, William, *The rural economy of Gloucestershire* (Gloucester, 1789).

Martin, Craig, *Renaissance Meteorology: Pomponazzi to Descartes* (Baltimore: Johns Hopkins University Press, 2011).

Mémoires de l'Académie royale des sciences depuis 1666 jusqu'en 1699 (Paris: par la Compagnie des libraires, 1733), Vol. 1.

Muzzarelli, Maria Giuseppina, *Nelle mani delle donne: nutrire, guarire, avvelenare dal Medioevo a oggi* (Roma-Bari: Laterza, 2013).

Nada Patrone, Anna Maria, *Il cibo del ricco e il cibo del povero. Contributo alla storia qualitativa dell'alimentazione* (Torino: Centro Studi piemontesi, 1981).

Nardi, Gabriele, *Lactis physica analysis* (Florence: typis Petri Nestij, sub signo Solis, 1634).

Naso, Irma, *Università e sapere medico nel Quattrocento: Pantaleone da Confienza e le sue opere* (Cuneo: Società per gli studi storici, archeologici ed artistici della provincia di Cuneo, 2000).

Newman, William R., *Atoms and Alchemy: Chymistry and the Experimental Origins of the Scientific Revolution* (Chicago: The University of Chicago Press, 2006).

Opitz, Donald L., Staffan Bergwik, and Brigitte Van Tiggelen, ed., *Domesticity in the Making of Modern Science* (New York: Palgrave, 2016).

Orland, Barbara, "Enlightened Milk: Reshaping a Bodily Substance into a Chemical Object," in Ursula Klein, Emma Spary, ed., *Materials and Expertise in Early Modern Europe: Between Market and Laboratory* (Chicago: The University of Chicago Press, 2010), 163–197.

Ottolini, Gerolamo *Prodromo dell'abate Gerolamo Ottolini, intorno alla maniera di migliorare la fabbrica de' formaggi* (Milan: Fratelli Pirola, 1785).

Partington, James Riddick, *A History of Chemistry* (London: MacMillan, 1969).

Plazio, Giancarlo, *La cera, il latte, l'uomo dei boschi: Mitologia e realtà sociale in una comunità prealpina* (Turin: Giappicchelli, 1979).

Pomata, Gianna, "Menstruating Men: Similarity and Differences of the Sexes in Early Modern Medicine," in Valeria Finucci and Kevin Brownlee, ed., *Generation and Degeneration: Tropes of Reproduction in Literature and History from Antiquity to Early Modern Europe* (Durham: Duke University Press, 2001), 109–152.

Ramazzini, Bernardino, *De morbum artificum diatriba* (Modena: typis Antonii Capponi, impressoris episcopalis, 1700).

Rankin, Alisha, *Panaceia's Daughters: Noblewomen as Healers in Early Modern Germany* (Chicago: The University of Chicago Press, 2013).

Sarti, Raffaella, *Europe at Home: Family and Material Culture, 1500–1800*, tr. Allan Cameron (Yale University Press, 2002).

Savoia, Paolo, "Cheesemakig in the Scientific Revolution: A Seventeenth-Century Royal Society Report on Dairy Products and the History of European Knowledge," *Nuncius*, 34, 2 (2019): 427–455.

Schook, Martin, *Tractatus de butyro, accessit eiusdem diatriba De aversatione casei* (Groningen: typis Johannis Colleni, 1664).

Sennert, Daniel, *Hypomnemata physica* (Frankfurt: sumptibus Clementis Schleichii et consortum, 1636).

Shapin, Steven, "The Invisible Technician," *American Scientist*, 77, 6 (1989): 554–563.

Smith, Pamela, *The Body of the Artisan: Art and Experience in the Scientific Revolution* (Chicago: The University of Chicago Press, 2004).

Snook, Edith, "'The Women Know': Children's Diseases, Recipes and Women's Knowledge in Early Modern Medical Publications," *Social History of Medicine*, 30, 1 (2017): 1–21.

Spary, Emma, and Anya Zilberstein, "On the Virtues of Historical Entomophagy," in Emma Spary and Anya Zilberstein, ed., *Food Matters: Critical Histories of Food and the Sciences, Osiris*, vol. 35 (2020), pp. 1–19.

Spary, Emma, *Feeding France: New Sciences of Food, 1760–1815* (Cambridge: Cambridge University Press, 2014).

Strocchia, Sharon, *Forgotten Healers: Women and the Pursuit of Health in Late Renaissance Italy* (Cambridge: Harvard University Press, 2019).

Struhal, Eva, "Who Can Read the Book of Nature? Early Modern Artists and Scientists in Dialogue," *Nuncius*, 32, 3 (2017): 501–513.

Valenze, Deborah, "The Art of Women and the Business of Men: Women's Work and the Dairy Industry *c.*1740–1840," *Past & Present*, 130, 1 (1991): 142–169.

Von Hoffmann, Viktoria, *From Gluttony to Enlightenment. The World of Taste in Early Modern Europe* (Urbana: University of Illinois Press, 2016).

Walker-Bynum, Caroline, *Holy Feast and Holy Fast: The Religious Significance of Food to Medieval Women* (Berkeley: University of California Press, 1987).

Whaley, Leigh Ann, *Women and the Practice of Medical Cure in Early Modern Europe, 1400–1800* (Palgrave Macmillan: Basingstoke, 2011).

Zilsel, Edgar, "The Sociological Roots of Science," in Id., *The Social Origins of Modern Science*, ed. by Diederick Raven, Wolfgang Krohn, and Robert S. Cohen (Dordrecht: Springer, 2003), pp. 7–21.

CHAPTER 4

Making Marmalade and Conserving Fruit within the Architecture of Seventeenth-Century Courtly Entertainment

Juliet Claxton

1 Introduction

> Last Autumne, I had leave to spend a whole day in observeing the Sweete Chymistry of a Lady, Who is most perfect in the Arte of making Marmalads, & conserving fruite in all their approved colors; And she hath taught me to confesse The Miracles of Heate; Tis sure the philosophers stone that can doe & undo all things.[1]

During the autumn of 1665 the Reverend John Beale (1608–1683) closely observed a woman while she prepared marmalade. In a letter to Robert Boyle (1627–1691) his term "sweet chemistry" implies that it had been a pleasant experience. His purpose, however, was not voyeuristic but to acquire an empirical understanding of the properties of heat in the transformation of fruit into conserve. Beale's writings reveal that he frequently watched the quotidian kitchen activities of his servants as they made bread, cheese or brewed ale.[2] In the very first volume of *Philosophical Transactions,* for example, he described how his cook generated a shiny residue from placing mackerel in a pot of water.[3] His servants became participants in his experiments, even if this was unintentional, at least on their behalf. But in his letter, Beale's use of the word 'Lady' implies a certain lightness of touch, inferring that this batch of marmalade was not made by a servant, but a woman of rank.

1 Cit. in Simon Werrett, *Thrifty Science: Making the Most of Materials in the History of Experiment* (Chicago: University of Chicago Press, 2019), p. 47.

2 Beale was noted for his writings on agriculture and natural philosophy and was an active correspondent of Samuel Hartlib and his circle. Beale's letters to Boyle (in the 1660s) detailed his annotations of household experimentation and urged the systematic order of empirical data in the service of memory and hypotheses. See: Richard Yeo, "Memory and Empirical Information: Samuel Hartlib, John Beale and Robert Boyle," in Charles T. Wolfe, Ofer Gal, ed., *The Body as Object and Instrument of Knowledge: Embodied Empiricism in Early Modern Science* (New York: Springer, 2010) pp. 185–210.

3 Werrett, *Thrifty Science*, p. 53.

 | DOI:10.1163/9789004512610_006

Marmalado or marmalade was first imported into England from Portugal during the late fifteenth century, suggesting that the English received the word direct from the Portuguese *marmelo* or 'quince'. As its name indicates, marmalade originally consisted of a preserve of quince jelly made with sugar and perfumed spices such as rosewater and musk that could be cut into squares for eating. The term marmalade came to denote several varieties of preserved fruit jelly and pippin apples were often used as a substitute for quince, particularly in northern Europe where they were more abundant.[4] It proved to be extremely popular in England as evidenced by the numerous recipes for marmalade in both manuscript collections and published volumes throughout the early modern period.[5] One of the first recipes for marmalade in English is found in *The Secrets of Alexis of Piedmont*, published in 1562, for a conserve of quinces as made in "Valencia or Genoa" that notes "we call it in English, Marmelade".[6] Its popularity arose from the belief that it contained medicinal properties, which created a warming, digestive palliative for the stomach. Sir Thomas Elyot in *The Castel of Helth* noted that quinces: "rosted or sodden, the core taken out and mixed with hony clarified, or sugar, then they cause good appetite, and preserueth the head from drunkennes: taken after meate, it closeth and draweth the stomacke together, and helpeth il digestion, and mollifieth the belly". He recommended "a quince rosted, or in marmalade, and after reste, to amende the lack of nature with selpe".[7] Largely due to the belief in its stomach calming properties, marmalade was incorporated into a food group known in England as "banquetting-stuff" – a selection of spiced wine and sweet-meats specifically chosen for their warming, medicinal properties, which aided digestion after a large meal or banquet.[8]

Preserving fruit using honey, sugar or alcohol was a practical necessity and has a long history in human society. Like most food preparation – including dairy, brewing and basic herbal remedies – preserving chiefly fell within the domestic remit and was commonly (but not exclusively) performed by

4 The addition of rosewater in many early recipes suggests that marmalade may have originated from the Moorish cuisine of the Iberian Peninsula.

5 A recipe 'To make marmalade of Appricoks' appears in a manuscript collection of recipes dated 1672 and attributed by the Floger Shakespear Library to Constance Hall. See: Kristine Kowlachuk, *Preserving on Paper* (Toronto and London: University of Toronto Press, 2019), p. 211. Recipes for true orange marmalade do not appear in English until the very end of the seventeenth century.

6 *The Secrets of Alexis of Piedmont* (London: 1652), p. 60: https://eebo.chadwyck.com.

7 Sir Thomas Elyot Knight, *The Castell of Health* (London: 1539), p. 31 and p. 46: https://eebo.chadwyck.com.

8 Philippa Glanville, Hilary Young, *Elegant Eating: Four Hundred Years of Dining in Style* (London: V&A Publications, 2005), p. 82.

women.[9] The locations and rooms where these domestic activities took place, such as kitchens, dairies and still-rooms were largely the domain of women, even if many early modern households functioned in a multiplicity of ways. There were male cooks, cheese makers and distillers, and while we should not fixate on a single prescriptive framework, the inevitable gendering of many domestic spaces both expresses and works to define the cultural meanings of cooking and physic in this period. In elite establishments where the management of a large and complex household required both hands-on knowledge and lived experience, most domestic food preparation was performed by servants. The lady of the house, however, was expected to have the knowledge and ability to manage and oversee the day-to-day running of her home.

As noted above, recipes for making quince marmalade appear in both household literature and recipe collections compiled from the seventeenth century onwards. Early modern recipes, however, are often frustratingly reticent about the exact steps involved, so observation was an important part of the process. Making marmalade and conserving fruit was and remains a complex task that requires intuition, an ability to judge temperature and familiarity with visual signals to achieve the correct setting point. Observing an experienced cook (as noted by John Beale) was the chief method of learning, and women initiated each other into the techniques. On November 2 1663, for example, Samuel Pepys (1633–1703) noted in his diary that "after a good dinner left Mrs. Hunt and my wife making marmalett of quinces", and just two days later Elizabeth Pepys appears to have mastered the skill as on November 4 Pepys wrote: "Home to dinner, and very pleasant with my wife, who is this day also herself making of marmalett of quince, which she now do very well herself".[10]

The knowledge transfer and the societal networks generated by seventeenth-century culinary and pharmacological literature have been extensively studied, but much less critical discussion has been concerned with the siting of knowledge production or the equipment specified within recipe collections.[11] Preserving and distilling was an experimental process, classified as natural

9 The extensive bibliography of early modern household management describing the skill-set and comportment expected from a householder is almost exclusively addressed to women. In Gervase Markham, *The English Huswife*, that was reprinted numerous times from 1615 to 1683, he encouraged women to: "have a perfect skill and knowledge in cookery, together with all the secrets belonging to the same, because it is a duty really belonging to a woman". See: *The English Housewife*, ed. Michael R. Best (Montreal and London: McGill-Queen's University Press, 1986), p. 60.

10 *The diary of Samuel Pepys. Daily entries from the 17th century London diary*, https://www.pepysdiary.com/diary/1663/11/02 & 04/.

11 Recent studies include: Michelle DiMeo, Sara Penell, *Reading and Writing Recipe Books 1550–1800* (Manchester: Manchester University Press, 2013); and Elaine Leong, *Recipes and Everyday Knowledge* (Chicago: Chicago University Press, 2018).

philosophy and often termed "kitchen physic" that needed its own space and technology. Where did this type of culinary experimentation and knowledge transfer happen, what equipment was used, and was it a 'lady' or her servant who stirred the pan of marmalade? This paper seeks to respond to these questions by exploring the culinary and medicinal experimentation that has been associated with Alathea Talbot, Countess of Arundel (*c.*1585–1654). The Countess has the rare distinction in that both her literary output, and household inventories are extant.[12] Linking her recipe collections to the 1641 inventory of the Countess' Westminster home Tart Hall, sites this activity within contemporary courtly architecture and illustrates where and how the Countess' own, uniquely gendered touch can be identified within different domestic locations.

2 The Countess of Arundel

Alathea Talbot was the youngest and eventually the sole survivor of five children of Gilbert Talbot, seventh earl of Shrewsbury (1552–1616) and his wife, Mary Cavendish (1556–1632). In 1606 she married Thomas Howard, Earl of Arundel, a member of one of England's premier, but relatively impoverished families.[13] The substantial dowry she brought to her marriage enabled her husband to reclaim much of the material status that the Howard family had lost during the sixteenth century, exemplifying the ability of early modern elite women to determine the socio-economic status of their marital families.[14]

Although she remained a life-long recusant Catholic, by birth and marriage Lady Arundel was one of the premier women at the early Stuart court. She was also intelligent and highly accomplished. She spoke Italian, took a lively interest in both art and architecture, and undertook extensive European travels both with her husband and on her own. During 1620–1623 the Countess travelled independently in Italy and the Netherlands, and after going into exile in 1641 she lived separately from her husband until her death in Amsterdam

12 Two seventeenth-century volumes of medicinal and herbal remedies have been linked to the Countess: Wellcome Collection, London, MS 213 (1606); and a sophisticated book of medicines and domestic science entitled *Natura Exenterate* published in 1655. Her household inventories are: the 1641 inventory of Tart Hall, including a unique room called the 'Dutch Pranketing Room'; and the 1655 probate inventory for 292 Singel, in Amsterdam, where the Countess live until her death in 1654.

13 Mary F.S. Hervey, *The Life Correspondence and Collections of Thomas Howard, Earl of Arundel* (Artemis Publishing: Chichester, 2005) p. 24.

14 Elizabeth V. Chew, "The Countess of Arundel and Tart Hall," in Edward Chaney, ed., *The Evolution of English Collecting: Receptions of Italian Art in the Tudor and Stuart Periods* (New Haven & London: Yale University Press, 2003), p. 287.

FIGURE 4.1 The Countess of Arundel, Daniel Mytens, *c.*1618. National Portrait Gallery, London

in 1654. Her behaviour throughout her life suggests that her wealth gave her a high degree of autonomy.

The first recipe collection attributed to the Countess is a handwritten manuscript held by the Wellcome Collection titled: *A Booke of diuers Medecines, Broothes, Salues, Waters, Syroppes and Oyntementes of which many or the most part haue been experienced and tryed by the speciall practize of Mrs Corlyon* (hereafter MS 213).[15] A second, anonymously printed book *Natura Exenterata* published in 1655 has also been accredited to the Countess.[16] There are very strong similarities between the recipes, structure and layout of the manuscript collection and the printed book, and unlike many contemporary recipe collections and household manuals that included recipes for puddings and drying meats alongside herbal remedies, both texts are primarily concerned with ailments and medicine. This suggests Lady Arundel developed extensive and widely recognised medical practices that went far beyond the localised parameters of her own society. Indeed, a palliative for smallpox credited to 'Countess Arundel' appears in another anonymous, seventeenth-century recipe collection:

> Against the small pox to wash ye face
> Take one handfull of Mallow leaves one handfull of hollyoock leaves, and shred
> ym very small, and boyle them in some milke then straine them, and put a little
> butter into it and dip them.[17]

A willingness to embrace this activity invites comparison with the activities of other contemporary gentlewomen who were renowned as medical practitioners. Lady Anne Clifford (1590–1676), was noted as a "distiller of medicinal waters, a dabbler in alchemy, and expert in the property of plants, flowers

15 Wellcome Collection, London, MS 213 is the earliest manuscript of this type held by the Wellcome Collection. The volume is dated 1606 and retains its original binding of calf skin with a gilt-stamped central arabesque ornament flanked by a double 'A', and the flyleaf is inscribed 'Liber Comitissae Arundeliae'.

16 The publication displays a copy of Wenceslaus Hollar's portrait of the Countess on the flyleaf and opens with a prefatory letter written by 'Philatros' (or 'friend of medicine'). The Countess' participation in the compilation of this book is unconfirmed, but her contemporaries, including Samuel Hartlib, attributed it to her, and the London bookseller Gabriel Bedell was clear in characterizing the volume as by the Countess, listing the work as 'The Countess of Arundel Experiments'.

17 Wellcome Collection, London, MS 7391/41, *Anonymous Recipe Book: Mid-17th Century*.

and herbs".[18] Another example is Anna of Saxony (1532–1585), wife of August of Saxony (1526–1586), who was a distinguished herbalist and medical practitioner, whose advice and medical distillations were sought after at several European courts.

The evidence from the Countess' surviving recipe collections and household inventories suggests that she had an interest in preserving and distillating, both for medicinal purposes and as an accompaniment to food. Her final probate inventory for 292 Singel, Amsterdam recorded numerous jars and pots of preserves and syrups including: "thirty potts of severall sorte of Sirups Conserves and preserves' and 'A Box with foure potts of Jelly of Pippin' in the 'uppermost Garrett".[19] Recipes for fruit preserves and syrups appear in both of the Countess' volumes, and she was familiar with exotic fruit such pomegranates as well as apples and sloe-berries found in English hedgerows. Quince and marmalade feature in *Natura Exenterata*, including the Countess' own recipe for marmalade:

> Take quinces and pare them; and pick out the cores and seeth them very tender in water, and strain them through canvas of eight pence the elle, and let them stand in the vessel wherein they be strained, then take of the water that they are sodden in, and take the white of two eggs, beat them, and put them into the same water, then take sugar and put thereto, and let them seeth together halfe a walme, then strain it through a blanket linnen, then put the syrup and the Quinces together, and boil them till they be very thick, and stirre it well for burning, and put it into a box. For two pound of quinces one pound of sugar.[20]

The recipe is very similar to other contemporary marmalade recipes, and it was customary to store the finished marmalade jelly in a wooden box usually lined

---

18 Ann Leigh Whaley, *Women and the Practice of Medical Care in Early modern Europe, 1400–1800* (Palgrave Macmillan: Basingstoke, 2011), p. 153.

19 *Inventory of pictures, etc. in the possession of Alathea, Countess of Arundel, at the time of her death at Amsterdam in 1654*, National Archives, Public Record Office, London (hereafter PRO), Delegates' processes, vol. VIII, NA DEL 1/7 f.687r. Listing foodstuff in a probate inventory was highly unusual unless they were very expensive or classified as commercial stock, their inclusion in this inventory implies they were deemed to be of some value. The inventory of household goods for 292 Singel, Amsterdam is dated 10 April 1655 and attested 25 June 1655. It is preserved in a volume of proceedings in the Court of Delegates in the case of Arundel v. Stafford and includes an important list of the remaining Arundel pictures and collection. It was first drawn up in Italian and then copied into English, probably by a scribe who was unfamiliar with both languages – several words are in Dutch.

20 For examples see: Wellcome Collection, London, MS 213/100 and *Natura Exenterata* (London: 1655), pp. 90–91: https://eebo.chadwyck.com/.

with sugared paper. The Countess advocated using marmalade in some surprising remedies. A recipe "To draw a bullet out of a wound" instructs: "Make a tent of a piece of a quince, or for lack of quince, of Marmalade dipped in oyle of eggs put it in the wound, and it wil draw out the bullet."[21] She also used the term 'marmalade' to denote a paste or jelly as found in a recipe for "Electuary or a Paste, or Marmalade for the Stone in the Reines to break the same".[22]

Both the manuscript and the printed volume note that recipes have been "approved by my selfe and others" implying that they had been tested and experimented upon by the Countess herself.[23] There is no current evidence stating that the Countess prepared any of the recipes credited to her, but Lynette Hunter has argued that female experimentations in domestic medicine was viewed as a social activity within the communities in which the Countess participated.[24] The cooking instructions in both MS 213 and *Natura Exenterata* are implicit rather than explicit and assume a prior knowledge of cooking, which suggests that Lady Arundel met with others to discuss and exchange their recipes or to demonstrate and observe the cooking process. Moreover, it was an established practice for noble women to send their daughters to learn from expert practitioners. In a letter from Anna of Saxony to her daughter Elizabeth, she complained:

> We are more than a little bothered that Your Dearest does not distill the water herself, as we know that the Countess of Hohenlohe personally taught Your Dearest to distill the aqua vitae as the old Countess of Mansfeld once did.[25]

Clearly, Anna was disappointed that her daughter appeared to show little interest in overseeing the preparation of her household's medicine, revealing the importance placed on personal involvement by elite women. To understand where and how these empirical activities took place, this paper turns to the 1641 inventory of the Countess' home at Tart Hall and its Dutch Pranketing Room.

---

21 *Natura Exenterata*, https://eebo.chadwyck.com/.

22 *Natura Exenterata*, https://eebo.chadwyck.com/.

23 Wellcome Collection, London, MS 213/123.

24 The Countess' sister, Elizabeth Grey, Countess of Kent (1582–1651) was also known for her medical experimentation. *A Choice Manuall, or Rare and Select Secrets in Physick and Chyrurgery: Collected, and Practised by the Right Honourable, the Countesse of Kent, Late Deceased*, attributed to Elizabeth Grey was published in 1653. See: Lynette Hunter, "Women and Domestic Medicine: Lady Experimenters, 1570–1620", in Lynette Hunter Sarah Hutton, ed., *Women, Science and Medicine 1500–1700* (Stroud: Sutton Publishing, 1997), p. 90.

25 Alisha Rankin, "Becoming and Expert Practitioner: Court Exerimentalism and the Medical Skills of Anna of Saxony (1532–1585)," *Isis*, 98 (2007): 23–53, p. 36.

3 Tart Hall

An existing house known as Tart Hall was purchased by the Earl and Countess of Arundel in 1633. It was located on the western edge of St James's park, (close to the mulberry gardens) just south of the present site of Buckingham Palace.[26] Although the Arundels purchased the property jointly, it is clear that the Countess herself paid and directed the architect Nicholas Stone to make alterations to the property, suggesting that it was purpose built to her own specifications.[27] No definitive drawing remains of Tart Hall, but it has been associated with a building in the background of an etching titled *Spring* by Wenceslaus Hollar (1607–1677).[28] In this image Tart Hall shows strong Dutch influence, and a surviving building of a similar style is Kew Palace in south-west London, originally built for the wealthy London silk merchant, Samuel Fortrey in 1631.

At the time the inventory was taken in 1641 the "old house" and the "new house" appear to have existed side by side, and included facilities that would have enabled a wide scope for producing the recipes that are recorded in MS 213 and *Natura Exenterata*.[29] The inventory itself was taken just before Lady Arundel's departure into exile and is, therefore, a "household" rather than a "probate" inventory. In noble establishments inventories were usually taken when there was a major change, such as the owner moving or a change in domestic administration. It was a useful way of keeping a check on the location of household items. The house was tidied, and items returned to their correct location; if things were missing, needed repair or were in the charge of a

26 The inventory for the main house was transcribed by Lionel Cust and published in three parts by *The Burlington Magazine for Connoisseurs* in 1911 and 1912. Tart Hall's architecture, inventories and the Arundel's collection have been the subjects of several previous studies See: Chew, "The Countess of Arundel and Tart Hall"; Diane Duggan, "A Rather Fascinating Huybrid", *The British Art Journal*, 4 (2003): 54–64 and Juliet Claxton, "The Countess of Arundel's Dutch Pranketing Room", *The Journal of the History of Collecting*, 22 (2010): 187–96.

27 Stone had spent six years in Holland and was married to the daughter of Hendrick de Deyser, a master mason from Amsterdam. See Lisa Jardine, *Going Dutch* (London, Harper Press, 2008), p. 205. Dutch influence extended to the interior of the house, and several items and pictures are labelled as 'Dutch' throughout the inventories of the house and its Dutch Pranketing Room.

28 Hollar entered Arundel's service in 1636 and Tart Hall would have been familiar to him. See Duggan, "A Rather Fascinating Hybrid."

29 The 1658 map of London by Richard Newcourt and William Fairthorne shows Tart Hall as a single building with two west facing wings. On the south side of the main house is a smaller structure, which may have included stables, and other utility areas. British Library, London, Maps Crace Port. 1.37.

FIGURE 4.2 Wenceslaus Hollar, Tart Hall, *c.*1640, Rijksmuseum, Amsterdam

servant this too was noted.[30] Thus, it is possible to be reasonably sure that the culinary apparatus recorded at Tart Hall was in the place where it was customarily used or stored. Tart Hall's inventory reveals four functional locations that may have been used either by servants or the Countess for preparing food, preserving and medicinal remedies. These include: the kitchens of both the old and the new houses; the servant's hall; and the Dutch Pranketing Room.

Tart Hall's inventory gives no indication of the configuration of these utilitarian locations. The milkhouse, larder and kitchen of the old house, however, are listed in sequence and were grouped together either on the lower floor of the old house or formed part of a separate block just visible in the far-right corner of Hollar's image.[31] Service areas were a common feature of elite households and in older houses they were often grouped together in a free-standing building, although in larger establishments this was sometimes connected to the main house by a passage or covered walkway. Detached kitchens frequently comprised a multi-storey, complex structure containing a number of chambers catering to a range of functions in addition to that of cooking.[32] By the seventeenth century changes in chimney and hearth technology reduced the fire risk, and along with the buttery and pantry kitchens elided into the main building.[33] In the "new house" there is a second kitchen. Again, the inventory gives no clues regarding the room's configuration within the house, but on this occasion its sequencing within the inventory implies that it was integral to the main building.

30 For example, a note stating that the 'snicke' or short corridor leading to the kitchen 'lyeth Broken up, & is very noisome, fit to be mended which as sayd will Cost at least 40s..wch Mr Marshe hath undertaken to se repayred forthwith'. See: Lionel Cust, "Notes on the Collections Formed by Thomas Howard, Early of Arundel and Surrey", *The Burlington Magazine for Connoisseurs*, 20, 104 (1911): 97–100, p. 98.

31 The contents of Tart Hall's milkhouse reveal a very well-furnished diary, equipped for supplying butter, cheese and other dairy products to the household. Butter was useful emollient and appears as a regular ingredient in both Wellcome Collection, London, MS 213 and *Natura Exenterata*. For example, a recipe for 'A Salue to heale a Canker Soore' uses 'Butter out of a Chorne not salted nor washed' and 'freshe Cheese curdes, wringe out the whaye'. See: MS 213/144. It was common practice to employ a dairy maid to produce these ingredients, and it is unlikely that the Countess spent much time in the milkhouse.

32 Tara Hamling, *A Day at Home in Early Modern England: Material Culture and Domestic Life, 1500–1700* (New Haven & London: Yale University Press, 2017), p. 71.

33 Mark Girouard, *Life in the English Country House: A Social and Architectural History* (Penguin: London, 1980), p. 66.

4 The Old and New Kitchens

The Countess' kitchens were well equipped and references to the paraphernalia recorded in the kitchens of the old and new houses are found in both MS 213 and *Natura Exenterata*. The larder and kitchen of the old house contained a significant quantity of items that were necessary for feeding a large, elite household maintained by an extensive corpus of servants. Amongst other items recorded in the inventory there were: mortars of brass and marble; a pastry board; a rolling-pin; an iron jack; bird spits; kettles; skillets; a beef fork; an apple press; a cleaver and chopping board; and numerous other culinary utensils.[34] As the evidence from the main inventory of Tart Hall implies that entertainment and recreation were the chief purposes of the house, it can be assumed that the kitchens regularly produced splendid banquets, such as a great feast in the Italian style given at the Arundels' house in Highgate in 1618:

> In Whitsun week the Countess of Arundel made a grand feast at Highgate to the Lord Keeper (Bacon), the two Lords Justices, the Master of the Rolls (Sir Julius Caesar), and I know not who else. It was after the Italian manner with four courses and four table cloths one under another: and when the first course and table-cloth were taken away, the Master of the Rolls, thinking all had been done, said grace, as his manner is when no Divines are present, and was afterwards well laughed at for his labour.[35]

The Countess obviously enjoyed entertaining as her contemporaries remarked upon her flair for elaborate dining, and most of the rooms at Tart Hall were dedicated to dining, conversation, display, the enjoyment of art and board games.[36]

Adjacent to this kitchen was a small room recorded as "the withdrawing Roome from the kitchen", which was richly furnished with guilt leather wall hangings, green and red cushions, and a leather-covered box (with a lock) containing 12 oyster knives. Over the mantlepiece were four pictures: "the first being of many Figures of Men woemen & Children: second a Doctore with a Rochett; the third of Erasmus, The 4th of a woeman in blacke with a kerchiefe

34 Cust, "Notes on the Collections," p. 341.

35 John Nichols, *The Progresses, Processions, and Magnificent Festivities of King James the First, His Royal Consort Family, and Court* (London: The Society of Antiquaries, 1828), p. 344.

36 Chew, "The Countess of Arundel and Tart Hall," p. 299.

on her head".[37] There is little to indicate the purpose of this room, it contains two muskets but no cooking utensils, nevertheless given its label and proximity to the old kitchen it may have been a gentile setting for the Countess to observe and oversee her servants' activities. Here she could watch or instruct without entering the hot and smoky environment of the main kitchen.

By contrast to the well-equipped kitchen in the old house the second kitchen in the "new house" appears sparsely furnished. Having two kitchens was not an uncommon feature in this period; several elite women are known to have had locations where they prepared special food stuff and medicines. Anna of Saxony, for example, had both distilling houses and a small private kitchen for medical work in her lodge at Augustusberg.[38] A second or "upper" kitchen was also a feature of wealthy merchant homes and two kitchens were commonplace in Dutch houses of the period and, as previously noted, Tart Hall shows substantial Dutch influence.[39]

The new kitchen was clearly designed to hold ingredients of some value. The main doors of both the old and the new kitchens were fitted with a lock and key, and as risk of robbery was very real locks were routinely fitted to exterior doors.[40] The new kitchen, however, also had a "greate Wooden Cubbard with 5. Shelues to putt sweete meates in with a lock & key", and alongside this was a "safe to putt could Meate in", again furnished with a lock and key.[41] The room also contained an "oyster table", which was a small, utilitarian table, with a drawer at the front. An example of this style of table can be seen in Zurbaran's domestic interior, "The House at Nazareth".[42] Listed alongside the table are "three little Table wayters standing vpon Iron feete", which were small salvers, used for serving food. The rest of the items in the room were all used

---

37 Cust, "Notes on the Collections," p. 98. A Rochett was a notched cane sometimes carried by physicians. For an example see: Wellcome Collection, Images: https://wellcomecollection.org/works/vkbdqrzg.

38 Rankin, "Becoming and Expert Practitioner," p. 24.

39 See Edward Roberts, Karen Parker, *Southampton Probate Inventories, 1447–1575*, Vol. 35 (Southampton: Southampton Records Series, 1992), p. 350. The configuration of a second kitchen is found in seventeenth-century Dutch dolls' houses. For examples in the Rijksmuseum, Amsterdam, see: https://www.rijksmuseum.nl/en/rijksstudio/works-of-art/dolls-houses.

40 Amanda Flather, *Gender and Space in Early Modern England* (Woodbridge: Boydell Press, 2007), p. 47.

41 Cust, "Notes on the Collections", p. 98.

42 Oyster tables were designed to be placed against a 'pier wall' or the wall space between two windows. For an example, see: Victoria and Albert Museum, London, W.53.1, 2–1948, http://collections.vam.ac.uk/item/O8474/table-unknown/.

FIGURE 4.3 Francisco Zurbaran, *The House in Nazareth*, c.1630, Cleveland Museum of Art

for brewing. Further paraphernalia for beer making are found in the "Seller of this Kitchen", and a pantry over the cellar had a "blankett to couer woort" and a "Bing" or bread oven.[43]

Expensive items such as sweet meats were frequently the responsibility of the mistress of the house, and she may have been unwilling to hand them over to the main kitchen where they might have been subject to petty pilfering. The Countess was clearly well informed regarding the cost of ingredients and in MS 213 expensive ingredients were often expressed in pecuniary terms: for example, "A Medecine food for those that are apte often to caste through weakeness of the Stomacke' uses 'two penny worthe of Saffron".[44] "A Medicine for the Collick and the Stone" uses "a pennyworth of cloves and mace an halfepennye worth of longe pepper, and two pennyworth of Turmarick", while

43 This is the only time that an oven is recorded in either kitchen, but as they were integral to the fabric of the building they rarely appear in inventories. Woort is a sweet liquid produced by steeping ground malt or other grain in hot water, which is then fermented to produce beer and distilled malt liquors or unfermented beer.

44 Wellcome Collection, London, MS 213/62.

another has "the weight of eyghte pence in Parmacetye, two pennyworthe of cloues" and "half a crowne of the powder of Mastick" or "the weighte of 3 shillinges of the best reysons".[45]

Lacking from either the old or new kitchens are any items for calibration, but ingredients were often measured by volume or sight, and scales do not habitually appear in kitchen inventories before the mid-eighteenth century.[46] Both MS 213 and *Nature Exenterate*, however, issue instructions to calibrate ingredients using hands, fingers or quotidian, household objects. Several measurements indicate the "breadth of your two fingers" or "as long and brode as a large hand" or a "mans hand".[47] A cure "to take away the pynn and webb in the eye" calls for "fyne white sugar as much as a walnut and a piece of Sanguis Draconis as bigg as a Beane".[48] Nuts and beans are effective units of measurement as they are usually consistent in size: for example, a "Salue for any Soore" suggests the using a pinch the size of "a greate walnut". Shellfish or birds also feature as a comparable unit, presumably as they too are normally of a similar size: for example, "as much Allome as a crabb" and "no bigger than a young cockes combes".[49]

Furthermore both MS 213 and *Natura Exenterata* refer to alternatives and substitutions for ingredients: for example, "Rennish wine and milke, if not rennish of the smallest white wine", while another recipe notes that "if turmerick is not available powder of stages horne" is an appropriate alternative.[50] A cure for shingles calls for "the blood of a catt or of a pigeon" and another recipe needed "deares sewett or for want thereof as much of sheepes sewett".[51] These careful notes on calibration and substitution would allow for recipes to be replicated with a degree of accuracy, allowing the Countess to instruct servants to recreate recipes on her behalf.

Elite women of this period were often responsible for large establishments with many dependents and the Countess must have relied upon senior servants, both male and female to produce food, brew beer and distil medicines. Like many of her contemporaries the Countess almost certainly issued her

---

45 Wellcome Collection, London, MS 213/68/70/76/110.

46 Sara Pennell, *The Birth of the English Kitchen, 1600–1800* (London: Bloomsbury, 2016), p. 90.

47 Wellcome Collection, London, MS 213 60/71/102, and MS 213/144.

48 Wellcome Collection, London, MS 213/12/61/72.

49 Wellcome Collection, London, MS 213/144/151/155/146.

50 Wellcome Collection, London, MS 213/70/78.

51 Wellcome Collection, London, MS 213/95/151. 'Sewett' refers to the fat from the thymus gland of an animal.

retainers with detailed instructions regarding the preparation of food, but it was also common practice to keep expensive ingredients away from the main kitchen for adding at the final cooking stage. The inventory of another seventeenth-century courtier and herblist, Lady Johanna St. John (1631–1705), reveals that she kept amber and musk amongst her personal items presumably to be added at the latter stages of a recipe's preparation by her own hand.[52] MS 213 contains a recipe for "A srype of Horehound good for the Consumption of the Lunges" which requires that "when it is almost boiled take a graine of muske well mixed with two spoonefulles of Rose water and putt it therein", implying that the Countess was familiar with this procedure.[53]

5 The Servants Hall

A final utilitarian location at Tart Hall was the detached servant's hall and the bulk of the items recorded were kept there for distilling. The equipment recorded adhered closely to the guidelines set out in Sir Hugh Platt's treatise on distilling *The jewel house of art and nature* published in 1594.[54] Tart Hall had two brass stills with pewter covers, three iron preserving pans with brass bottoms, a copper kettle, and a marble cistern. Still rooms were an important feature of early modern households. Melford Hall in Suffolk, for example, had a still house with "foure stilles & a limbecked" and "18 fruit dishes and 24 fruite glasses".[55] But, home distillation was not a preserve of the nobility and stills appear in merchants' inventories, particularly in those belonging to apothecaries.[56] Several recipes in MS 213 and *Natura Exenterata* call for a still or "glass sillitory": for example, "An excellent water food for a Surfett" instructed "to distill all in a lymbicke, wth a lymbicke, wth a very softe fyer".[57] Flowers and fruit distilled into waters, conserves and syrups all needed a dependable heat resource – so a good wood supply was a necessary requirement for distillation. The still room at Melford Hall contained "lumber", while at Tart Hall there was

52 Leong, *Recipes and Everyday Knowledge*, p. 62.

53 Wellcome Collection, London, MS 213/135.

54 Sir Hugh Platt, *The jewel house of art and nature containing divers rare and profitable inventions, together with sundry new experiments in the art of husbandry: with divers chimical conclusions concerning the art of distillation* (London: 1594), https://eebo.chadwyck.com/.

55 Lyn Boothman, Richard Hyde Parker, *Savage Fortune: An Aristocratic Family in the Early Seventeenth Century* (Woodbridge: Boydell Press, 2006), p. 94.

56 Roberts, Parker, *Southampton Probate Inventories*, p. 351.

57 Wellcome Collection, London, MS 213/123.

"an ordinary sawe for Billetts".[58] The fire risk in still houses was considerable, hence their frequently detached status; an arrangement that also kept any unpleasant cooking odours away from the main house.

Distilling fruit and flowers implies an aromatic, sweet-smelling activity, but without the convenience of modern, heat-regulated hotplates and extractor fans it is time consuming and produces a sticky residue. Some of the Countess' recipes for distillation would have produced pungent smells and called for very extended cooking times. In MS 213 the recipe for "The distillation of a Pigg very good for those that ar weake and faynte" recommends to "sett it in a kettle of seething water upp to the necke and let it boile continuallye the space of 14 howers".[59] The pre-modern kitchen relied on aural signals to record time, such as church bells, or used smell or visual indicators to judge if cooking was completed.[60] It was hot, dirty work and often required several pairs of hands. Producing good results from distillation needed both patience and experience and was mostly left to experts. Leong notes that Lady Johanna St. John relied on the wife of her senior steward, Elizabeth Hardyman, to distil household medicines and at least twice used the services of her local apothecary, Mr Goram, to distil citron water, although she supplied him with eight citrons and twelve lemons to be distilled to her own recipe.[61]

FIGURE 4.4 Two women and a man work at distilling plants and herbs, Wellcome Collection

58 'Billets' was a term used for wood suitable for fuel.

59 Wellcome Collection, London, MS 213/108.

60 Pennell, *The Birth of the English Kitchen*, p. 988.

61 Leong, *Recipes and Everyday Knowledge*, pp. 61–62.

6 The Dutch Pranketing Room

One area within Tart Hall that has previously been associated with the Countess' medicinal experimentation is the Dutch Pranketing Room.[62] In many ways the room was representative of a typical Jacobean banqueting house, but it also housed the formal display of the Countess' impressive collection of 496 pieces of Chinese porcelain arranged alongside items of brass, glass, and exotica.[63] The layout of the Pranketing Room, which is carefully recorded in the inventory, demonstrates that the Countess used the arrangement of objects within the room to maximise the visual display of artworks, but the contents of the room also reveals 36 items of high-quality brass and iron cookware. Such as: "a Milke strayer of brass & two funnels of Brass", and an "Eight square Barzeele to putt fire in".[64]

Another small collection of culinary equipment was grouped together under a dresser on the fourth side of the Pranketing room, adjacent to a door leading towards the kitchen yard:

> One Brass Pan to Boyle fish in with a Couer, one skellett of brass tinned & one greate Brass ladle very deepe, one little Brasse baking pann, two brass panns tinned for frying of Eggs, one Brass Morter but noe Pestell belonging there unto, three Iron Potlides, two Marble Kettles hooped with Iron, Two salters of Torteaux shell.[65]

These items are identifiable from kitchen inventories and seventeenth-century images of domestic interiors. A pan to cook eggs, for example, can be seen in Diego Velasquez's *An Old Woman Cooking Eggs* and eggs appear numerous times in MS 213; for example, "A medicine for winde in the Stomake" uses "new laid egg".[66]

62 Jennifer Rabe, "Mediating between Art and Nature: The Countess of Arundel at Tart Hall," in Susanna Burghartz, Lucas Burkart, Christine Gottler, ed., *Sites of Mediation* (Leiden: Brill, 2016), pp. 183–210.

63 The porcelain, artworks and collections listed in the Pranketing Room have been the subject of a previous study. See: Claxton, "The Countess of Arundel's Dutch Pranketing Room."

64 Arundel Castle, Arundel, IN 1, *An Inventory of the Parcells of Purselin, Glasses & Other Goods now remayning in the Pranketing Room*, f. 10.

65 Arundel Castle, Arundel, IN 1, f. 24.

66 Wellcome Collection, London, MS 213/61.

FIGURE 4.5 Diego Velazquez, *Old Woman Cooking Eggs*, c.1618, National Galleries Scotland

On the first side of the room next to the chimney the inventory recorded "a very large Deepe vessell of Purselin And therein a Turned sticke with barre at the end: gagged of wood. to make some sorte of liquor with".[67] The description of this porcelain vessel suggests that it had a functional, rather than a decorative purpose, and the qualities of porcelain's glaze, combined with its ability to withstand heat would have made it ideal for producing medicinal preparations or banqueting-stuff. There is no manuscript evidence supporting porcelain's use in the production of food or herbal remedies, but porcelain was being used by apothecaries and there are extant porcelain apothecary jars from this period.[68] The qualities of porcelain's glaze would have been understood and

67 Arundel Castle, Arundel, IN 1, f. 5. Wooden sticks often appear in images and recipes, sometimes even a species of tree is specified, for example, a recipe in Wellcome Collection, London, MS 1793/7 *Seventeenth-Century Recipe Collection* calls for 'a green ash stick'. Elder and ash were the most popular as these trees were pollarded to produce long straight lengths of wood, which could easily be turned into utensils.

68 For an example of a porcelain apothecary jar see: British Museum, London, item no. 1963,0422.6: https://research.britishmuseum.org/research/collection_online/collection_object_details.aspx?images=true&objectId=255300&partId=1.

appreciated as it was well known that metallic vessels could affect both the taste and the efficacy of the medicines they held. Recipes often specified using an earthen dish. In a recipe for the 'ague' taken from a mid-seventeenth century recipe collection it states: "You must boyle the drink in an earthen pipkin because brasse will make it taste".[69]

Previous readings of this inventory have understood the presence of this cookware as evidence that the Pranketing Room acted as a multi-functional space allowing the Countess to prepare delicate banqueting-stuff or experiment with medicinal recipes.[70] This conclusion is not implausible. A deeper reading of the Pranketing Room inventory, however, reveals that while some limited cooking and food preparation may have taken place in the room, it remained primarily a space for courtly entertainment.

A display of *artificialia, naturalia* with dining and cooking facilities was not unique to the Pranketing Room, and other contemporary banqueting houses contained similar items. The 1604 inventory for Wardour Castle's Posslyn House had an impressive collection of tableware, porcelain and Venetian glass, while the banqueting house at Melford Hall contained a marble table, and items of plate, brass and pewter.[71] The Grand Pavilion in the Hortus Palatinus, Heidelberg, designed by Salomon de Caus (1576–1626) for the Elector Palatine, Frederick V (1596–1632), housed a collection of art and natural wonders and some limited cooking facilities, suggesting that it functioned as a dining facility away from the formalities of the court.[72] Moreover, the Countess may have viewed this banqueting house when she accompanied Elizabeth of Bohemia (1596–1662) to Heidelberg in 1613. Banqueting houses were designed for display and exhibition, and the presence of high-quality cookware in the Pranketing Room suggests that some food preparation was integral to an evening's entertainment in banqueting houses.

Some of the items of cookware may even have been part of the display as they were evenly spaced around the room. On a single shelf on the third side of the room were:

> Two long round potts with Eares of Purselin: And on that shelfe a Pot lid, two chafing dishes and two little Candlestickes of Brass, and four little Purselin Dishes, & for ornament a Frogg of Purselin.[73]

---

69 Wellcome Collection, London, MS. 774/36 *Townshend Family Cookery and Medical Receipts 1636–47*.

70 Rabe, "Mediating between Art and Nature", p. 186.

71 Wiltshire and Swindon Archives, MS 266/22/2/2, *Wardour Castle Inventory*.

72 Luke Morgan, *Nature as Model: Salomon de Caus and Early Seventeeth-Century Landscape Design*, (Philadelphia: University of Pennsylvania Press, 2006), pp. 152–197.

73 Arundel Castle, Arundel, IN 1, f.13.

FIGURE 4.6 Brass chafing dish, *c.*17th century, Victoria & Albert Museum

Chafing dishes were used for heating ingredients and instructions to use a 'chafing dish of coals' appears numerous times in both MS 213 and *Natura Exenterata* as well as many other contemporary recipe collections. But as these two brass chafing dishes were placed in amongst a decorative arrangement, they may have been part of the display of objects and artworks. They might have resembled a small, high-quality seventeenth-century chafing dish in the collection of the Victoria and Albert museum.

Any cooking performed in the Pranketing Room must have been constrained. Taking into consideration both the contents of the Pranketing Room and the Countess' recipes reveals that the preparation and cooking processes for many of the recipes found in MS 213 and *Natura Exenterata* were not compatible with a room filled with expensive porcelain and other objects of art. Several recipes found in these volumes needed to be cooked for extended periods of time creating a lot of grease and smoke, and although the room had a chimney the only other sources of heat were two braziers. The initial preparation of the ingredients too was both messy and smelly. A cure for a cough required the lungs of a fox to be marinated in white wine for 12 hours, although as the inventory recorded "a fox tayle to keepe all neate" the more decorative parts of the animal were deemed appropriate for the room.[74] Other recipes are even more problematic: "An other Brothe good for the same purposes" (i.e. weakness) takes "a red Cocke of an yeare and an halfe olde. Pluck hym aliue and beate hym with a rodd until the bloode do appeare through the skinn, then breake his necke and drawe hym".[75] While a broth for those afflicted with

74 Wellcome Collection, London, MS 213/47; Arundel Castle, Arundel, IN 1, f.22.
75 Wellcome Collection, London, MS 213/105.

"melancholye" sets mutton knuckles in a pipkin over a fire with the instruction to "lett them boile until such tyme as the brothe doth smell of the moate".[76] The putrid smell of moats was a common complaint, as they usually consisted of stagnant water and were used as an overflow for kitchen and household waste, including effluent from the privies.

In fact, the corpus of recipes from MS 213 or *Natura Exenterata* suitable for preparation or cooking in the Pranketing Room are limited, but it is plausible that the final, arrangement or addition of an expensive or fragrant ingredient was performed in the room. An example from MS 213 is "A syrupe of Damaske Roses good to purge choler", which was prepared in two stages. The first required that a preparation of roses and water in a pipkin were steeped overnight, then boiled until only a pint and a half of liquid remained. During the second stage the remaining liquid was re-boiled with sugar to produce thick syrup before being poured into a glass.[77] Unless the room was extremely well-ventilated boiling sugar for a long time would have left sticky deposit on the Countess' porcelain. One explanation, however, is that the first stage of this recipe was performed by a servant in Tart Hall's old kitchen or servant's hall, before being brought to the Pranketing Room for a final short reduction, which was decanted into a glass and served to a guest by the Countess herself.

The Countess' recipe for marmalade notes that it needed to be reduced to a solid jelly before being boxed. Quince is high in pectin (the setting agent) but even in a modern kitchen this procedure takes extended cooking at a high temperature. The final addition of flavourings such as rosewater, however, needs to be done when cooking is completed, or the essence will be lost. While there are no records of this activity in the Pranketing Room, is very likely that marmalade was amongst the foodstuffs regularly served at the Countess' banquets. Markham's *The English Housewife* lists the common ingredients for a banquet and the order in which they should be served.

> I will now proceed to the ordering or setting forth of a banquet; wherein you shall observe that march panes have the first place, the middle place, and last place; your preserved fruits shall be dished up first, your pastes next, your wet suckets after them, then your dried suckets, then your marmalades and goodinyakes, then your comfits of all kinds; next, your pears, apples, wardens baked, raw or roasted, and your oranges and lemons sliced; and lastly your wafer cakes.[78]

76 Wellcome Collection, London, MS 213/111.

77 Wellcome Collection, London, MS 213/133/134.

78 Best, *The English Housewife*, p. 121.

The serving and arrangement of banqueting-stuff was extremely important, as many of the ingredients were expensive, particularly the 'suckets' that formed an important element of the banquet.[79] Both wet and dry suckets were sweetmeats composed of fruit, frequently citrus such as lemons or oranges that had been preserved in sugar and spices.[80] Wet suckets were presented in their sugar syrup and both cream and butter were served as accompanying sauces to these delicacies. The expensive nature of these items makes it likely that they were stored under lock and key – at Tart Hall they were probably kept in the lockable cupboards recorded in the new house's kitchen.

As most of the porcelain in the Pranketing Room was ornamental (the display is symmetrical and would have been difficult to replace accurately if the porcelain was being regularly used) there is an apparent lack of dishes to serve the banquet course. But Tart Hall's inventory records a small room, one wall of which was sited "next to the Pranketing Room", which stored a substantial quantity of tableware including "eleauen ouall Dishes of White Earth, 9 fruite Dishes carued of Earth, 2 Bason and Eures of White Earth, & three earthen Dishes more & 4 Poringers of earth white", while another cupboard in the same room held "ten dozen & 7 glasses great & little".[81] These items were stored adjacent to the Pranketing Room so that they could be easily transported into the room when the Countess entertained guests to a banquet. Moreover, as the bulk of the cookware itemized in the inventory is on the fourth side of the room next to a door leading to the kitchen it is credible to conclude that the banquetting-stuff was primarily prepared in the kitchens, before being brought to the Pranketing Room, allowing the Countess to personally add final, expensive ingredients or finishing touches to the sweetmeats, confections or medicinal preparations before they were served to her guests.

7 Conclusions

Each one of the utilitarian spaces within Tart Hall may have acted as a site for knowledge transfer, but the nature of empirical information that was either presented or acquired differed between rooms, companions and social situations. Culinary knowledge and experience were passed between the Countess, her

79 'Suckets' were fruits preserved in sugar, either candied or in a syrup.

80 Peter Brears, "Rare Conceites and Strange Delights," in C. Anne Wilson, ed., *Banquetting Stuffe* (Edinburgh: Edinburgh University Press, 1991), pp. 60–114.

81 Cust, "Notes on Colletions," p. 99.

companions and servants, and amongst the servants themselves. Deciphering Lady Arundel's 'touch' both from her recipes and from the Tart Hall inventory remains circumstantial, but presuming that she conducted herself in a similar fashion to other contemporary elite women (and there is no evidence to the contrary) it is reasonable to conclude that the first stages of food or medicinal preparation were performed in one of Tart Hall's two kitchens or servant's hall. The old house kitchen was the domain of the servants, but almost certainly the Countess, her ladies or a senior servant oversaw culinary activities from the small withdrawing room next to the kitchen. The room acted as a gentile location allowing for observation to occur between the Countess and her servants. This arrangement would have allowed the Countess to retain the kitchen of the new house as a space for experimenting with recipes and storing expensive comestibles, allowing her to make luxury or exotic additions at the final stages of preparation and cooking. The detached status and distilling equipment recorded in the servant's hall implies that complicated distillations and preserves, which required extended cooking periods took place in this location.

The Dutch Pranketing Room remained primarily a space for entertainment, but it is conceivable that the room also included a performative aspect where the final elements of banquetting-stuff became part of courtly entertainment. From the room's viewing gallery, a guest could both admire the magnificent display of porcelain, brass, *naturalia* and works of art, and watch as the Countess added the final flourish to the banquet dishes brought into the room by her servants. It is noteworthy that the room contained a painting of "the monkey and cat" that is recorded next to the chimney of the Pranketing Room. The picture interprets one of Aesop's ancient Greek fables, where a monkey flatters a cat and persuades it to pull scalding chestnuts out of the fire, which the monkey quickly eats. It is traditionally seen as a warning against the evils of sycophancy, but in this case perhaps the visual joke may have been to remind the Countess to use a cloth to protect her 'touch' when receiving and handling hot dishes.

The probate inventory for 292 Singel, Amsterdam, where the Countess lived until her death in 1654 reveals that she carried her interest in preserving and distilling fruit with her into exile.[82] Unfortunately, her financial situation, particularly towards the end of her life was severely constricted. Numerous items in her probate inventory were judged "oulde" [old]; for example, an "ould table with a leather cover" and "one Ould Straw Bedd of my Ladyes".[83] Lord

82 PRO, Delegates' processes, vol. VIII, NA DEL 1/7 f. 686r and f.687r.

83 PRO, Delegates' processes, vol. VIII, NA DEL 1/7 f.687r.

FIGURE 4.7 Abraham Hondius, *The Monkey and the Cat*, c.17th century, Cleveland Museum of Art

Hyde, first Earl of Clarendon (1609–174) commented that at her demise she was 'so clogged with her husband's debts that [even] if her age would permit, she cannot return to her native country.[84] The culinary equipment recorded at 292 Singel, however, indicates that the Countess remained actively involved in experimenting, preserving and creating medicinal remedies. It is even conceivable that substantial quantity of preserves, syrups and jams recorded in her final inventory (far in excess of the amount required for her personal consumption) were distributed amongst her fellow Royalist exiles, following the Catholic tradition of providing charitable sustenance to the unfortunate and impoverished.

84 Geoffrey Smith, *The Cavaliers in Exile, 1640–1660* (London: Palgrave Macmillan, 2003), p. 83.

References

Alexis of Piedmont, *The Secrets of Alexis of Piedmont* (London: 1652).

Boothman, Lynn, Richard Hyde Parker, *Savage Fortune: An Aristocratic Family in the Early Seventeenth Century*, (Woodbridge: Boydell Press, 2006).

Brears, Peter, "Rare Conceites and Strange Delights," in Anne Wilson, ed., *Banquetting Stuffe* (Edinburgh: Edinburgh University Press, 1991).

Chew, Elizabeth V., "The Countess of Arundel and Tart Hall," in Edward Chaney, ed., *The Evolution of English Collecting: Receptions of Italian Art in the Tudor and Stuart Periods* (New Haven & London: Yale University Press, 2003).

Claxton, Juliet, "The Countess of Arundel's Dutch Pranketing Room," *The Journal of the History of Collecting*, 22 (2010): 187–96.

Cust, Lionel, "Notes on the Collections Formed by Thomas Howard, Early of Arundel and Surrey", *The Burlington Magazine for Connoisseurs*, 20, 104 (1911): 97–100.

DiMeo, Michelle, Sara Penell, *Reading and Writing Recipe Books 1550–1800* (Manchester: Manchester University Press, 2013).

Duggan, Diane, "A Rather Fascinating Hybrid", *The British Art Journal*, 4 (2003): 54–64.

Elyot Knight, Thomas, *The Castell of Health* (London: 1539).

Flather, Amanda, *Gender and Space in Early Modern England* (Woodbridge: Boydell Press, 2007).

Girouard, Mark, *Life in the English country house: A Social and Architectural History* (London: Penguin, 1980).

Glanville, Philippa, Hilary Young, *Elegant Eating: Four Hundred Years of Dining in Style* (v&A Publications: London, 2005).

Hamling, Tara, *A Day at Home in Early Modern England: Material Culture and Domestic Life, 1500–1700* (New Haven & London: Yale University Press, 2017).

Hervey, Mary F.S., *The Life Correspondence and Collections of Thomas Howard, Earl of Arundel* (Artemis Publishing: Chichester, 2005).

The Diary of Samuel Pepis. Daily entries from the 17th century London diary, https://www.pepysdiary.com/diary/1663/11/02 & 04/.

Hunter, Lynette, "Women and Domestic Medicine: Lady Experimenters, 1570–1620," in Lynette Hunter and Sarah Hutton, ed., *Women, Science and Medicine 1500–1700* (Stroud: Sutton Publishing, 1997).

Jardine, Lisa, *Going Dutch* (Harper Press: London, 2008).

Kowlachuk, Kristine, *Preserving on Paper* (Toronto & London: University of Toronto Press, 2019).

Leong, Elaine, *Recipes and Everyday Knowledge* (Chicago: Chicago University Press, 2018).

Markham, Gervase, *The English Housewife*, Michael R. Best ed. (McGill-Queen's University Press: Montreal & London, 1986).

Morgan, Luke, *Nature as Model: Salomon de Caus and Early Seventeeth-Century Landscape Design* (Philadelphia, University of Pennsylvania Press, 2006).

Natura Exenterata (London: 1655): https://eebo.chadwyck.com/.

Nichols, John, *The Progresses, Processions, and Magnificent Festivities of King James the First, His Royal Consort Family, and Court* (London: The Society of Antiquaries, 1828).

Pennell, Sara, *The Birth of the English Kitchen, 1600–1800* (Bloomsbury: London, 2016).

Platt, Hugh, *The jewel house of art and nature containing divers rare and profitable inventions, together with sundry new experiments in the art of husbandry: with divers chimical conclusions concerning the art of distillation* (London: 1594).

Rabe, Jennifer, "Mediating between Art and Nature: The Countess of Arundel at Tart Hall," in Susanna Burghartz, Lucas Burkart, Christine Gottler, ed., *Sites of Mediation* (Leiden: Brill, 2016), pp. 183–210.

Rankin, Alisha, "Becoming and Expert Practitioner: Court Exerimentalism and the Medical Skills of Anna of Saxony (1532–1585)," *Isis*, 98 (2007): 23–53.

Roberts, Edward, Karen Parker, *Southampton Probate Inventories, 1447–1575*, vol. 35, (Southampton: Southampton Records Series, 1992).

Smith, Geoffrey, *The Cavaliers in Exile, 1640–1660* (London: Palgrave Macmillan, 2003).

Werrett, Simon, *Thrifty Science: Making the Most of Materials in the History of Experiment* (Chicago: University of Chicago Press, 2019).

Whaley Ann Leigh, *Women and the Practice of Medical Care in Early Modern Europe, 1400–1800* (Basingstoke: Palgrave Macmillan, 2011).

Yeo, Richard, "Memory and Empirical Information: Samuel Hartlib, John Beale and Robert Boyle," in Charles T. Wolfe, Ofer Gal, ed., *The Body as Object and Instrument of Knowledge: Embodied Empiricism in Early Modern Science* (New York: Springer, 2010), pp. 185–210.

CHAPTER 5

Women in Secrets: Drug Inventions between Household, Guilds and Small Scale-Economy

Sabrina Minuzzi

1 From the Margins to Ubiquity? The Connective Tissue of Early Modern Medicine

Over the last few decades, we have realised that the idea of women being marginalized in the history of medicine is only true on the surface, however, in actuality, it's a rather outdated concept.

Since the 1960s, the reconstruction of the history of medicine has begun to go beyond academic texts alone along with professional sources, and has started to reflect on the silence of the sources usually examined.[1] Excluded from academic career tracks as well as from formal professional training, women hardly ever had professional titles nor were they usually represented in guilds, though this varied from country to country and over time.[2] In actual fact, in the medical field, they were often penalised even though they actually made a great contribution on an everyday basis.[3] Expanding on the sources examined has led to progressively going beyond the idea of dividing the profession into three major roles – physicians, surgeons and apothecaries – professional labels that were at times used fluidly by the key players themselves.[4] At the same time, this has led to focusing on the concept of the *medical agent*, which expands on

1 Monica H. Green, "Gendering the History of Women's Healthcare," *Gender & History* 20, 3 (2008): 487–518; Mary E. Fissell, "Introduction: Women, Health, and Healing in Early Modern Europe," *Bulletin of the History of Medicine* 82, 1 (2008): 1–17; Sharon T. Strocchia, "Introduction" to *Women and Healthcare in early modern Europe*, special issue of *Renaissance Studies* 28, 4 (2014).

2 A recent comparative European framework is provided by Anna Bellavitis, *Il lavoro delle donne nelle città dell'Europa moderna* (Rome: Viella, 2016).

3 Monica H. Green, "Documenting Medieval Women's Medical Practice", in Luis García-Ballester, Roger French, Jon Arrizabalaga, Andrew Cunningham, ed., *Practical Medicine from Salerno to the Black Death* (Cambridge, MA: Cambridge University Press, 1994), pp. 322–351; Monserrat Cabré, "Women or Healers? Household Practices and the Categories of Health Care in Late Medieval Iberia", *Bulletin of the History of Medicine* 82, 1 (2008):18–51.

4 Sandra Cavallo, *Artisans of the Body in Early Modern Italy: Identities, Families and Masculinities* (Manchester: Manchester University Press, 2010), pp. 49–54.

 | DOI:10.1163/9789004512610_007

the range of activities the female figure took part in, and the equally inclusive concept of *bodywork* practiced by a variety of individuals that dealt with "care" in addition to "curing" the body.[5] Numerous figures – both male and female – have thus come to light. They previously were not considered or believed to only be found in the shadows populated by empiricists, women and charlatans.[6] This is especially true when looking at the Anglo-Saxon environment. Women increasingly played greater roles in various forms of healthcare provision, from the domestic environment to working for organizations meant to help those in weaker social strata.[7]

Studies have looked at new settings where medical knowledge was generated, utilised and shared, and the most important has turned out to be the household, with a continuous exchange of information among the empirical, professional and commercial spheres.[8] Another important area where knowledge of medical/pharmaceutical practices came about was in lay caregiving institutions and in religious institutes, whose boundaries were equally permeable, allowing for constant communication with domestic, professional and commercial medicine.[9]

Research carried out on multiple sources, despite being fragmented at times, has allowed us to identify a varied array of female figures, found in such great numbers as to definitively challenge the marginal role attributed to them until a few decades ago. Or rather, marginality in terms of the organised type of work found in guilds and the medicine taught at the universities coexists

---

5 For the codification of women as *medical agents* see Monica H. Green, "Bodies, Gender, Health, Disease: recent Work on Medieval Women's Medicine," *Studies in Medieval and Renaissance History* 2 (2005): 1–46; by *bodywork*, common denominator for extremely varied medical-hygienic and treatment activities, and exemplified by the research of Sandra Cavallo, see Fissell, "Women, Health, and Healing," p. 10.

6 A binomial that is still prolific, e.g. Margaret Carlyle, "From Practice to Print: Women Crafting Authority at the Margins of Orthodox Medicine," *Mémoires du livre*, 1 (2014). For the medical *penumbra* cf. Laurence Brockliss, Colin Jones, *The Medical World of Early Modern France* (Oxford: Clarendon Press, 1997), pp. 262–273.

7 Deborah E. Harkness, "A View from the Streets: Women and Medical Work in Elizabethan London," *Bulletin of the History of Medicine*, 82, 1 (2008): 52–85.

8 Enlightening is the European framing provided by Sandra Cavallo, "The Domestic Culture of Health in Early Modern Europe," in J. Eibach, M. Lanzinger, ed., *The Domestic Sphere in Europe (16th to 19th century)* (London: Routledge, 2020), pp. 455–74.

9 Exemplary for the plurality of sources and female figures intercepted is the recent research by Sharon T. Strocchia, *Forgotten Healers: Women and the Pursuit of Health in Late Renaissance Italy* (Cambridge (Mass.): Harvard University Press, 2019); a further broadening of the range of sources, which goes so far as to embrace the aspects of religious healing is in Sara Ritchey, Sharon T. Strocchia, ed., *Gender, Health, and Healing, 1250–1550* (Amsterdam: Amsterdam University Press, 2020).

with the wide variety of women active in many areas of medical practice; this included exchanging knowledge even with the academic and professional, lay and religious, as well as commercial worlds. Keeping all of this in mind, women in the history of medicine could actually be seen as the *connective tissues* of a living organism. We speak of connective tissues in the plural because they have highly differentiated morphological features and varying functions: filling, supporting, protecting and always essential in the exchange of nutrients among the various tissues and organs with which they interact.[10] Women, similar to connective tissues, are far less identifiable than individual organs – with men still playing that dominant role – yet they are numerous and equally indispensable to the life of the organism, in other words, the history of medicine, thanks to their versatility and omnipresence.

From a quantitative point of view, women of secrets did not have a large impact in terms of the history of female medicine, but they certainly did from a qualitative viewpoint, bringing quite a bit to the great deal of differentiation of the connective tissues that they embody.

In the following pages we shall start from a reflection on the main source taken into examination, relating to the authorisations for the handling of medicinal secrets, essential in order to correctly interpret the historical and social context in which the practice can be found.[11] This is a source that, similar to professional ones, favours male representation. Hence, in reviewing the few female figures that emerge from the survey, we shall intersect, when possible, the biographical sources available and, in some cases, print books and manuscripts, which can partially make up for the silence or the omissions of the main serial source. The intersection of sources – along with the broadening of their range – remains one of the most effective antidotes to correcting the invisibility suffered by the female component in the history of medicine.[12] In some cases we know of the existence of these women only because we have dug deeper into the biography of spouses or male acquaintances who appeared in the principal serial source.

We shall try to put together clues to reconstruct the socio-cultural profiles of the women who took part in the invention and/or handling of medicinal secrets within a a normative system that allowed them to begin or continue selling medicine on a small-scale level. The name of each one of them opens a

10 *Dizionario di Medicina Treccani* (Rome: Enciclopedia Treccani, 2010), ad vocem.

11 In the course of time, the process would be managed by a single magistrature, the *Provveditori alla Sanità*.

12 Crucial was the biographical perspective adopted by Sandra Cavallo in her *Artisans of the Body*, which has made otherwise ignored figures and ties emerge among artisans who dealt with the wellbeing and curing of the body.

short section, as a way of making up for the invisibility that most of them may have otherwise been destined to.

2 The Context: Secrets by "Particular Persons" or *particolari*

Here we look at the female figures that emerge when analysing the privileges granted first by the Senate and later by the Venetian Provveditori alla Sanità for the handling and sale of medicinal secrets invented by individuals.[13]

This documentation is part of a broader series of practices begun by the Republic at the end of the fifteenth century, which was created as a full-fledged patent system unprecedented in Italy and Europe at the time.[14] Provided they were innovative, the most varied technologies – astronomical/astrological and musical, *agrimensor* techniques, medicinal techniques, medicinal secrets, etc., but also more immaterial inventions, such as printed works of invention (books and etchings) and teaching methods – could be the subject of privilege. Petitions for privileges put forward by artisans, engineers, glassworkers, master calligraphers, agrimensors, inventors of medicinal secrets, etc. – more numerous – can be found in the files of the Senate along with the first petitions by authors of print works, emphasizing the material origins of early forms of safeguarding intellectual property and related economic gains.[15]

For the medical-pharmaceutical sector, the inventors/handlers of medicinal secrets could ask for a simple licence or a privilege, depending on the degree of originality that the drug presented and the ambitions of the applicant.[16]

---

13 Alongside the reflections induced by be research, the paragraph summarises passages from Chap. 1 of Sabrina Minuzzi, *Sul filo dei segreti: farmacopea, libri e pratiche terapeutiche a Venezia in età moderna* (Milan: Unicopli, 2016), where further information can be found.

14 Roberto Berveglieri, *Inventori stranieri a Venezia (1474–1788). Importazione di tecnologia e circolazione di tecnici artigiani inventori. Repertorio* (Venezia: Istituto Veneto di Scienze, Lettere ed Arti, 1995); for Italy Luca Molà, "Inventors, Patents, and the Market for Innovations in Renaissance Italy," *History of Technology*, 32 (2014): 7–34.

15 A selection of documents that illustrate the range of privileged innovations besides the print works, amongst which the medicinal secrets, is in Sabrina Minuzzi, *The Invention of the Author. The Privilegio di stampa in Renaissance Venice* (Venice: Marsilio, 2017). Fundamental reflections on the contribution of the patenting system to the construction of intellectual property in Pamela O. Long, "Invention, authorship, 'intellectual property' and the origin of patents: notes towards a conceptual history", *Technology and Culture* 32, 4 (1991): 846–884.

16 While the licence was an authorisation to handle and sell a secret, always managed by the Provveditori alla Sanità, the privilege protected the originality of the invention from the attempts at imitation, frequent in the case of secrets that turned out to be particularly

After consulting the medical College and at times, in the case of privileges, after a phase of experimentation (often at the lazarettes and/or prisons), the Senate granted the privilege and the Provveditori alla Sanità the licences. At first unevenly, in the seventeenth century the Provveditori began to grant both on its own.[17] The practice of offering privileges for an invention is unique to Venice, a city with an urban fabric made up of an extraordinarily lively array of artisans and businesses, which reflected itself in the sophisticated articulation of the Arts.[18] Contrary to what one might think, Venetian guilds were quite receptive to the privileged innovations, which, in some cases, were just what was needed in times of economic crisis.[19] The apothecary guilds were mostly favourable to making pharmaceutical offerings more competitive throughout the city and country: in addition to requesting privileges for secrets of their own invention, the apothecaries always aimed to obtain the secrets that turned out to be most effective in therapeutic practice once the exclusive had expired. Another Italian city that showed a marked inclination towards alchemical-pharmaceutical experimentation and invention was Florence, where, however, the *signoria* always sought to centralise most experimental activities. The Medici dedicated a wing of the Casino of St Mark's (Casino Mediceo erected in 1574) to these experimentations. This structure became a symbol and tool of the family's scientific – as well as artistic and political – patronage.[20] Throughout the seventeenth century and beyond, the court also acted as catalyst for the push for innovation locally thanks to friends and relatives of the Medici family and the professionals they surrounded themselves with.[21] Unsurprisingly, in the Florentine socio-political context, the organisation into guilds remained at a rather embryonal stage also for the pharmaceutical matter, nor was there

efficacious; at times the privilege was indeed requested several years after the granting of a simple licence, that is after experience had made it appear to be efficacious.

17 The Provveditori alla Sanità made the originality of the formula a mandatory requisite, first *de facto*, and much later *de iure* (1760s).

18 Venetian Arts had a strong economic component but also served purposes of social representation and mutual assistance. Antonio Manno, *I mestieri di Venezia. Storia, arte e devozione delle corporazioni dal XIII al XVIII secolo* (Cittadella: Biblos, 1995); Richard Mackenney, *Tradesmen and Traders: The World of the Guilds in Venice and Europe, c. 1250–c. 1650* (London-Sydney: Croom Helm, 1987).

19 Francesca Trivellato, "Guilds, Technology, and Economic Change in Early Modern Venice," in S.R. Epstein, Maarten Prak, ed., *Guilds, Innovation and the European Economy, 1400–1800* (Cambridge: Cambridge University Press, 2008), pp. 199–231.

20 Marco Beretta, "Material and Temporal Powers at the Casino di San Marco (1574–1621)," in Sven Dupré, ed., *Alchemy and Art Technology from Antiquity to the 18th Century* (Cham: Springer, 2014), pp. 129–156.

21 See the studies of Sheila Barker and Sharon Strocchia cited in the conclusion.

ever a system of licenses and privileges like that of Venice, created to respond to a wide-ranging – and – decentralised – tendency to experiment and come up with inventions.[22]

After the fundamental study carried out by David Gentilcore on the charlatans in Italy between the mid-sixteenth century and 1800,[23] a closer analysis looking only to the Venetian archival records has revealed that, during the same time frame, the local Health Board (Provveditori alla Sanità) and the Senate allowed and sometimes privileged 375 individuals to manufacture and sell one or more medicinal secrets.[24] It emerged that, over the years, two different regulatory paths for licensing medicinal secrets were established in Venice – one for the quacks and another for the so-called (by law) "particular persons" or *particolari*. The former implied the public assessment of the formula, before the whole Medical College to which the Provveditori or the Senate requested an opinion, whereas the latter was carried out in the presence of only two senior physicians together with the *Protomedico* (first physician) – in such a way that allowed the secrecy of ingredients and production techniques to be much more protected.

An analysis of licences and privileges combined have allowed us to establish that, out of the 375 individuals authorized in Venice, charlatans made up 6.9%, while "particular persons" were apothecaries (16.8%), physicians (8%), surgeons (6.7%), clergymen (8%), distillers, barbers, grocers, dye sellers, and ordinary people with no specific – or, better yet, no specified – profession. The category of "particular persons" with no declared profession is the biggest section in the graph; they represent 43.3% of the total.[25] Through a thorough biographical study – resorting to parish, notarial and census records – we can sometimes collect enough evidence to draw up a profile of at least some of these *particolari* with no occupational label. They include among their ranks: noblemen, notaries, civil servants, dye sellers, musicians, choir singers in St Mark's cathedral, etc.

This grey area includes only a few women who invented and handled, or apparently only handled, medicinal secrets: there are 13, which is 3.4% of the total.

22 Renato Ciasca, *L'arte dei medici e speziali nella storia e nel commercio fiorentino dal secolo XII al XV* (Firenze: Olschki, 1927), Katharine Park, *Doctors and medicine in Early Renaissance Florence* (Princeton: Princeton University Press, 1985).

23 David Gentilcore, *Medical Charlatanism in Early Modern Italy* (Oxford: Oxford University Press, 2006).

24 Minuzzi, *Sul filo dei segreti*. The negative responses, preserved only for the 18th century, do not fall within the calculations.

25 All the data are also processed by century, in Minuzzi, *Sul filo dei segreti*.

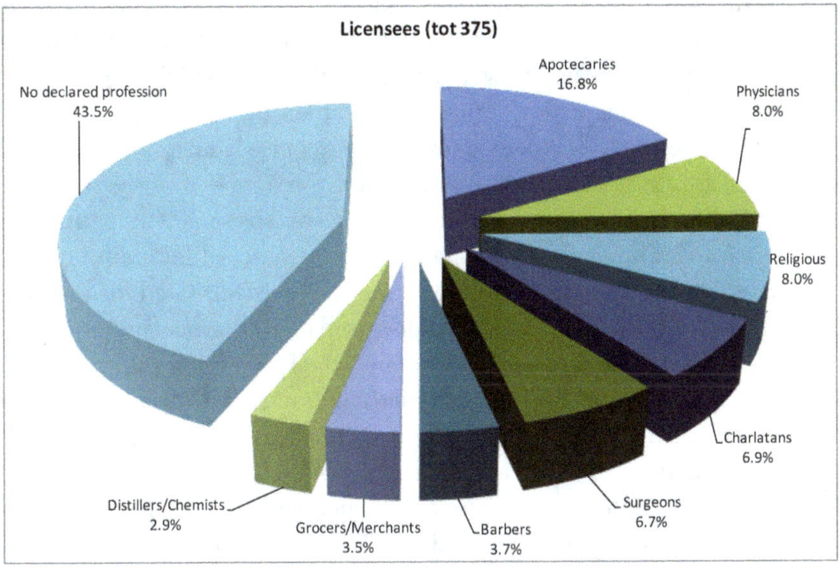

FIGURE 5.1 Venetian apothecaries

3 Women between Artisans of Secrets and the Apothecary Guild

The "particular persons," even those without a stated profession, are figures that can be likened to artisans who created something new with their own hands thanks to knowledge accumulated over years of experience. Pamela Long's definition is quite fitting: "Skilled workers who learned through formal or informal apprenticeships and oral instruction [sharing] a common culture that values handwork and hands-on skill that accompany them".[26] In terms of women, these are always figures who carried out informal apprenticeships, though they were often family members of those working in the guild of the apothecaries and other areas of medicine. It is important to point out, in fact, that the largest swath of experimenters of secrets came from the ranks of medicinal apothecaries, which, in Venice, were clearly separate from wholesale apothecaries (selling confections, kitchen spices, waxes, etc.).[27] In the space of two and a half centuries, up to the fall of the Republic, the apothecary category saw the greatest growth among inventors of medicinal secrets, reaching 22%

26 Pamela O. Long, *Artisan/Practitioners and the Rise of New Sciences, 1400–1600* (Corvallis: Oregon State University Press, 2011), p. 132, n. 1.

27 Manno, *I mestieri di Venezia*, pp. 47–48.

in the eighteenth century, when privileges alone were granted.[28] In the various petitions, apothecaries and others, seemed to be proud of their own medicinal inventions, which they hoped would be helpful for humanity, in addition to being a source of income for themselves and their own families. Some were driven by the desire to their own innovations recognised and protected materially as well as intellectually, because they were the source of social prestige.[29] Numerous declarations went well beyond what was customary for petitions.

The point can be made, thanks, above all, to sources that mention them indirectly, that the few women of the secrets that existed shared the dignified attitude and innovative spirit that characterised many "particulars."

Requesting a privilege or a licence for medicinal secrets fell within the individual's "public activities," so to speak, in Italy transferred to the male figure according to an ancient juridical tradition that became an everyday custom.[30] It was not like trying to register with the guild of the apothecaries as a master (*maestro*), but it came very close to it. In Venice, women were never formally prevented from joining guilds most in line with the work carried out by inventors/handlers of secrets, but, in practice, this never happened even when the tasks in the workshop involved the entire family.

As it can be deduced from a census of all the types of workshops opened in Venice between 1567 and 1568, in rare cases women managed the sale of medicinal remedies: the names of only two women who managed the retail sale of medicinal remedies on a daily basis in the family apothecary shop do appear – without being registered with the *mariegola* as master apothecaries.[31]

28 Minuzzi, *Sul filo dei segreti*, Grafico 6, p. 316.

29 It is not unusual to come across references to their "sleepless nights spent in experimenting on *Materia Medica* ... to invent secrets whose good Reputation our Lord liked to be much greater than our Profit": Archivio di Stato di Venezia (hereafter ASV), *Sanità, Terminazioni*, b. 91, petition of 1st April 1727 of the Morossi brothers, apothecaries at the Orso.

30 Marina Graziosi, "Fragilitas sexus: alle origini della costruzione giuridica dell'inferiorità delle donne", in Nadia Maria Filippini, Anna Scattigno, and Tiziana Plebani, ed., *Corpi e storia* (Roma: Viella, 2011), pp. 19–38.

31 The *mariegola* was the register containing the list of the registered in an Art (from the Latin *matricula*) together with the main resolutions and the minutes of the sessions. The secondary literature originated from Mackenney, p. 88; Girolamo Dian, *Cenni storici sulla farmacia veneta* (Venezia, 1900 [Rist. Anast.: Filippi, 1983]), p. 167, confounded some scholars who wrote about the female apothecaries in Venice. In actual fact, no woman was a master apothecary in Venice, not even in the 16th century, as is shown by the direct examination of the cited census (ASV, *Milizia da Mar*, b. 446). See the recent article by Emilie Fiorucci, which reveals a female presence in 2 apothecaries and in 3 *da grosso* (wholesale apothecaries), Emilie Fiorucci, "Women at work in a southern European town: women, guilds and commercial partnerships in Venice in the sixteenth century", in Anna Bellavitis, Beatrice Zucca Micheletto, ed., *Gender, Law and Economic Well-being in Europe*

Conversely, looking at the authorised/privileged secrets, it would appear that women played far more important roles than merely selling, such as in coming up with and sharing medical and scientific knowledge as well as handling *materia medica*.

4 Angela Sala: A Written Licence to Cure

In March of 1591, Angela Sala, who described herself simply as the wife of a certain Venetian Teodoro di Franceschi, requested a written licence to protect the secrets of her invention, after receiving several verbal authorisations. Angela's petition shows how the Provveditori, at times, only gave licences verbally – it seems, in comparing with other cases, in situations that looked more straightforward; this custom extended until the early years of the eighteenth century:[32]

> ... Although many times I, Anzola di Sala wife of Thodaro di Franceschi from Venice, appearing before you most Honourable Lords, have had been verbally granted the licence to be able to medicate and cure various ailments with my secrets and the practice acquired, nonetheless, to be safe and to live more cautiously in compliance with the good laws of this *Serenissima Repubblica*, I intend, by this letter, now to appear before you to ask that you give me licence to medicate now kindly *on paper*, as you have given me verbally, so that I may, for the benefit of many people, exercise these gifts that Our Lord has given me and cure the infirmities herein described ...[33]

Angela Sala treated fevers, syphilis, pulmonary disorders and congestion. She produced certified testimonials from the Venetians she had cured, which, together with the experience of the past years, allowed her to immediately obtain what she had requested. Angela did not appeal based on a situation of economic hardship, as often occurred in the request for secrets. Her petition was level-headed, firm, an descriptive. The people she had treated were "authoritative" Venetians, as defined by the Provveditori, which certainly knew who they were. Angela Sala wanted written authorisation to act within the rules and, above all, to be "safer" in carrying out her medical and pharmaceutical

---

from the Fifteenth to the Nineteenth Century: North versus South? (London: Routledge, 2019), pp. 215–241.

32 As regards the licences granted only orally and traced thanks to references within the documentation, cf. Minuzzi, *Sul filo dei segreti*, pp. 29–30 and chart 1.

33 ASV, *Sanità, Notatori*, reg. 736, c. 50r–51r, 1st March 1591.

activities. Certainly, a written document would increase the curer's prestige with potential patients (and she didn't seem to lack in those), but it was also a good card to play before the professionals in order to get round the lack of qualifications via regular medical training, which was reserved for men. Angela's attitude leads us to think that she consciously used the written authorisation as a substitute for the qualification that she could never have obtained.

5 Marietta Colochi Salaried Physician[34]

A women of secrets who has already attracted the attention of modern scholars is Marietta Colochi.[35] She was the wife of Nicolò Colochizario (m. 1552), a physician of Greek origin at the service of the Republic who had invented an anti-plague secret recipe based on *smartella*, an herb in the myrtaceous family. The spouses would often work side by side in Venice and in *Terraferma* (on the mainland) with the official task of medicating the infected with the anti-plague secret. In 1558, Marietta put forth her own candidature for the role of physician of the *Lazaretto Vecchio*, which is the role that had been occupied by her husband in the past. She had the expertise she'd gained working on the field and could boast previous paid healthcare positions. But it was a public post, up until then filled only by physicians with traditional training, and, therefore, was highly unusual for a woman. Unsurprisingly, her candidature was rejected.[36]

However, the *Provveditori alla Sanità* did not intend in any way to give up her collaboration. In September 1559, the Senate assigned to her and her son Giacomo the task of medicating patients with a secret described as a family remedy.[37]

---

34 BMC, *Ms. Donà dalle Rose*, 181, cc. 4r–8r, a private agreement of 1598 describes the "secret of the excellent late sir Nicolò Colocchi salaried physician of the excellent Dominio Veneto, and then upon his death used for the quondam madonna Marieta his spouse herself salaried physician"; secret donated to the writer by Giovan Battista Padavino, secretary of the Council of the Ten (c. 4r).

35 Jane Stevens Crawshaw, "Families, medical secrets and public health in early modern Venice", *Renaissance Studies*, 28, 4 (2014): 597–618; Minuzzi, *Sul filo dei segreti*, pp. 119–128.

36 In Italy very few women obtained a title following an official training in the trades of medicine, between 13th–15th centuries, although then as later, they could be authorised to medicate or were called 'medica' or 'medichessa': 24 in the Kingdom of Naples, 5 in Florence, none in Venice cf. Katharine Park, "Medicine and the Magic: The Healing Arts", in Judith C. Brown, Robert C. Davis, ed., *Gender and Society in Renaissance Italy* (London; New York: Longman, 1998), pp. 129–149: 134–136.

37 ASV, *Senato Terra*, reg. 42, c. 50r, 7 September 1559: "by whose wife, with said secret, many have been cured, and as witnesses, examined for the Health Office and in regard to every case that could arise, act upon this family having such secret".

Giacomo died prematurely but Marietta continued on with her activities until 1568, with an assistant-servant in tow to teach techniques for sanitising places and objects (making it unnecessary to burn them) and medicating patients with the family secret.[38] A print recipe also survives bearing witness to her tireless work: *Queste son le vere et esperimentate ricette per guarir giandusse et carboni con molta facilità. Insegnate da M. Marieta Colochi, fu di M. Nicolò Colochi.*[39]

Once again, upon invitation of the *Provveditori alla Sanità*, in 1567, Marietta educated Sicily-born Ascanio Olivieri – her elder daughter Paola's husband and the newly elected physician for the lazarettes – on the handling and dosage of the secret anti-plague remedy. Although the vulgate states that Marietta bequeathed the secret to her elder daughter Paola, neither her will dated 29 July 1568 nor her daughter's dowry mention it.[40] Certainly she arranged for Olivieri spouses to allow Daria, the unmarried sister/daughter-in-law, to live in their home. Ascanio was to transfer to Daria an annuity of 3 ducats a month, which Marietta had begun to receive in 1559 in exchange for her medical services. This occurred following Daria's petition to the *Provveditori alla Sanità* in the autumn of 1568.[41] In the petition-memorial Daria supplies copious details and dates concerning her parents' Nicolò and Marietta's collaboration with the Health Board, but it is clear that she had never accompanied her widowed mother on the missions nor did she ever handle the remedy – like her sister Paola. Daria's tone is the heartfelt one of an "orphan" asking for a source of sustenance.[42] Looking beyond the rhetoric, the words are those of a person with few resources, far from the self-confidence of Marietta's petition in 1559 or from the tone seen in Angela Sala's petition.

After Marietta's death, the secret returned to being of exclusively male pertinence. The formula underwent variations that were then channelled into the official print collections of 1630 and 1682 with only the names of the male key players: Nicolò Colochi, Ascanio Olivieri and Scipione Paragatta, who had later married Daria.[43]

38 We can read about the assistant that Maria took with her in ASV, *Sanità, Notatori*, reg. 733, c. 4r (petition of her son-in-law Paragatta, 1576).

39 Biblioteca Apostolica Vaticana, Roma, R.I.IV.1551 (int. 86).

40 ASV, *Notarile, Atti*, reg. 2576, cc. cc. 183r–185v, 222v (Paola's dowry).

41 ASV, *Sanità, Notatori*, reg. 731, cc. 83r–84.

42 Ivi, "Whereupon I finding myself poor Daria her daughter [...] reduced to poverty at a young age, endowed only with the secret left to me by my mother, without father and mother to tend to me".

43 *Raccolta di avvertimenti & raccordi per conoscer la peste ... presentata al magistrato ill.mo della Sanità di Venetia, & di ordine di quello mandata alla stampa per curarsi, & preservarsi* (Venice, 1630 and subsequently in 1682). On 27 July 1577 the son-in-law Ascanio Olivieri ceded to the Republic the formula of Colochi's secret in exchange for a cash reward

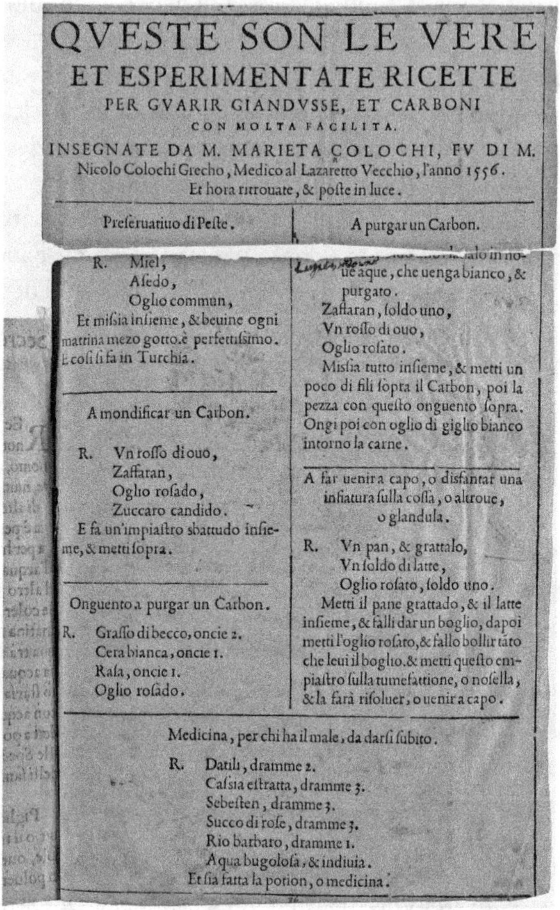

QVESTE SON LE VERE
ET ESPERIMENTATE RICETTE
PER GVARIR GIANDVSSE, ET CARBONI
CON MOLTA FACILITA.
INSEGNATE DA M. MARIETA ÇOLOCHI, FV DI M.
Nicolo Colochi Grecho, Medico al Lazaretto Vecchio, l'anno 1556.
Et hora ritrouate, & poste in luce.

Preseruatiuo di Peste.

R. Miel,
Asedo,
Oglio commun,
Et missia insieme, & beuine ogni
mattina mezo gotto.è perfettissimo.
E cosi si fa in Turchia.

A mondificar un Carbon.

R. Vn rosso di ouo,
Zaffaran,
Oglio rosado,
Zuccaro candido.
E fa un'impiastro sbattudo insie-
me, & metti sopra.

Onguento a purgar un Carbon.

R. Grasso di becco, oncie 2.
Cera bianca, oncie 1.
Rasa, oncie 1.
Oglio rosado.

A purgar un Carbon.

[illegible] lasalo in no-
ue aque, che uenga bianco, &
purgato.
Zaffaran, soldo uno,
Vn rosso di ouo,
Oglio rosato.
Missia tutto insieme, & metti un
poco di fili sopra il Carbon, poi la
pezza con questo onguento sopra.
Ongi poi con oglio di giglio bianco
intorno la carne.

A far uenir a capo, o disfantar una
infiatura sulla cossa, o altroue,
o glandula.

R. Vn pan, & grattalo,
Vn soldo di latte,
Oglio rosato, soldo uno.
Metti il pane grattado, & il latte
insieme, & falli dar un boglio, dapoi
metti l'oglio rosato,& fallo bollir tato
che leui il boglio.& metti questo em-
piastro sulla tumefattione, o nosella,
& la farà risoluer, o uenir a capo.

Medicina, per chi ha il male, da darsi subito.

R. Datili, dramme 2.
Cassia estratta, dramme 3.
Sebesten, dramme 3.
Succo di rose, dramme 3.
Rio barbaro, dramme 1.
Aqua bugolosa, & indiuia.
Et sia fatta la porion, o medicina.

FIGURE 5.2 Rome, Biblioteca Apostolica Vaticana, R.I.IV.1551 (int. 86), Marietta Colochi's anti-plague recipes

6 Inventor Giulia Tartara and Her Daughter Orsetta

A female inventor of secrets is mentioned in the petition of an heir. In 1638, Giulia Tartara, wife to "domino messer Antonio Tartaro," was authorised by the Medical College of Padua to handle and sell pills of her own invention to treat syphilis.[44] Perhaps her husband Antonio Tartaro was the public officer

---

(which he would have liked to be larger), cfr. ASV, *Secreta*, Materie Miste Notabili, cc. 312 et seq.

44 ASV, *Sanità, Notatori*, reg. 740, c. 142r–v, [26 August 1649].

identified as the author of a *Raccordo* (i.e. *Proposal*) on the subject of numismatics published in Padua in 1617.[45] For ten years, the secret offered up good results, so much so that, a few months after Giulia's death in February 1649, her son-in-law Carlo Negri appeared before the *Provveditori alla Sanità* to request a licence for it in Venice as well. Carlo had married Orsetta, the only daughter of the Tartaros, who now possessed Giulia's secret.[46] In line with the social conventions of the time, it was the son-in-law and not the daughter who made the request to continue handling and dispensing the secret along with Orsetta, mentioned only here. Carlo's petition took a calm tone explaining the request for trade on a domestic scale not out of need, but because "it is not right that such a distinguished secret, offering such public relief and well-known to many, should die out due to our taciturnity."[47]

7 Maria Vezzali and Her Grandchildren, Handlers of a Chemical Secret

Far more often than might be imagined, behind the petitions put forth by men, there were not only women but also the youngsters of an household: all involved in the handling and sale of a medicinal secret.

Domenico Vezzali had trained by working eleven years, between 1657 and 1668, in the Struzzo apothecary owned by a wealthy man named Antonio de Sgobbis.[48] There he had learned many chemical techniques, subsequently opening his own apothecary shop named Tre Santi at S. Giacomo dall'Orio. After retiring in 1715, by then in his eighties, Vezzali asked to be able to continue to handle and sell his "natural caustic sweetener" from home.[49] This was

45 Biblioteca Nazionale Marciana, Venezia (hereafter BNM), Misc. 259.17: *Dichiaratione del raccordo … In materia di Monete raccordate, & publicate a Sua Serenità, Con le resolutioni raccontate in Dialogo delle oppositioni, et quesiti, che potessero esser fatti contra d'esso raccordo* (In Padova, per il Crivellari, 1617). The work is cited by Emmanuele Antonio Cicogna as a comment to the various members of the Tartaro family buried in the church of San Severo (demolished in 1829) in the *sestiere* of Castello, *Delle Inscrizioni Veneziane*, III (Bologna: Forni, 1969), p. 101. "Messer", without further specifications, is a term widely used at the time in the petitions, for physicians (es. Nicolò Colochi, Ascanio Olivieri), apothecaries, notaries (e.g. Scipione Paragatta), nobles.

46 ASV, *Sanità, Notatori*, reg. 740, c. 142r.

47 Ivi.

48 The apothecary Antonio de Sgobbis, author of the *Nuovo et universale Theatro Farmaceutico* (Venice 1667) played a leading role in the affirmation of chemistry in Venice, cf. Minuzzi, *Sul filo dei segreti*, pp. 77–81.

49 ASV, *Sanità, Terminazioni*, b. 89, n. 30, June 1715.

a chemical remedy with a formula that had been approved in the past by the late *protomedico* (first physician) Iseppo Ton, and which, for years, Venetian surgeons had used to medicate cutaneous and oral ulcerations, callosity and scrofula.[50] Unlike his old master de Sgobbis, Vezzali had not amassed a large fortune but was well-known in professional circles for his secret. In fact, he was granted the licence immediately given the local fame of the remedy. After a few years of selling and handling the secret in a private capacity, Vezzali dictated his last wishes on 17 February 1718, and this is a precious document for understanding who had contributed to his small artisanal operation. First among the tangible goods he mentioned was his own "great secret:"

> The largest amount of capital in my limited fortune is the great secret, the natural caustic made by me for a long time with the utmost universal benefit, and honored with a privilege granted exclusively to my household alone by the most honourable Magistrate of Health. This I leave to my *delightful spouse* [Maria], and to my *beloved grandchildren, who for many years have worked wonderfully alongside me to create and handle it.*[51]

Without Domenico's will we would have never understood the family component of Vezzali's pharmaceutical activities, which involved his wife Maria and their young grandchildren, sons of their only daughter Caterina who had married the count Giuseppe Mandricardi. The history of the secret continued within an exclusively family environment until 1760, with the grandson Ruggero, who was given a certification signed off on by the most famous Venetian surgeons:

> We attest [...] that, over the course of many years, we have used and still use fruitfully and with admirable effect the ancient natural caustic remedy called "*of the Vezzalis*" dispensed here in Venice only by the Abbot d. Ruggiero Mandricardi Vezzali from his highly reputable home. And we state that, in our most difficult cures, we cannot do without it, being the surest remedy and experimented by us, superior to certain other remedies that are called caustic.[52]

50 The first authorisation, dateable between 1698 and 1711, years of the medical officer Ton, was not registered in the Notatori, cfr. ASV, *Sanità, Notatori*, reg. 747, cc. 175r–176r.

51 ASV, *Notarile, Testamenti*, b. 23, n. 55.

52 ASV, *Sanità, Rapporti medici*, b. 588, 10 September 1760, the underscore is mine.

8 Caterina Pasetti and Elisabetta Bonetti, between Distillers and Apothecaries

By studying various archival documents, we can find references to wholly unknown licences, either because they were issued verbally or because the documentation was lost, especially in the seventeenth century. In the will dictated in 1656 by then 35-year old Giacomo Pasetti, producer and seller of spirits and distilled waters in the neighbourhood of Santa Maria Formosa, we learn that he had inherited his shop near St. Mark's Square from his grandfather.[53] Giacomo explains he had taught his wife Caterina to handle all of his secrets for distilled waters, and urges the latter, in the event of his untimely death, to teach the secrets in turn solely to the children that would be born from the marriage and never to a potential second husband.

Thanks to a notarial document, we learn about a cosmetic secret created by Elisabetta Bonetti, married to a man named Giacomo who was an apothecary at the Novizza shop in Campo S. Maria Zobenigo. After being widowed, in 1685 Elisabetta gave her apothecary shop on rent for 100 ducats a year, "reserving for herself Isabetta the secret of the hair oil, *as her particular secret*, which, however, the Bernardinis [tenants] are content to sell in the above-mentioned shop free of charge."[54]

9 Caterina Pontano and a Formula Worth 200 Gold Ducats

The *Electuary of the Pontano* has quite an exemplary story, telling of the continuous comings and goings between the apothecary shop and the domestic environment, with women handling as well as administering secrets. Caterina Moretti Pontano is mentioned in the registers of the Provveditori alla Sanità only in 1618, one year after her death. Eleven years before, her second husband Gentile Pontano, a worker at the apothecary, had obtained a licence for a vermifuge he had invented, known as the *Electuary of the Pontano*.[55] After the untimely death of Gentile, who in the meantime had become a master apothecary and ran the shop at the Cedro in Campo S. Luca, Caterina cashed in her

53 ASV, *Notarile, Testamenti*, b. 908, n. 72, 10 June 1656.

54 ASV, *Notarile, Atti*, reg. 3926, contract on 13 April 1685, c. 58v.

55 ASV, *Sanità, Notatori*, reg. 738, c. 109r–v, 22 May 1618; Ivi, reg. 737, c. 212r, 22 October 1607. Further details and documentation on family and secret in Minuzzi, *Sul filo dei segreti*, pp. 161–163.

dowry together with the Gentile's licence – estimated at 200 gold ducats – to continue to produce the secret at home, with her four children from her first marriage, two boys and two girls.[56]

For a year after, Caterina maintained herself and her children by producing and selling the medicinal secret. When she died as well, her four children asked for and obtained that their stepfather's license would be transformed into an exclusive privilege, given the great market demand and the risk of counterfeiting. None of the Pontano heirs ever became an apothecary, but the sales of the secret went so well that, in 1646, they were forced to pay a tax of 2 ducats on the revenues, like any other member of the guild of apothecaries.[57]

The electuary resurfaces in documents that testify to its new entrance in the apothecary shop in 1681, while, just a few years later, in 1686, it returned to being handled in the domestic environment.[58] The *Provveditori* willingly seconded all the passages, as long as the secret continued to circulate in a very recognisable guise, in jars marked by the same label featuring a bridge [Ponte-Pontano] and the name of the inventor.[59]

10 Caterina Melchiori, Francesca Mascotela and Two Anna Marias: Provincial Aristocrats and Female Apothecaries

Only twice is Francesca Mascotela, wife of the Trentino nobleman Giuseppe Felice Maria Scutellio (1666–1731) mentioned in the registers of the *Provveditori alla Sanità*, in 1730 and later in 1760, and the names of the other two women in the title of this section never are.[60] Parish and notarial documents found in Trento and Venice, printed recipes and a couple of printed books published by Giuseppe, have allowed us to reconstruct the facts and to bring these female figures to life.[61]

In this section, we'd like to point out how a series of authorised and (later) privileged secrets involved all of the members of two households, the Scutellios

56 ASV, *Giudici del Proprio, Mobili*, b. 142, c. 43v, 27 September 1616.

57 Dian, *Cenni storici*, p. 157.

58 ASV, *Sanità, Terminazioni*, b. 85, n. 149: a grandson of Caterina ceded the privilege to the apothecary Pietro Giorgio Castelli, who continued to produce it successfully in his shop alla Madonna at the feet of the Rialto Bridge; Castelli then transferred it to his children who were not apothecaries.

59 Ibid.

60 ASV, *Sanità, Terminazioni*, b. 92, n. 254; *Sanità, Rapporti medici*, 1° sett. 1760.

61 The story is reconstructed in detail in Sabrina Minuzzi, "Quick to say quack. Medicinal secrets from the household to the apothecary's shop in early eighteenth-century Venice", *Social History of Medicine*, 32, 1 (2019): 1–33.

and the Cimolin-Scutellios, friends who inherited the surname of the inventor of the secrets, Giuseppe, along with the privilege of dealing in them. The Scutellios settled in Venice at the end of the seventeenth century. The family included, in addition to Giuseppe: the mother Caterina Melchiori Scutellio (d. 1710 ca., daughter of an esteemed physician from Trento and skilled in the treatment of women's ailments and disorders); Caterina's daughter-in-law Francesca Mascotela (d. 1754 ca.) who learned "Great Rarities" from Caterina and was the trusted custodian of husband Giuseppe's knowledge and secrets; and lastly the couple's son, Carlo Antonio (d. 1759). A subsequent nuclear family, that of the Cimolin-Scutellios, included the herbalist father Antonio (d. 1799), the mother and handler of the secrets Anna Maria (d. 1808), who also dealt in selling them from the home, and her son from her first marriage Girolamo (d. 1794), a surgeon involved in making the secrets who also used them in his profession. After two generations of being handled from the home, with Anna Maria Ferramonti (d. 1788), an unmarried heiress from a family apothecary business and a close friend of Girolamo, the Scutellio's privileged secrets began to be part of the official therapeutic provision of Venice. Until the early years of the twentieth century, they were indeed sold by the pharmacies of the lagoon city.

Giuseppe Scutellio also bequeathed to his son Carlo Antonio his *Libro d'Arcani*, a manuscript of recipes and *curationes* which, handed down to the Cimolin-Scutellios, was finally sold in exchange for an annuity in 1795. Now no longer traceable, it is nevertheless possible to reconstruct its content: exactly like how Venetian artisan glassmakers passed down the secrets of their art through private "recipe books," Scutellio passed down his privileged pharmacological secrets that had never been shared. Though incredibly different in type from what the glassmakers left behind, the function of the *Libro d'Arcani* was still the same:[62] bequeathing the secrets from an artisanal endeavour in which women were inventors and colleagues in the creation and the subsequent modifications of the formulas.[63]

62 Tullio Toninato, Cesare Moretti, *Ricettari muranesi: XVI–XX secolo* (Venice, [1991]); Iidem, "Ricettari Muranesi", *Rivista della Stazione Sperimentale del Vetro*, 3–4, 1992: 197–206. The research of Moretti and Toninato have brought to light around seventy recipe books of glassmakers now in private collections, but retrievable in copy from the Biblioteca della Stazione Sperimentale del Vetro di Murano. Dating back mainly to the 18th and 19th century, they are nearly always synthetic, simply providing, for the various types of glass, the minerals necessary and their quantities. Trivellato, "Guilds, Technology", p. 224 and note 95.

63 This is inferred from the descriptions of the mother and the wife in the print books of Scutellio.

11 Margaret Paston and Her Venetian 'Foundry' or 'Laboratory': At the Intersection between Two Worlds

Margaret Paston (1652–1723), daughter of the first Earl of Yarmouth, Sir Robert Paston (1631–1683), was one of the female figures of a greater socio-economic status as compared to the others who presented themselves before the *Provveditori.*

On 30 April 1709, the *Provveditori alla Sanità* returned enthusiastically from a unique visit to S. Severo parish:

> Madam Margarita countess of Termauth [sic] of a noble and distinguished English family, wife of the notable Girolamo Alberti, secretary of the venerable Council of Ten, possesses the rarest knowledge in handling vegetal, mineral, semi-mineral, and animal substances, extracting from them various specifics, worthy of consideration by humans as well as beneficial for the curing of many infirmities. Therefore, Their Excellencies, having inspected her laboratory, all well seen and considered it, have commended such virtuous applications of the afore-said woman, and, in witness whereof, have left her the present declaration.[64]

It is the first and last time that the *Provveditori* authorised a "laboratory," defining it as such, and it is significant that they should do so for a foreign noblewoman. She is similar, in many ways, to other aristocratic English experimenters with a passion for secrets who left behind many handwritten records.[65]

On 29 December 1672 – though the Paston family was against it – Margaret was wed in London to Girolamo Alberti, who was living there as a secretary representing the Serenissima.[66] The union resulted in two children born in London: Caterina in 1674 and Girolamo in 1675; actually, just before the birth of the second child, in his last dispatch to the Senate, Alberti declared his intention to come home, with "the wife pregnant, who soon will give birth to her

---

64 ASV, *Sanità, Notatori*, reg. 746, c. 111r; the relevant termination, without supplementary documentation attached, is in ASV, *Sanità, Terminazioni*, b. 88, n. 1139.

65 Sara Pennell, Elaine Leong, "Recipe Collections and the Currency of Medical Knowledge in the Early Modern Marketplace", in Mark S. Jenner, Patrick Willis, ed., *Medicine and the Market in England and its Colonies. c.1450–c.1850* (Basingstoke: Palgrave Macmillian, 2007), pp. 133–152.

66 Girolamo Alberti was resident in London between 5 December 1670 and 21 June 1675, 353 dispatches, cf. Rawdon Brown, *L'archivio di Venezia con riguardo speciale alla storia inglese* (Venezia; Torino: Antonelli e Basadonna, 1865), p. 268. The marriage was recorded by the London notary Anthony Wright on 8th May 1676, cf. ASV, *Avogaria di Comun*, b. 237, n. 96. Margaret's dowry amounted to 4,000 pounds, that is about 22,588 current ducats.

second child in the forthcoming season."[67] The family continued to grow in Venice, with the birth of other children, while Girolamo Alberti carried on with his diplomatic missions, as secretary resident in Poland and then as secretary of the Council of Ten.[68] By the time her children were grown, Margaret returned to dedicating herself to pharmaceutical experimentation, for which her grandfather William Paston was famous, and to the chemical-alchemical passion that she had shared in youth with her father Robert. Sir Robert Paston had attended meetings at the Hartlib Circle in mid-seventeenth century London, and in 1661 he became member of the Royal Society. In the meanwhile, sir Robert had set up a chemical laboratory in the family residence – Paston or Oxnead Hall,[69] where young Margaret embarked on an apprenticeship that she later perfected in Italy – enabling her to create the "virtuous applications" that the Venetian magistrature admired and officially authorized in 1709.

Like other English, French, German and Italian noblewomen, Margaret also left a collection of medicinal secrets, now in the Wellcome Library, Ms. 3777.[70] It consists of an autograph manuscript that she started to compile in 1708. The date can be gleaned from a passage in which Margaret says that for the first time she had prepared the "polvere belzuardica" when her husband was about to leave for Poland and that she had obtained magnificent results over 24 years. Girolamo Alberti left in November 1684, so the writing must have begun at the end of 1708.[71]

In the manuscript, Margaret Paston Alberti always spoke of her laboratory as the "foundry," putting it on the same level as the celebrated Fonderia dei Medici

67 BNM, Mss. It. VII, 1671–1672, *Dispacci al Senato*, n. 353. The dates of birth of the children are in copy in ASV, *Avogaria di Comun*, b. 237.

68 It is hard to trace the birth of the successive children given the numerous transfers of the Alberti family after its arrival from England. The contrà (parishes) verified are at least three: S. Trovaso, S. Vio and S. Severo.

69 Francesca Vanke, Paula Findlen, "Imperfect Reds: Robert Paston's Alchemical Interests", in A. Moore, N. Flis, F. Vanke, ed., *The Paston Treasure: Microcosm of the Known World* (New Haven: Yale Center of British Art, 2018), pp. 218–231.

70 Wellcome Library, London (=hereafter WL), Ms. 3777, acquired in 1911, is described by Nathan Flis in *The Paston Treasure*, pp. 526–528, no. 194.

71 WL, Ms 3777, c. 3r. Alberti was sent by the Serenissima to Poland as resident from November 1684 to 2 August 1699. In the *The Paston Treasure*, pp. 526–528, moreover full of information and identifications on the English side, the drafting is erroneously considered allograph and attributed to the 1680s. In actual fact is was written by Margaret, who had learned the language perfectly and had assimilated graphically the structure of the italics. A similar case is that of Robert Dudley, resident in Italy for several decades, whose missives do not betray his English origins at all, cf. https://italian-paleography.library.utoronto.ca/islandora/object/italianpaleography%3AIP_059 (thanks to Laura Nuvoloni for the lead).

in Florence, one of the liveliest centres for chemical-alchemical experimentation in Italy in the sixteenth and seventeenth centuries.[72] From the very first pages, a number of recipes refer to handling methods utilized by the Fonderia Medici and show that she had direct contact with the professionals working there; some recipes are "of the Serenissimo Gran Duca," with whom Margaret exchanged not only works of art but also medicinal formulas.[73] Calling her laboratory a "foundry" was justified, because it was much more than the average *urban kitchen* equipped with a few supplementary tools.[74] The English noblewoman used (and described) a range of varied, sometimes sophisticated, tools: presses and "torcoleti" in various sizes, mortars, alembics and volumetric flasks (glass spheres with a flat bottom and elongated neck), circulatory vessels, filling sand, sponges and all kinds of gauze, burners with curved or blind chimneys, etc. Often, when she did not deem her equipment to be sufficient, she would turn to that of the specialised artisans, suggesting she made use of the kilns of glassmakers, *pignateri* (pot makers) and *scaletteri* (bakers) because they were larger or could reach and maintain different temperatures.[75] The proximity and the collaboration with the world of Venetian artisans is indeed a distinctive trait of Margaret's manuscript as compared to the other experimenters from the European aristocracy.

She left the recipe book to her heirs, whom she hoped might continue with the production activities and research in the foundry: "I have decided to leave this secret in memory to my heirs, so that they may find some other virtues that I still have not discovered;" and again: "I write this distinctly so that my

72 See Giovanni Piccardi, "La Fonderia medicea di Firenze," in *Memorie di Scienze Fisiche e Naturali, Rendiconti della Accademia Nazionale delle Scienze detta dei XL*, serie V, vol. XXIX, part II, tome II, vol. 123 (2005): 197–210.

73 WL, Ms. 3777, cc. 5v ("Sua Altezza Reale il Gran Duca fa in questa forma, che dalli suoi medici è stato stimato ancora più valido"), 7r ("come fanno sempre i giulebbi del Ser.mo Gran Duca"); c. 46r (La preciosissima acqua cordiale del SS. G[ran]. D[uca]. di Fiorenza) etc. For the gifts of art paintings see Flis, *The Paston Treasure*, p. 526.

74 On the simple equipment required for the Galenic handling of the medicinal secrets see the results of the study conducted on over 1,300 domestic inventories and 28 apothecary shops of the country of Glamorgan between 1600 and 1750 in Withey, *Physick and the Family*, pp. 99–120.

75 WL, Ms. 3777. Some examples: "and place it for six days in the fire of glare of the glassmakers and the pot-makers ... in a glass chamber pot with its lid shut" (c. 6v); "cook it in the mouth of the bakers' oven" (c. 7v); "you will need a low stove on the ground ... with a large flask that has a long neck at least three arms in length with its thin hose with the beak to distil" (c. 90v).

heirs shall be informed of the differences of the salts, which is something very important to know for a foundry."[76]

So the manuscript was meant for the family, and Margaret, who had compiled it alone, shared the fruits of her labour and personal experience, which otherwise would have gone lost. She categorically excluded the recipes circulating in various printed collections, unlike what many contemporary noblewomen often did: "I shall not describe in this book the virtues of those specifics that are already printed, and this shall suffice as warning to those who will survive me."[77] It is no surprise that the collection of recipes opened her the two most precious formulas, which may have been useful to her descendants as gifts to give to courts and their peers, further strengthening the network of favours that she herself had always cultivated:[78] from the *drinkable gold* (c. 1r) to the "balle di polvere belzuardica" (c. 1v) ("balls of bezoar powder") based on ingredients like saffron, precious stones, oriental vegetables and golden leaves: "These balls are gifts only for royalty, owing to their high cost and their great virtues" (c. 3r). But most of the secrets were naturally much less precious remedies, which in some cases she herself dispensed freely to the poor and that she hoped her heirs would continue to produce and distribute, as was the custom of aristocratic ladies.[79]

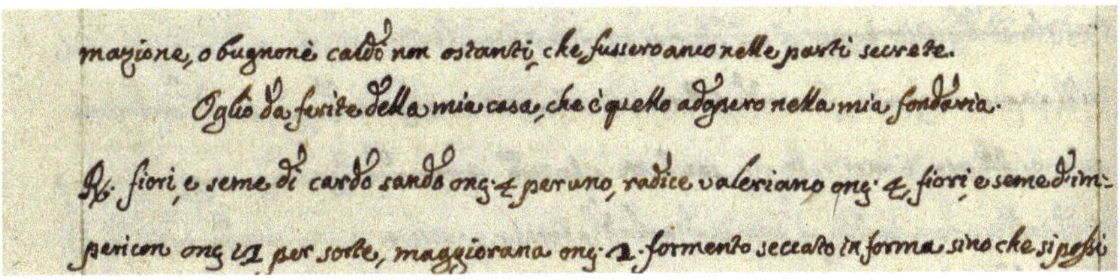
mazione, o bugnone caldo non ostanti che fussero anco nelle parti secrete.
Oglio da ferite della mia casa, che è quello adopero nella mia fondaria.
R. fiori e seme di cardo santo onz 4 per uno, radice valeriana onz 4, fiori e seme d'im-
pericon onz 12 per sorte, maggiorana onz 1 formento seccato in forma sino che si pesti

FIGURE 5.3 Wellcome Library, Ms. 3777, *Margaret Paston's Recipes Book*, f. 26v. One of the many occasions on which Margaret speaks of her 'fondaria' or 'foundry'

76 WL, Ms. 3777, cc. 29v, 36r: "I believe that for me both of them are excellent operations, and thus it is necessary to leave these again to my heirs".

77 Ibid., c. 7r. Another claim to not having drawn from print recipes is in the manuscript of the wife of a merchant from Augsburg, Regina Zangmeister (d. 1597), in Leong, Rankin, "Secrets and Knowledge. Introduction", in Elaine Leong, Alisha Rankin, ed., *Secrets and Knowledge in Medicine and Science, 1500–1800* (Aldershot: Ashgate, 2011), p. 15.

78 The culture of medical gift exchanges brings together courts and convents, cf. Strocchia, *Forgotten Healers*.

79 WL, Ms. 3777: "Other favourite balsam of mine, it being made from great experiences with the poor" (c. 29r); "My form of nectar and curing the wounds that is more than necessary to be described here so that my descendants can make charity for the poor."

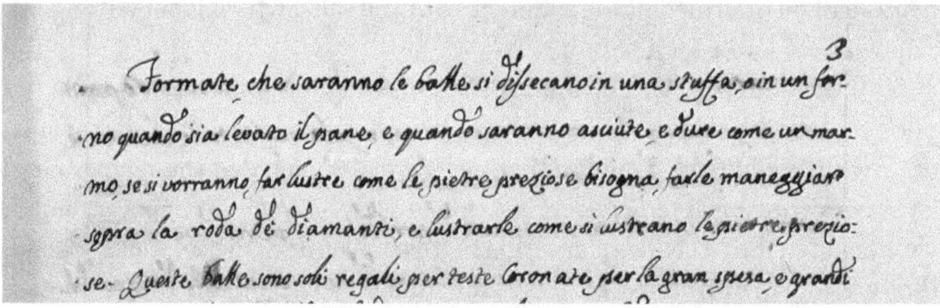
Formate che saranno le balle si disecano in una stuffa, o in un for-
no quando sia levato il pane, e quando saranno asciute e dure come un mar-
mo, se si vorranno far lustre come le pietre preziose bisogna farle maneggiare
sopra la roda de' diamanti, e lustrarle come si lustrano le pietre prezio-
se. Queste balle sono soli regali per teste Coronate per la gran spesa, e grandi

FIGURE 5.4 Wellcome Library, Ms. 3777, c. 3r. Reference to the baker's oven and to the gemstone grinding wheel used in the preparation of bezuard powder balls

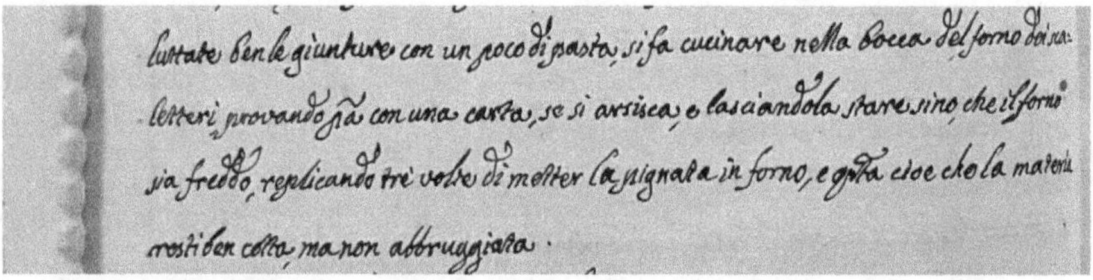
lutate ben le giunture con un poco di pasta, si fa cucinare nella bocca del forno dei pa-
letteri, provando già con una carta, se si arsisca, e lasciandola stare sino che il forno
sia freddo, replicando tre volte di metter la pignata in forno, e questa cioè che la materia
resti ben cotta, ma non abbrugiata.

FIGURE 5.5 Wellcome Library, Ms. 3777, c. 7v. Reference to the use of the baker's oven

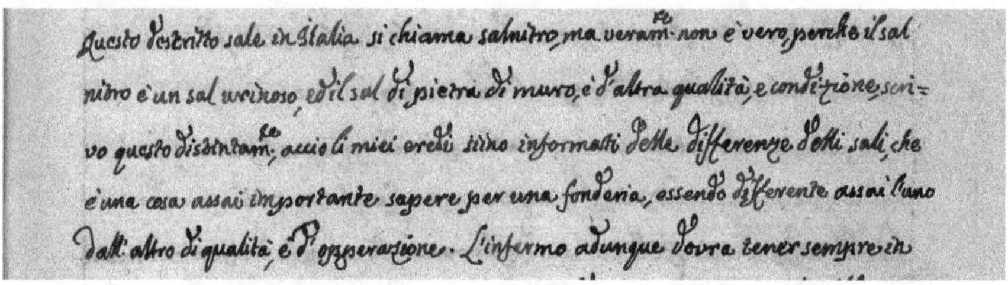
Questo descritto sale in Italia si chiama salnitro, ma veram^te^ non è vero, perche il sal
nitro è un sal urinoso, ed il sal di pietra di muro, è d'altra qualità, e condizione, scri-
vo questo distintam^te^ acciò li miei eredi siino informati delle differenze delli sali, che
è una cosa assai importante sapere per una fonderia, essendo differente assai l'uno
dall'altro di qualità, e d'opperazione. L'infermo adunque dovrà tener sempre in

FIGURE 5.6 Wellcome Library, Ms. 3777, c. 36r. Transmission of important knowledge to the heirs who, in Margaret's hopes, will run her foundry

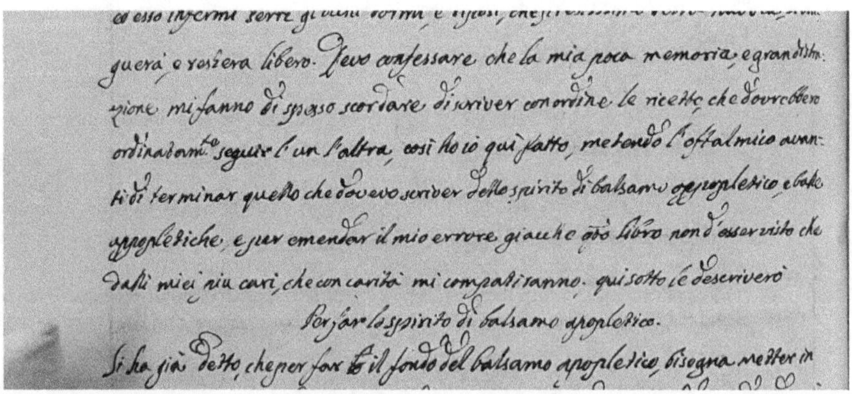
guerà, e resterà libero. Devo confessare che la mia poca memoria, e grande dista-
zione mi fanno di spesso scordare di scriver con ordine le ricette, che dovrebbero
ordinatam^te^ seguir l'un l'altra, così ho io qui fatto, mettendo l'oftalmico avan-
ti di terminar quello che dovevo scriver dello spirito di balsamo apopletico, e balle
appopletiche, e per emendar il mio errore, giache questo libro non d'esser visto che
dalli miei più cari, che con carità mi compatiranno, qui sotto le descriverò
Per far lo spirito di balsamo apopletico.
Si ha già detto, che per far il fondo del balsamo apopletico, bisogna metter in

FIGURE 5.7 Wellcome Library, Ms. 3777, c. 64v. Margaret says she is confused and distracted while her writing becomes more cursive

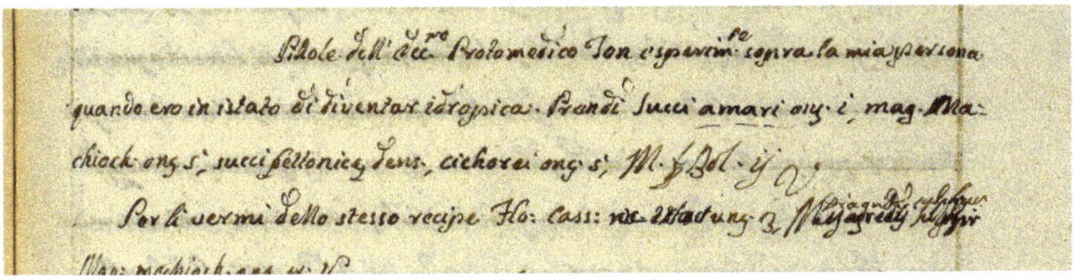

FIGURE 5.8 Wellcome Library, Ms. 3777, c. 91r. Medicinal remedies from the *protomedico* Iseppo Ton

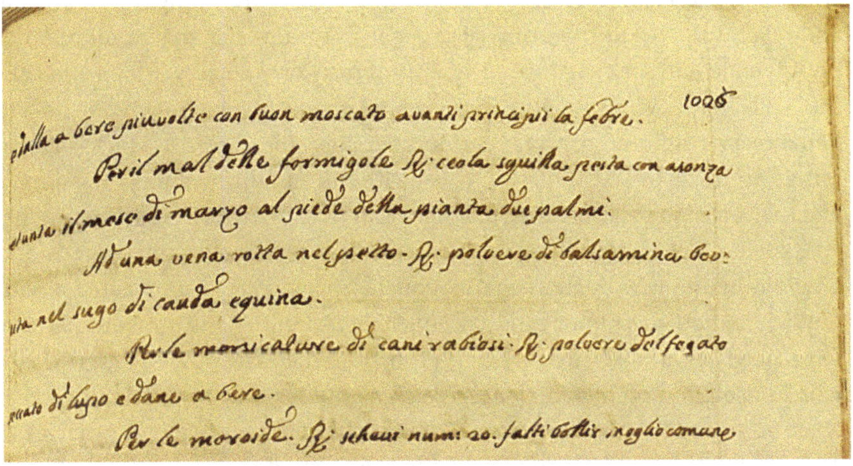

FIGURE 5.9 Wellcome Library, Ms. 3777, c. 126r. The recipes become shorter and shorter, the cartulation is misspelled (from *c.*1010 [but 110] to the end of the remedies, *c.*1027 [but 127])

The secrets that emerged from Margaret's foundry were also destined for sale, implicit in the authorisation of the Provveditori: from the manuscript, it can be deduced that she had been producing them for over twenty years, but evidently, around 1709, they were being circulated so widely as to make necessary the request for a marketing licence. In a passage she cites, incidentally, a bowl of "fortissimo spirito di nitro senza flemma" that she had prepared to send to the Fair of Senigallia, a market square in the Mediterranean Basin which was of extraordinary importance for Venice as well.[80] The Fair took place every

80 Ibid., cc. 88v–89r. On the Fair of Senigallia, see the various publications from the early 20th century by Roberto Marcucci, Giovanni Pagani, "Venezia e la fiera di Senigallia", *Atti e Memorie della R. Deputazione di Storia Patria per le Marche*, ser. IV, vol. 6, 1 (1929); Paola Lanaro, ed., *La pratica dello scambio: sistemi di fiere, mercanti e città in Europa, 1400–1700* (Venice: Marsilio, 2003), *ad indicem*.

year in the historic centre of the Adriatic city near Ancona, reaching its peak in popularity in the eighteenth century: sending one's secrets to the Fair of Senigallia meant that Margaret's preparations enjoyed a wide-ranging distribution network.

What makes Margaret Paston Alberti's collection of secrets so unique is that she lived between two different worlds, geographically and culturally: England and Italy, aristocracy and professionals/artisans. The debt to English and European scientific knowledge is evident in the declared provenance of many recipes: from her grandfather William and from her father Robert, numerous ones from the laboratory of "my Paston home" and from English aristocrats and royalty, as well as from other European countries. She makes mention of both female and male figures, several from the Royal Society, which she calls the "collegio dei Virtuosi," and physicians of great repute with whom she had had direct or indirect contact.[81] As for the Italian side, the aristocracy is represented above all by the Medicean court with the entourage of its professionals and the Venetian petty nobility, to whom Margaret often dispensed treatments and advice. She had many scientific exchanges with local professionals and religious figures who dealt in pharmacopeia. She was in constant contact with physicians, both as a patient and as a colleague, asking for and providing advice. She placed her utmost trust on Iseppo Ton, protomedico between 1682 and 1711, who himself authored medicinal secrets patented by way of the same process reserved to the "particular persons" who requested a licence or a privilege.[82] Thanks to Iseppo Ton's pills, she avoided the risk of dropsy ("anasarca") and she relied on him for help in establishing the correct doses in administering a remedy.[83] Physicians and surgeons from the front lines and from the Stato da

81 For the reconstruction of the network of English relations, refer to Vanke, Findlen, "Imperfect Reds".

82 ASV, *Sanità, Terminazioni*, b. 86, n. 578, 14 May 1696: "Possessing the excellent dominus Iseppo Ton medical examiner of this excellent magistrate specific for fluxions, gallic disease, and worms authenticated as such by many experiences, that with all distinction have made known their particular activity, while therefore he implores the benignity of this same most excellent magistrate and allows him and the excellent d. Gio. Antonio his doctor collegiate, to be they themselves the ones to compose and administer for the relief of the oppressed, its excellences made from the above instance their own considerations, have commendable regard for the petitioning subject, have conceded to the excellent medical examiner and said Giovanni Antonio his son to perform the above-said."

83 WL, Ms. 3777, "Pills of the excellent medical examiner Ton tested on myself when I was in the state of becoming dropsical, which is followed by the remedy for worms of the same (c. 91r); the collaboration is implied when he writes: "I do not have the dose for the powder but I will have it dosed by the medical examiner of health and will write in in the margins.", c. 58r.

Mar sent her recipes for treating wounds – Castelli, part of the retinue of Alvise Mocenigo, who oversaw Venice's maritime territories, and Van der Weyden "famous physician from the Levant" etc.[84] Even her husband Girolamo Alberti sent her recipes found during his diplomatic missions, even if there is no evidence as to his own personal medical or scientific interests.[85] At times, it was Margaret who assisted the local physicians in treating patients. She tells of an episode from 1697, when, even before having opened her foundry, she helped the "surgeon Francesco Bon" to cure someone named Tiepolo who, to avoid being sent to France as an ambassador, had severed his jugular vein and was bleeding to death.[86] Often Margaret took turns with other professionals at the bedsides of the sick: in the space of four days, she completely cured the wife of a boatman with one of her syrups, "to my and the physicians' amazement," she confessed modestly.[87] Margaret's secrets served to cure herself as well, like the common people of Venice and of Selva di Cadore in the Belluno area, where the Alberti family owned a country home.[88] Margaret, therefore, interacted with physicians, both as a patient and as a colleague-healer.

Among the professional figures with whom she exchanged recipes, there were plenty of chemists, some more famous than others. She explains that the *Specifico dolcificato della fonderia detto febrifugo* was taught to her by a Roman chemist who stopped in Venice, the same one who had also taught it to two local apothecaries; the recipe for the *nitro dolcificato* had been given to her by the chemist-alchemist Francesco Travagino, who was very well-known both in

84 Ibid., c. 99r. Alvise Mocenigo (1662–1732) was general Provveditore from March 1695–, 1707–1711 and 1718–1720; then elected doge, cf. Gino Benzoni, DBI, *ad vocem*.

85 Ibid., c. 85r: "Recipes sent to me by my husband from the Levant having been given to him by a virtuous Jewish doctor"; c. 91r: "For the pleurisy / recipe given to me by mister Girolamo my husband").

86 Ibid., c. 73r–v. This is Giovanni Domenico Tiepolo, who was elected on 20 June 1697 ordinary ambassador in France but refused to leave, BMC, *Ms. Venier, Consegi*, 87, at the time. In another passage of the Ms. 3777, c. 104r, Margaret describes a recipe stones of the physician Bon, who signed three curations of the *Pratica chirurgica* by Sebastiano Melli (Venice: Antonio Bortoli, 1724).

87 WL, Ms. 3777, c. 29v.

88 "Plaster to fortify the nerves of a member that has been dislocated, tested on my own life (c. 73r); Rinse for every sort of wound and sore with which I have cured at the villa of Selva many poor creatures without other balsams or plasters that here I described for the benefit of the poor (c. 72v); this balsam is my favourite more than any other possessing the same virtues as the vulnerary balsam of the Fondaria [of the Medici of Florence]: With this I have dome many miraculous things, both in the villa of Selva, and in Venice", c. 29v).

Italy and in England, where he was a member of the Royal Society and had a close relationship with Boyle.[89]

Another constant presence in the collection of Margaret's secrets were the nurse-apothecaries of the religious orders, who reflect what was happening in a country in which pharmacopeia was much practiced within the regular orders. The recipe for the drinkable gold that opens the manuscript was donated to Margaret by the "father Petronio from Verona, nurse of the Capuchins, truly a great man of the profession."[90] This was Petronio da Verona (formerly Rocco Domenico Mastagni, 1660–1744), botanist and naturalist, who completed the work of his master Fortunato da Rovigo, an herbarium of the flora of Monte Baldo in 7 volumes.[91] A Jesuit, Prince Spinola di Genova missionary of the Indies went to visit Margaret Paston in 1684 and taught her a variant of the expensive *balls of bezoar powder*.[92] The "virtuoso signore Antonio," apothecary of the Benedictines of San Giorgio taught her how to distil cordial waters based on herbs.[93] A male monastery nurse gave her the recipe to make the celebrated decorated metallic cups containing a laxative/emetic solution, often distributed as a gift: "The true recipe for making the antimonial cups, taught to me by brother Filippo Capuchin, nurse, several times made by him for his religion."[94] Margaret almost certainly referred to the Capuchins of the Redentore in Giudecca, who were famous in the late sixteenth century for their excellent nursing traditions.[95] While she includes recipes given to her by religious figures, Margaret can still be quite critical of some of them. For example, she provides recipes for the balsam for sores from Bolognese Carthusian Doroteo Spinelli, who was stationed in Padua, and that of the Benedictines

---

89 WL, Ms. 3777, cc. 88r, 118v. On Travagino and his role in English and above all Boylean science in Veneto culture see C. Pighetti, *L'influsso scientifico di Robert Boyle nel tardo '600 italiano* (Milano: FrancoAngeli, 1988); Ivano Dal Prete, *Scienza e società nel Settecento veneto. Il caso veronese, 1680–1796* (Milano: FrancoAngeli, 2008), pp. 75–76.

90 WL, Ms. 3777, c. 1r: Arcane physicist-chemist | Preparation of drinkable gold.

91 This and other news in the item DBI dedicated to Fortunato da Rovigo, ed. Pamela Anastasio. See also the modern edition of a manuscript on fish by fra Petronio: Giuseppe Sandrini, "Il libro della natura e il sogno del mare", in *Produzioni marine* (Verona: Alba Pratalia, 2008), pp. 9–32, 97–125.

92 WL, Ms. 3777, c. 3r. A panacea for 'extreme maladies': pestilential fevers, stones, epidemic maladies, urine retention.

93 Ibid., c. 48r–v.

94 Ibid., c. 86v.

95 During the plague epidemic in 1630 the Capuchins of the Giudecca were also tasked with dealing with the plague sufferers admitted to the Venetian Lazarettes. The 17th century furnishings of their apothecary are reproduced in Renato Vecchiato, *Gli speziali a Venezia. Pagine di storia*, (Venezia 2013), pp. 74–75.

of San Giorgio Maggiore di Venezia,[96] though, in line with the evolution of eighteenth-century pharmacopeia, she takes issue with the jumble of ingredients: "Although to say it with my usual earnestness I find both of them to be great mixtures of many herbs and drugs, when I shall be able to create a balsam, whether it is a medication or not, with fewer ingredients and greater virtues, I believe that, for me, honestly, it will be even better."[97] What is surprising, in this short series of examples, is that the religious people with whom she came into contact were exclusively male figures.

Scattered clues reveal Margaret Paston's scientific-experimental attitude. She told of her successes but was honest in saying when she had not yet tested the recipe, or that she had tested it several times without an outcome.[98] In a delightful diary passage, she explains how she learned – the hard way – how to correctly handle an antipyretic preparation based on nitro, after having provoked an explosive reaction with thick yellow smoke that shattered the glass bowl she was working with and caused a loud bang.[99] Margaret possessed and shared a type of sensorial knowledge that was exquisitely artisanal, a quality common for those who make and experiment and don't only have theoretical knowledge.[100] She had direct experience with the matter she was handling and was extremely precise in describing the correct consistency to be obtained as well as the processes to achieve it: "the consistency of honey," "the tenacity of dough ready for breadmaking," "as hard as marble," "the jar must not be closed too tightly so that it does not crack, but breathes a little,:" and so on.

In many circumstances, Margaret shows a confident awareness of diverse European therapeutic practices and of how a medicinal remedy ought to be. She came of age in an era that valued the specific; as we saw in the case of the two balsams for wounds, she did not believe in combining too many

96 Doroteo Spinelli published, using the name of his grandchildren Giovanni Battista and Francesca, *L'economia delle fabbriche* (Bologna, 1708) and *La Marescalca in Villa* (Bologna, 1714).

97 WL, Ms. 3777, cc. 31r–34r.

98 Ibid., c. 74r: "To heal every ulcer in the mouth that might be cancer or rotten tongue secret of mine several times approved without ever managing to get satisfaction from it". She does not say who was the inventor.

99 Ibid., c. 89r: "I want to teach another antipyretic preparation as well using the nitro that a great accident allowed me to learn and that almost cost my life."

100 Pamela H. Smith, *The Body of the Artisan: Art and Experience in the Scientific Revolution* (Chicago-London: The University of Chicago Press, 2004); Pamela O. Long, *Artisan/Practitioners and the Rise of the New Sciences, 1400–1600* (Oregon State University Press, 2011); Pamela H. Smith, Amy R.W. Meyers, Harold J. Cook, ed., *Ways of Making and Knowing: The Material Culture of Empirical Knowledge* (Ann Arbor: University of Michigan Press, 2014).

substances to obtain a medicine.[101] She shunned remedies that caused very violent reactions in favour of more gentle therapies that were more popular in some areas of Europe as opposed to others: with regards to the antipyretic preparation gone in a bang, she wrote that she had included it because the medication could be requested in Germany and in Turkey, where harsher therapies were also appreciated.[102] She possessed a sound knowledge of dietary habits and the various diseases that ran rampant through different European countries. She said she had given a Venetian nobleman a little of her excellent *Poppy Water*, a digestive that all the English families used, given the diet based on heavy meat pies, which, unlike the Italian diet, create "surfetts, che vol dir carghezza di stomaco" (heaviness in the stomach).[103] Again, while she illustrates the recipe for *Acqua che guarisce il mal scorbutico* (*Water that cures scurvy*), she makes it known that scurvy in England is a lot more widespread than in Italy, so she describes its symptoms.[104]

The English noblewoman leaves much more than a recipe book to her heirs. It is a reference book in which she shares wide-ranging scientific knowledge and not only the chemical-pharmaceutical kind. At a certain point, however, the regular and narrative rhythm of the manuscript is interrupted. Something seems to have thrown Margaret a bit off course:

> I must confess that my scarce memory and great distraction often make me forget to write the recipes that should follow one another in an orderly fashion; so I have done here, putting the ophthalmic before finishing what I had to write of the apoplectic balsam and the apoplectic balls, and to make amends for my error as this book must be seen by none than my nearest and dearest who will forgive me charitably, here below I will describe them.[105]

From the beginning of the manuscript, it is clear that it was not structured for the recipes to follow one another in an orderly fashion. It has somewhat

101 Harold J. Cook, "Markets and Cultures: Medical Specifics and the Reconfiguration of the Body in Early Modern Europe", *Transactions of the Royal Historical Society*, 21 (2011): 123–145.

102 WL, Ms. 3777, c. 89r: "I put it aside not liking the medicaments that cause the vomiting, however for Germany and Turkey where they seek vomit medications I think that this would be excellent, and I will teach the easiest way to make it without danger". On the search for therapeutic sweetness that was taking hold in Italy, see. Minuzzi, *Sul filo dei segreti*, pp. 215–222.

103 WL, Ms. 3777, c. 46v, underlined and English in the original.

104 Ibid., cc. 54v–56r.

105 Ibid., c. 64v.

systematic blocks that bring together remedies by pathology or type of remedy. Starting from c. 116r, however, there are even more groups of recipes explained in just a few lines and the narrative context used to enrich them is completely absent at times. Nor does she explain the source or whether she'd experimented with them or not.

Her husband Girolamo Alberti had passed away on 1 March 1716, with years of his salary yet to be collected, and this money was needed in part to settle the debts he'd accumulated over his lifetime, as well as to pay for funeral costs.[106] Sad family events definitively distracted Margaret from her chemical experiments. Her son Carlo Antonio, secretary of the Senate, had already been tried by the Inquisition in 1705 for occult practices along with a group of peers; but, in 1721, he committed what was considered to be a serious crime for the secular authorities, falsifying decrees and ducats. Notwithstanding Margaret's petition to the State Inquisitors, Carlo Antonio was put to death in secret on the night between 21 and 22 January 1722.[107]

The foundry-cum-laboratory, authorised by the *Provveditori* in 1709, was almost certainly closed in May 1723, when Margaret died. In 1721, she had nominated her son Giulio Cesare Alberti as legal representative for all of her interests. In the will, also dated 1721, she designated him universal heir to her properties in Venice and in England, having made sure her other children were taken care of financially.[108] But she made no mention of the secrets or of the foundry.

12 Lucietta Visomio, Elisabetta Amorosi Manini and the Legnaghi Sisters: The Evolution of Secrets

In the second half of the eighteenth century, when either "particular persons" or charlatans could only present authorities with secrets that were innovative and effective enough to obtain a privilege, there were three female figures

106 ASCPV, *Parrocchia di Sant'Agnese, Morti*, reg. 4. For the debts see ASV, *Notarile Atti*, reg. 12716, 19 March 1716, cc. 11r–12v. Girolamo's uncle, Giulio Cesare Alberti, who died in 1686 without descendants, had left his large inheritance bonded for 100 years to a testamentary obligation (*fedecommesso*) to avoid it being ceded to grandchildren amongst whom Girolamo because "of excessively liberal spirit, not to say prodigal, if I left in their will the control of it (= inheritance) I am sure that it would reduce their children to the state of poverty," cfr. ASV, *Notarile Testamenti*, b. 773–774, n. 124.

107 I refer to the well documented passage in Fabiana Veronese, "L'orrore del sacrilegio: abusi di sacramenti, pratiche magiche e condanne a morte a Venezia nel primo ventennio del Settecento", *Studi Veneziani*, 52, 2 (2006): 265–342: pp. 314–315.

108 ASV, *Notarile, Testamenti*, b. 949, n. 216, opened on 7 May 1723.

whose work saw them going back and forth between the apothecary and the domestic environment in what was then a mature pharmaceutical market.

In 1746, Nicola Corradi sold his “al Lupo” apothecary in campo SS. Filippo e Giacomo, close to Saint Mark’s, to Francesco Montagna, on one precise condition: “with explicit pact, that in said sale and purchase neither the *ceroto emmoliente detto del signor Antonio* normally sold in the aforementioned shop by the said Messrs Corradi, nor the credits apparent from the books of the aforementioned shop are included.”[109] That was a secret based on resins, oils, *litargirio, tremetina* and a few other ingredients, to treat fistulas, superficial tumours and cutaneous lesions in general, including deep ones. Normally prescribed by the Venetian surgeons and physicians and even used in Epirus, Macedonia and Albania, since the end of the seventeenth century, the *cerotto* (plaster) *ammolliente omogeneo* was associated with the “al Lupo” apothecary, even if none of the Corradi family members had ever requested the privilege for it.[110] According to documents, starting in 1746, Nicola continued to deal in this *cerotto* from home, managing to maintain his family – a wife and two daughters – until 1758. In actual fact the women of the home participated in the processing of the *cerotto* long before Nicolò died in 1758, at the age of just 55, after two years of being completely disabled.[111]

Between 1758 and 1760, his wife Lucietta Visomio and their two daughters pursued its handling and sale, until, in August 1760, his widow presented herself before the *Provveditori* asking that she be personally granted the privilege to produce and market the *cerotto*. She performed a test before the *protomedico* Pietro Santorini, obtaining a certification of compliance and very positive comments, from another physician who had witnessed it as well and who claimed he had already been using it in his profession for many years with excellent results.[112]

109 ASV, *Notarile, Atti*, b. 1143, 1st October 1746.

110 See the documentation attached to the file dated 15th September 1760 in ASV, *Sanità, Rapporti medici*, b. 588, among which a declaration dated 1st August 1743 of the physician Carlo Antonio Grazzi.

111 ASV, *Sanità, Necrologi*, reg. 947, 11 May 1760: “Nicola Coradi quondam Marco d’anni 55 da caduta plopletica [sic] anni 2, sorpreso da un accidente gagliardo dopo un’ora morì. Medico Biancini, prete zago. S. Samuel”.

112 Thus, Lotario Giuseppe Lotti, collegiate physician: “as I have for many years sold its excellent effect, having used it in many cases,” in ASV, *Sanità, Rapporti medici*, b. 588. Symptom of the fame and success of the Plaster are the counterfeits that in the wake of Nicolò Corradi’s death started to multiply, the first of which by Francesco Montagna who in 1746 had taken over the apothecary.

Mercoledì addì 4. Giugno 1760.

IPSE ALIMENTO SIBI. Hor.

N.° 35.

GAZZETTA VENETA

Che contiene

Quello, ch' è da vendere, da comperare, da darsi a fitto, le cose ricercate, le perdute, le trovate, in Venezia, o fuori di Venezia, il prezzo delle merci, il valore de' cambj, ed altre notizie, parte dilettevoli, e parte utili al Pubblico.

FIGURE 5.10 *Gazzetta veneta* of 4 June 1760, p. 4. Advertisement in which Lucietta Visomio announces that she is producing and keeping for sale at home the well-known plaster *cerotto del Sig. Antonio*

Lucietta Visomio and her daughters continued to produce and sell the secret from their home, using the same type of advertising the apothecaries typically relied on. So she got Lucia to write in the *Avvisi* of the *Gazzetta veneta*:

> It is made known to the public that the secret of the handling of the true and legitimate supreme homogeneous emollient bandage, commonly called *Cerotto del Sig. Antonio*, which was always made by Niccola Corradi and his masters at the Lupo apothecary in campo SS. Filippo e Giacomo, now that Corradi has passed away, is now solely owned by the widow Mrs. Lucia Visomio, not having at any other time been demonstrated to others, but currently produced and dispensed only by her in her home in *piscina S. Samuele* at the Galante house.[113]

113 *Gazzetta veneta*, n. 35, 4 giugno 1760. In the *Gazzetta*, as in other periodicals, there are several advertisements of Venetian apothecaries and authorised secret sellers, cf. Minuzzi, *Sul filo dei segreti*.

The secret continued to stay in the hands of the widow and then of her daughters Caterina and Margherita Corradi. When the new formula came out in 1777, *the protomedico* offered an enthusiastic opinion on it.[114]

Elisabetta Amorosi Manini from Rovigo – whose double-barrelled surname perhaps shows she may have been related to the aristocratic Manini family of Vicenza[115] – was the inventor of a medicinal secret based on apples and juice extracted from vine shoots, effective against scurvy and rickets. The *specifico antiscorbutico di Rovigo*, with an emphasis on the word "specific," reflects that the eighteenth-century incarnation of the drug is simple in its formula yet not at all ordinary. In evaluating it, the *protomedico* compared it to *cinchona*:

> The specific anti-scurvy of Rovigo, from the ingredients used to make it, turns out to be innocuous and homogenous, soothing. As it consists of apples and *pampani* juice from white grapevines, you can immediately see that it must give energy and strength to a mechanism that has been worn out. But then it is reflected in the particular effects that can be seen daily from it, it must be said that it is also effective and distinctly specific for scurvy, and even more distinctly for rickets. The woman making such a secret is thus worthy of all the praise and the protection for the benefit of the people by preserving intact and pure the result of a remedy of which, after that of cinchona, I have no doubt to assert frankly that we do not have anything more powerful and safe in the medical arts.[116]

The Legnaghi sisters brought the century to a close. In 1785, their brother Giuseppe, the owner of the "al Tempo" apothecary, had left them the formula of an ancient family secret: *acqua da spasimo e vermi*. The sisters turned to the Provveditori to have the privilege to continue producing the secret themselves as an artisanal family activity, "imploring the grace to be able to keep in the family the work and the sale of the water itself."[117] And they soon obtained it, while the sales point remained the apothecary Al Tempo, as had happened for the Bonetti widow's balsam.

---

114 ASV, *Sanità, Rapporti medici*, b. 588, report of the medical examiner Paitoni dated 14 May 1777. For the granting of the *jus privativo* cfr. ASV, *Sanità, Notatori*, reg. 768, cc. 38r, 50r, 16 May 1777; the petition of the Corradi daughters is in ASV, *Sanità, Suppliche*, b. 172, 22 April 1777.

115 Francesco Tommasini, Illustre serie dell'antichissima, e nobilissima famiglia Manini (Vicenza: Angelo Bontognale, 1690).

116 ASV, *Sanità, Rapporti medici*, b. 588, 24 August 1771.

117 Underlining in the original, in ASV, *Sanità, Notatori*, reg. 776, c. 39r, 27 April 1785 e cc. 60v–61r, 1st June 1785. The family dimension is referred to in the license "for more than a century in *said family* the water itself was made and sold in the apothecary ..."

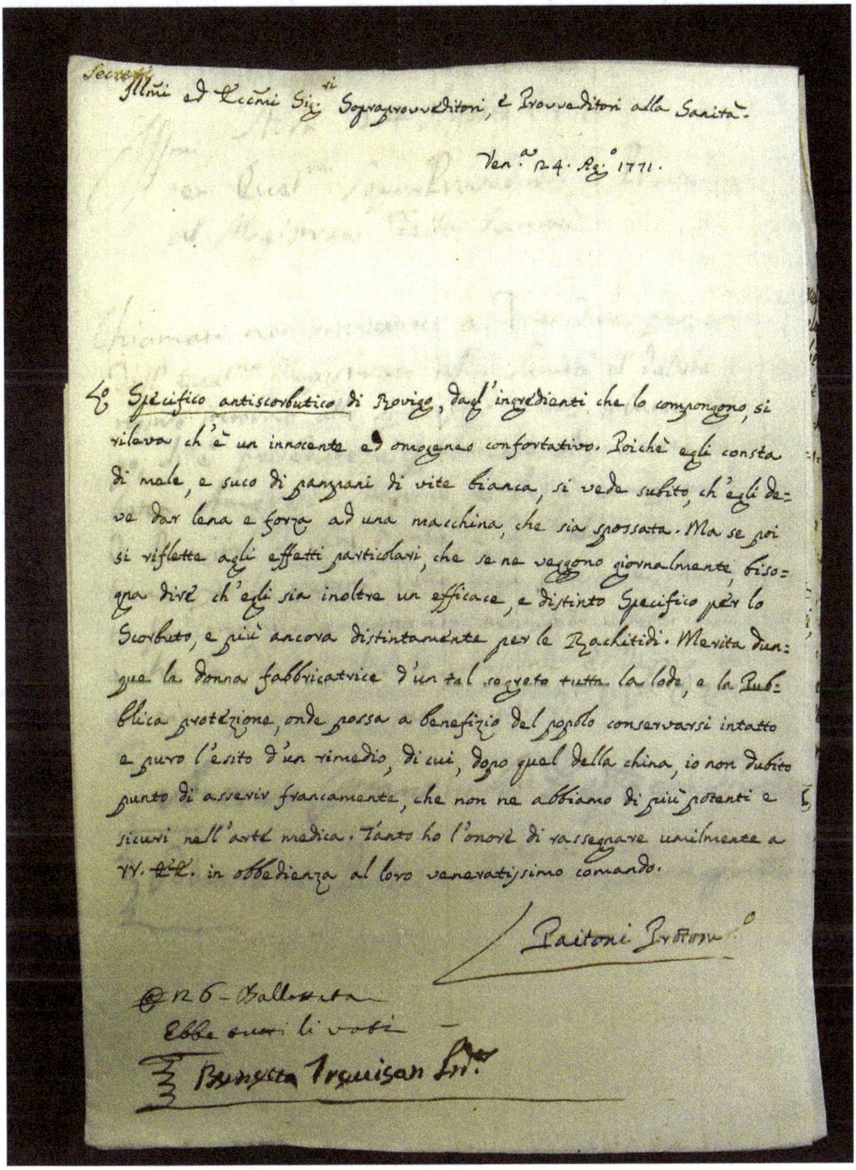
Secreti

Ill.mi ed Ecc.mi Sig.ri Sopraprovveditori, e Provveditori alla Sanità.

Ven.a 24. Ag.o 1771.

Lo Specifico antiscorbutico di Rovigo, dagl'ingredienti che lo compongono, si rilava ch'è un innocente ed omogeneo confortativo. Poichè egli consta di mele, e suco di ganzari di vite bianca, si vede subito, ch'egli deve dar lena e forza ad una macchina, che sia spossata. Ma se poi si rifletta agli effetti particolari, che se ne veggono giornalmente, bisogna dire ch'egli sia inoltre un efficace, e distinto Specifico per lo Scorbuto, e più ancora distintamente per le Rachitidi. Merita dunque la donna fabbricatrice d'un tal segreto tutta la lode, e la Publica protezione, onde possa a benefizio del popolo conservarsi intatto e puro l'esito d'un rimedio, di cui, dopo quel della china, io non dubito punto di asserir francamente, che non ne abbiamo di più potenti e sicuri nell'arte medica. Tanto ho l'onore di rassegnare umilmente a VV. EE. in obbedienza al loro veneratissimo comando.

Paitoni Protomedico

@ 26 - Ballottata

Ebbe tutti li voti —

Benetto Trevisan Prov.r

FIGURE 5.11 Archivio di Stato di Venezia, *Sanità, Rapporti medici*, b. 588, 24 agosto 1771. *Protomedico* Giambattista Paitoni enthusiastically evaluates Elisabetta Amorosi Manini's antiscorbutic remedy

13 Conclusions

The women in secrets discussed here are characterised by their belonging to a middling-sort milieu, where the lowest echelons are not represented and the highest ones include a foreign woman. There is no figure similar to Maddalena the weaver of Rome illustrated by Tessa Storey, who lived off an economy of makeshifts, she being a room-letter (when not working as a procuress), herbalist and maker of cosmetic waters and oils, and who was tried for magical practices.[118] On the contrary, the women in secrets working in Venice were all regularly authorised and were eager to have that recognition, as in the case of Angela Sala, who after repeatedly being given verbal licences, asked specifically for a written licence as a substitute for the professional title that she would have never been able to obtain.

In terms of those of the highest social standing, there was the Englishwoman Margaret Paston, educated in an elite scientific context. This noblewoman perfectly fits into the lively Venetian landscape of artisans, making for an extraordinarily fascinating hybrid: she brought together the traits of the court aristocracy with artisanal abilities used in working with chemical and natural substances as well as refined therapeutic knowledge with the marketing of her remedies at the Italian fairs. However, Paston was always unique in the panorama of the women of privileged secrets, and it's no surprise that she is the only woman to leave behind her own handwritten compilation.

The written record can naturally be traced back to her elevated social status, which made her similar to other noblewomen, mostly English, who compiled collections. Her manuscript reveals an attitude toward her work that distinguishes her a little from the Italian noblewomen with strong interests in medicine and science or in pharmacopeia. Take, for example, Cristina di Lorena, above all involved in the *patronage* of the work of physicians and high-level professionals, and, in general, the women of the Medicean court, active more in the sharing of knowledge and medicinal secrets than in the direct handling of the matter.[119] Similarly, the females figures of the Spada-Veralli nobility in Rome studied by Sandra Cavallo had developed solid knowledge relating to

118 Tessa Storey, "Face Waters, Oils, Love Magic and Poison: Making and Selling Secrets in Early Modern Rome," in Leong, Rankin, ed., *Secrets and Knowledge in Medicine and Science*, pp. 143–63.

119 This is clear in Barker's reconstructions: Sheila Barker, "The Contribution of Medici Women to Medicine in Grand Ducal Tuscany and Beyond", in Alessio Assonitis, Brian Sandberg, ed. *The Grand Ducal Medici and Their Archive (1537–1743)* (Turnhout: Brepols Publishers, 2016), pp. 101–116; Sheila Barker, "Christine de Lorraine and Medicine at the Medici Court", in J.C. Brown, G. Benadusi, ed., *Medici Women: The Making of a Dynasty in Grand Ducal Tuscany* (Toronto: Centre for Reformation and Renaissance Studies, 2015), pp. 154–180; Strocchia, *Forgotten Healers*. An exception was Caterina Sforza: see Sheila Barker, Sharon Strocchia, "Household Medicine for a Renaissance Court: Caterina Sforza's Ricettario Reconsidered," in Sara Ritchey, Sharon Strocchia, ed., *Gender, Health, and Healing, 1250–1550* (Amsterdam: Amsterdam University Press, 2020), pp. 139–166, p. 142.

healthy living. They were able to recommend the best medicinal remedies for ailments and disorders to family members, friends, servants and poor people, but ordered the cures and the compounds they needed from apothecaries.[120] Margaret Paston was also unique as compared to her English and European peers because she had settled in an Italian city characterised by a peculiar artisanal-experimental liveliness, a factor that further inspired her in collaborating with artisans. As has been correctly pointed out, there are socio-cultural and geographical differences that led to women in the medical and scientific practices in the different European contexts participating in a variety of ways.[121]

While Italian aristocratic ladies were less keen to handle secrets, this was not true for these women in secrets, who mostly came from a middling-sort milieu of professionals with strong family and artisanal ties. This is where the women/artisans of secrets existed – between the apothecary and the home or vice versa, between the shop/laboratory and urban kitchen (some better equipped than others). In the domestic environment, professional titles didn't matter, there were no gender differences in terms of knowledge and ability to work, and medicinal secrets continued to be created and sold. Families, women and men alike, conducted small economies of scale with the home – no longer the shop – as the centre of it, though sometimes they did have to pay some taxes to the apothecary guild (like the Pontano heirs, the only documented example, but presumably not the only one). Regardless of the differences in social class, it was the artisanal knowledge, the know-how, that brought together both women and men in secrets. There was the desire to use one's own hands, which is an important distinguishing factor from the sole knowledge of the therapeutic properties of *materia medica* and from the proper administration of prepared medicinal compounds.

These were women who acted within the rules and who, from what can glean from documents, enjoyed society's respect and deference. There is Giuseppe Scutellio who, in his print books, highlighted the medical and scientific capabilities of his mother, Caterina, and of his wife, Francesca, who worked alongside him as equals handling medical matter and giving advice to patients. *Protomedico* Paitoni, otherwise known for his strict evaluations, expressed respect and admiration for Elisabetta Amorosi.[122] Marietta Colochi was considered by her contemporaries to be a "medica" and Margaret Paston

120 Cavallo, "The Domestic Culture of Health".

121 Storey, "Face Waters, Oils, Love Magic and Poison"; Strocchia, *Forgotten Healers*, p. 24; Cavallo, "The Domestic Culture of Health."

122 Gentilcore, *Medical Charlatanism*.

was seen as a peer by local physicians and the professionals who came through Venice.

The great absence of religious female figures in this area should be pointed out. The men of the church, secular or not, on average, held 8% of the secrets licenced to "particular persons." The same could not be said for the women of the church. There was only one request for a privilege submitted collectively by the Converts of Treviso on 6 May 1765 for the *Balsamo dell'Avogadro*, which was already being used at the time. The formula had been given to them by the last descendent of the Avogadro family.[123] However, medical examiner Paitoni denied the licence because of the insufficient documentation provided: there were not enough ingredients listed to judge whether it merited secret status, and it was necessary to explain the method of processing it, because the effects could vary radically even if the substances were the same. Paitoni could certainly teach a thing or two to scholars who, still today, consider some medicinal secrets as being exactly the same due to having similar ingredients. But the fact remains that the absence of female religious figures in authorised secrets is striking when compared with the lively picture of religious women dedicated to apothecary in and around Florence as per Sharon Strocchia's research. Perhaps an explanation lies in the different typology of remedies: while in Venice it was necessary to demonstrate the innovative nature of medicinal secrets for which an exclusive permission was requested, Florentine religious organisations were mainly centres for dealing in non-experimental/non-innovative remedies as well as hubs for sharing knowledge pertaining to a more ordinary pharmacopoeia. Certainly, if we were to do research into the medical-scientific or pharmaceutical knowledge put into practice in the female religious institutes of Venice, we would find some surprises. In fact, it is well-known that some female religious institutes in Venice also sold medicinals that they had created, making for competition with the apothecaries – the Augustinians of Santa Lucia and the Franciscans of Santa Maria Maggiore and Santa Chiara, as well as the Conservatorio delle Penitenti –[124] though nobody asked, individually or collectively, for a privilege for secrets.

123 ASV, *Sanità, Rapporti*, b. 588, file on the date. The petition tells that in 1756 the last descendent of the Treviso lineage, Nestore degl'Azzoni Avogadro, decided to gift the secret of the "Ancient balsam belonging to his family donated to the reverend Converted mothers, licenced by the bishop Paolo Giustiniani to enter the cloistered convent every time there was the need to show how the remedy should be made."

124 Dian, *Cenni storici*, p. 153. In the *mariegola* of the apothecaries are listed in 1713 the "monasteries and hospitals that freely sell medicines," most of which male, which had to be more active in this sense. In ASV, *Milizia da Mar, Specieri medicinali*, b. 554 an *ad lites* booklet of 44 pages is preserved, which traces back over the initiatives taken by the

Therefore, there are plenty of new and exciting paths to be explored that would help complete the picture of those varied and wide-ranging connective tissues that we have been able to hone in on only in the last few decades.

References

Barker, Sheila, "Christine de Lorraine and Medicine at the Medici Court", in J.C. Brown, G. Benadusi, ed., *Medici Women: The Making of a Dynasty in Grand Ducal Tuscany* (Toronto: Centre for Reformation and Renaissance Studies, 2015), 154–180.

Barker, Sheila, "The Contribution of Medici Women to Medicine in Grand Ducal Tuscany and Beyond," in Alessio Assonitis and Brian Sandberg, ed., *The Grand Ducal Medici and Their Archive (1537–1743)* (Turnhout: Brepols Publishers, 2016), 101–116.

Barker, Sheila, Sharon Strocchia, "Household Medicine for a Renaissance Court: Caterina Sforza's Ricettario Reconsidered," in Sara Ritchey, Sharon Strocchia, ed., *Gender, Health, and Healing, 1250–1550* (Amsterdam: Amsterdam University Press, 2020), 139–166.

Bellavitis, Anna, *Il lavoro delle donne nelle città dell'Europa moderna* (Roma: Viella, 2016).

Beretta, Marco, "Material and Temporal Powers at the Casino di San Marco (1574–1621)," in Sven Dupré, ed., *Alchemy and Art Technology from Antiquity to the 18th Century* (Cham: Springer, 2014), pp. 129–156.

Berveglieri, Roberto, *Inventori stranieri a Venezia (1474–1788). Importazione di tecnologia e circolazione di tecnici artigiani inventori. Repertorio* (Venezia: Istituto Veneto di Scienze, Lettere ed Arti, 1995).

Brockliss, Laurence, Colin Jones, *The Medical World of Early Modern France* (Oxford: Clarendon Press, 1997).

Brown, Rawdon, *L'archivio di Venezia con riguardo speciale alla storia inglese* (Venezia; Torino: Antonelli e Basadonna, 1865).

Cabré, Monserrat, "Women or Healers? Household Practices and the Categories of Health Care in Late Medieval Iberia", *Bulletin of the History of Medicine*, 82, 1 (2008): 18–51.

Carlyle, Margaret, "From Practice to Print: Women Crafting Authority at the Margins of Orthodox Medicine", *Mémoires du livre*, 1 (2014).

Collegio degli Speziali between 1654 and 1731 to stem the drug market of the convents of SS. Giovanni e Paolo, S. Francesco della Vigna e di S. Giobbe, up to a final penalty of 50 ducats levied upon the latter two.

Cavallo, Sandra, "The Domestic Culture of Health in Early Modern Europe", in J. Eibach, M. Lanzinger, ed., *The Domestic Sphere in Europe (16th to 19th century)* (London: Routledge, 2020).

Cavallo, Sandra, *Artisans of the Body in Early Modern Italy: Identities, Families and Masculinities* (Manchester: Manchester University Press, 2010).

Ciasca, Renato, *L'arte dei medici e speziali nella storia e nel commercio fiorentino dal secolo XII al XV* (Firenze: Olschki, 1927).

Cicogna, Emmanuele Antonio, *Delle Inscrizioni Veneziane* (Bologna: Forni, 1969).

Cook, Harold J., "Markets and Cultures: Medical Specifics and the Reconfiguration of the Body in Early Modern Europe," *Transactions of the Royal Historical Society*, 21 (2011): 123–145.

Dal Prete, Ivano, *Scienza e società nel Settecento veneto. Il caso veronese, 1680–1796* (Milano: FrancoAngeli, 2008).

Dian, Girolamo, *Cenni storici sulla farmacia veneta* (Venezia, 1900).

Dizionario di Medicina Treccani (Rome: Enciclopedia Treccani, 2010).

Fiorucci, Emilie, "Women at Work in a Southern European Town: Women, Guilds and Commercial Partnerships in Venice in the Sixteenth century", in Anna Bellavitis, Beatrice Zucca Micheletto, ed., *Gender, Law and Economic Well-being in Europe from the Fifteenth to the Nineteenth Century: North versus South?* (London: Routledge, 2019), 215–241.

Fissell, Mary E., "Introduction: Women, Health, and Healing in Early Modern Europe", *Bulletin of the History of Medicine*, 82, 1 (2008): 1–17.

Gazzetta veneta, n. 35, 4 giugno 1760.

Gentilcore, David, *Medical Charlatanism in Early Modern Italy* (Oxford: Oxford University Press, 2006).

Graziosi, Marina, "Fragilitas sexus: alle origini della costruzione giuridica dell'inferiorità delle donne", in Nadia Maria Filippini, Anna Scattigno, and Tiziana Plebani, ed., *Corpi e storia* (Roma: Viella, 2011).

Green, Monica H., "Bodies, Gender, Health, Disease: Recent Work on Medieval Women's Medicine, *Studies in Medieval and Renaissance History*, 2 (2005): 1–46.

Green, Monica H., "Documenting Medieval Women's Medical Practice," in Luis García-Ballester, Roger French, Jon Arrizabalaga, Andrew Cunningham, ed., *Practical Medicine from Salerno to the Black Death* (Cambridge: Cambridge University Press, 1994), 322–351.

Green, Monica H., "Gendering the History of Women's Healthcare," *Gender & History*, 20, 3 (2008): 487–518.

Harkness, Deborah E., "A View from the Streets: Women and Medical Work in Elizabethan London", *Bulletin of the History of Medicine*, 82, 1 (2008): 52–85.

Lanaro, Paola, ed., *La pratica dello scambio: sistemi di fiere, mercanti e città in Europa, 1400–1700* (Venice: Marsilio, 2003).

Long, Pamela O., "Invention, authorship, 'intellectual property' and the origin of patents: notes towards a conceptual history," *Technology and Culture*, 32, 4 (1991): 846–884.

Long, Pamela O., *Artisan/Practitioners and the Rise of New Sciences, 1400–1600* (Corvallis: Oregon State University Press, 2011).

Mackenney, Richard, *Tradesmen and Traders: The World of the Guilds in Venice and Europe, c.1250–c.1650* (London-Sydney: Croom Helm, 1987).

Manno, Antonio, *I mestieri di Venezia. Storia, arte e devozione delle corporazioni dal XIII al XVIII secolo* (Cittadella: Biblos, 1995).

Marcucci, Roberto, Giovanni Pagani, "Venezia e la fiera di Senigallia", *Atti e Memorie della R. Deputazione di Storia Patria per le Marche*, ser. IV, vol. 6, 1 (1929).

Melli, Sebastiano, *Pratica chirurgica* (Venice: Antonio Bortoli, 1724).

Minuzzi, Sabrina, "Quick to Say Quack. Medicinal Secrets from the Household to the Apothecary's Shop in Early Eighteenth-century Venice", *Social History of Medicine*, 32, 1 (2019): 1–33.

Minuzzi, Sabrina, *Sul filo dei segreti: farmacopea, libri e pratiche terapeutiche a Venezia in età moderna* (Milan: Unicopli, 2016).

Minuzzi, Sabrina, *The Invention of the Author. The Privilegio di stampa in Renaissance Venice* (Venice: Marsilio, 2017).

Molà, Luca, "Inventors, Patents, and the Market for Innovations in Renaissance Italy," *History of Technology*, 32 (2014) 7–34.

Park, Katharine, "Medicine and the Magic: The Healing Arts", in Judith C. Brown and Robert C. Davis, ed., *Gender and Society in Renaissance Italy* (London; New York: Longman, 1998), pp. 129–149.

Park, Katharine, *Doctors and medicine in Early Renaissance Florence* (Princeton: Princeton University Press, 1985).

Pennell, Sara, Elaine Leong, "Recipe Collections and the Currency of Medical Knowledge in the Early Modern Marketplace", in Mark S. Jenner and Patrick Willis, ed., *Medicine and the Market in England and its Colonies. c.1450–c.1850* (Basingstoke: Palgrave Macmillan, 2007), 133–152.

Piccardi, Giovanni, "La Fonderia medicea di Firenze", *Memorie di Scienze Fisiche e Naturali, Rendiconti della Accademia Nazionale delle Scienze detta dei XL*, serie V, vol. XXIX, part II, tome II, vol. 123 (2005): 197–210.

Pighetti, C., *L'influsso scientifico di Robert Boyle nel tardo '600 italiano* (Milano: FrancoAngeli, 1988).

Ritchey, Sara, Sharon T. Strocchia, ed., *Gender, Health, and Healing, 1250–1550* (Amsterdam: Amsterdam University Press, 2020).

Sandrini, Giuseppe, *Produzioni marine* (Verona: Alba Pratalia, 2008).

Smith, Pamela H., Amy R. W. Meyers, Harold J. Cook, ed., *Ways of Making and Knowing: The Material Culture of Empirical Knowledge* (Ann Arbor: University of Michigan Press, 2014).

Smith, Pamela H., *The Body of the Artisan: Art and Experience in the Scientific Revolution* (Chicago-London: The University of Chicago Press, 2004).

Stevens Crawshaw, Jane, "Families, Medical Secrets and Public Health in Early Modern Venice," *Renaissance Studies*, 28, 4 (2014): 597–618.

Storey, Tessa, "Face Waters, Oils, Love Magic and Poison: Making and Selling Secrets in Early Modern Rome", in Elaine Leong and Alisha Rankin, ed., *Secrets and Knowledge in Medicine and Science, 1500–1800* (Farnham: Ashgate, 2011), 143–63.

Strocchia, Sharon T., "Introduction," *Renaissance Studies*, 28, 4 (2018).

Strocchia, Sharon T., *Forgotten Healers: Women and the Pursuit of Health in Late Renaissance Italy* (Cambridge: Harvard University Press, 2019).

Tommasini, Francesco, *Illustre serie dell'antichissima, e nobilissima famiglia Manini* (Vicenza: Angelo Bontognale, 1690).

Toninato, Tullio, Cesare Moretti, "Ricettari Muranesi", *Rivista della Stazione Sperimentale del Vetro*, 3–4 (1992): 197–206.

Toninato, Tullio, Cesare Moretti, *Ricettari muranesi: XVI–XX secolo* (Venice, 1991).

Trivellato, Francesca, "Guilds, Technology, and Economic Change in Early Modern Venice", in S.R. Epstein and Maarten Prak, ed., *Guilds, Innovation and the European Economy, 1400–1800* (Cambridge: Cambridge University Press, 2008), 199–231.

Vanke, Francesca, Paula Findlen, "Imperfect Reds: Robert Paston's Alchemical Interests", in A. Moore, N. Flis, F. Vanke, ed., *The Paston Treasure: microcosm of the known world* (New Haven: Yale Center of British Art, 2018), 218–231.

Vecchiato, Renato, *Gli speziali a Venezia. Pagine di storia* (Venezia 2013).

Veronese, Fabiana, "L'orrore del sacrilegio: abusi di sacramenti, pratiche magiche e condanne a morte a Venezia nel primo ventennio del Settecento", *Studi Veneziani*, 52, 2 (2006): 265–342.

PART 3

Eighteenth-century Spaces of Gendered Knowledge

CHAPTER 6

The "Anonymous Neapolitan": Faustina Pignatelli and the Bologna Academy of Sciences

Paula Findlen

... I am
by the side of the great Fausta, the diva
whose name is the sound of fame.
CORILLA OLIMPICA[1]

∵

Exactly eight months after the Bologna Academy of the Institute for Sciences made Laura Bassi (1711–78) its first female member on 20 March 1732, the Bolognese academicians met to deliberate whether to admit another woman into their ranks. On 20 November 1732, Faustina Pignatelli (1705–69), princess of Colubrano, became the second female member. Based on "the most definitive testimonials of this Signora's great and marvelous worth in mathematics and especially in algebra," they unanimously appointed Pignatelli "among the honorary academicians."[2] The basis for their enthusiasm rested primarily on the enthusiastic endorsement of a young Neapolitan mathematician and astronomer Pietro Di Martino (1707–46), recently arrived in Bologna in April 1732.[3]

---

1 Alessandro Ademollo, *Corilla Olimpica* (Florence: C. Ademollo, 1887), p. 62.

2 Archivio dell'Accademia delle Scienze, Bologna (hereafter AASB), *Verbali degli Atti* (20 November 1732); see Marta Cavazza, "Les femmes à l'Académie: le cas de Bologne," in Daniel-Odon Hurel, Gérard Laudin, ed., *Académies et sociétés savantes en Europe (1650–1800)*, (Paris: Honoré Champion, 2000), pp. 161–175, esp. pp. 162–163.

3 Pietro Nastasi, "Pietro De Martino (Di Martino), *Dizionario biografico degli italiani* (Rome: Enciclopedia Italiana, 1960–) [hereafter DBI], vol. 38, pp. 603–606. The most recent assessments of his work include Salvatore Serrapica, *Scienza e filosofia nella Napoli settecentesca. I fratelli Di Martino* (Sorrento, 2011); and Claudia Addabbo, "The *Philosophiae Naturalis Principia Mathematica* in Naples," in Elizabethanne Born, Mordechai Feingold, ed., *Reading Newton in Early Modern Europe* (Leiden: Brill, 2017), pp. 23–63.

 | DOI:10.1163/9789004512610_008

Pietro Di Martino belonged to a circle of young philosophers and mathematicians under the patronage of Monsignor Celestino Galiani (1681–1753). After a long career in Rome, Galiani returned to Naples in 1732 to become bishop of Taranto and take up his appointment as *cappellano maggiore* of the Kingdom. He had great plans for the reform of Neapolitan science, which included the creation of a Neapolitan Academy of Sciences in 1732, modelled in part on the Bologna Academy of which he was a member.[4] In 1731–32, three Neapolitan men of science – the physician and protomedico Francesco Bonocore, Antonio Caracciolo, Prince of Tonella, and Di Martino – became academicians, further cementing ties between the two cities. Di Martino had been sent to Bologna not only to deepen his knowledge of astronomy by working with Eustachio Manfredi in the Institute observatory, before taking up his appointment as the new professor of astronomy and nautical science at the University of Naples.[5] Undoubtedly inspired by the endless fêting of Bassi during the year in which she received a degree and professorship in addition to her academy membership, his praise of Pignatelli was, in part, a challenge to the Bolognese to acknowledge that there were women of science beyond their native city.

The delicate nature of this discussion revolved not only around the relative merits of these two women, but also the desirability of admitting more than one woman to the Bologna Academy of Sciences. The Bolognese academicians did not want their recognition of Pignatelli to set a worrisome precedent. Bassi had been a singular instance of a city acknowledging, indeed publicly embracing the demonstrable talent of a local prodigy, a young woman well versed in natural philosophy and capable of debating a wide variety of scientific topics in public to great acclaim. The decision to recognize her was not without controversy. The vociferous disagreements about whether to acknowledge Bassi's learning publicly were still fresh in everyone's memory, and yet the group in favor of rewarding her with institutional recognition of her accomplishments – a degree, an academy membership, and ultimately a salaried university

4 Fausto Nicolini, *Un grande educatore italiano. Celestino Galiani* (Naples: Giannini, 1951); Vincenzo Ferrone, *The Intellectual Roots of the Italian Enlightenment: Newtonian Science, Religion, and Politics in the Early Eighteenth Century*, tr. Sue Bretherton (Atlantic Highlands, NJ: Humanities Press, 1995).

5 Raffaele Mazzei, "Relazioni culturali fra Bologna e Napoli nel secolo XVIII," *Il Carobbio* 3 (1977): 285–297; and Aldo Brigagli, Pietro Nastasi, "Bologna e il Regno delle Due Sicilie: aspetti di un dialogo scientifico (1730–1760)," *Giornale critico della filosofia italiana*, 63 (1984): 145–178. I have used this longer version of their foundational study, which can also be found in Renzo Cremante, Walter Tega, ed., *Scienza e letteratura nella cultura italiana del Settecento*, (Bologna: Il Mulino, 1984), pp. 211–232.

professorship – prevailed.[6] The prospect of Pignatelli's admission raised concerns that the Bologna Academy of Sciences might now admit any and every learned woman who demonstrated aptitude in the sciences.

Perhaps this was part of the reason the academy secretary, Francesco Maria Zanotti (1692–1777), delayed sending news of "this illustrious and magnificent aggregation," since the first official correspondence informing Pignatelli of this singular honor dates to 11 March 1733.[7] There was indeed some concern that celebrating women of science, far and wide, rather than recognizing a single woman who had been nurtured and educated in the local environment, might send a far more radical message than the majority intended. Bassi was supposed to be exception, not the rule.

In the meantime, however, Pietro Di Martino began to involve Pignatelli in discussions with colleagues in Bologna. On 11 December 1732, in addition to reading aloud his own paper responding to an astronomical problem proposed by Manfredi, he offered Pignatelli's "Solution to a Numeric Problem." Assuring listeners that her reputation as a geometer was second to none, Di Martino displayed the letters that "Faustina Pignatelli wrote to me."[8] None of these letters survive, but Di Martino's description of them does. They were material demonstrations of the mathematical mind he described with such fervor and conviction that the academy admitted her. They had no other proof of her abilities, indeed did not subject her to the intensive private exams that proceeded Bassi's admission to the academy and her public examination by the university. No one besides Di Martino knew her at all.

In Pignatelli's case, her correspondence was crucial in the constitution of her identity at the center of a Neapolitan network and the making of her reputation abroad. She published only once, anonymously. Her near invisibility today serves in marked contrast to her reputation in her own lifetime.[9] Her scientific letters convinced a group of male academicians to honor and publicly acknowledge Pignatelli's learning. Di Martino's campaign to make her the first

6 The literature on Bassi is now considerable, but a great point of departure is Marta Cavazza, *Laura Bassi. Donne, genere e scienza nell'Italia del Settecento* (Milan: Editrice Bibliografica, 2020); and Beate Ceranski, *"Und sie fürchtet sich vor niemandem": Die Physikerin Laura Bassi (1711–1778)* (Frankfurt: Campus Verlag, 1996).

7 AASB, *Antica Accademia*, Corrispondenza Zanotti, n. 95 (11 March 1733).

8 AASB, *Memorie*, Titolo Quarto (Pietro De Martino, "Problema sulle annualità. Letta 11 Xbre 1732); Annarita Angelini, ed., *Anatomie accademiche*. Vol. 3. *L'Istituto delle Scienze e l'Accademia* (Bologna: Il Mulino, 1993), p. 328.

9 On a different form of invisibility, see Paola Bertucci, "The In/visible Woman: Mariangela Ardinghelli and the Circulation of Knowledge between Paris and Naples in the Eighteenth Century," *Isis*, 104 (2013): 226–249.

female foreign member succeeded because her words, diagrams, and solution to an unspecified mathematical problem were persuasive.

There was still the problem of setting a potentially dangerous precedent. Immediately after admitting Pignatelli, the academicians discussed a "proposal not to accept any more women into the Academy."[10] They kept this promise until the admission of the French mathematical philosopher and physicist Émilie du Châtelet (1706–49) in 1746.[11] Pignatelli became an important part of the symbolic capital of the Bologna Academy of Sciences, strengthening

FIGURE 6.1 The Bologna Academy of Sciences. *De Bononiensi Scientiarum et Artium Instituto atque Academia Commentarii* (Bologna, 1748). Courtesy of Special Collections, Stanford University Library

10 AASB, *Antica Accademia*, Corrispondenza Zanotti, n. 95 (11 March 1733).

11 Cavazza, "Les femmes"; Massimo Mazzotti, "Mme Du Châtelet, académicienne de Boulogne," in Olivier Courcelle, Ulla Kölving, ed., *Émilie Du Châtelet, éclairages et documents nouveaux*, ed. (Ferney-Voltaire: Centre international d'étude du XVIIIe siècle, 2008), pp. 121–126; and Paula Findlen, "Tra uomini. Laura Bassi all'Istituto delle Scienze (1732–78)"/"Always among Men: Laura Bassi at the Bologna Academy of Sciences (1732–78)," in Luisa Cifarelli, Raffaella Simili, ed., *Laura Bassi. Emblema e primato nella scienza del Settecento* (Bologna: Società Italiana di Fisica, 2012), pp. 71–88 (English pp. 189–206).

its ties with the Kingdom of Naples. For almost forty years, she corresponded with the academy, especially Zanotti, who was secretary at the time of her admission and later president. While Pignatelli never visited Bologna, Zanotti made a point of finally meeting her in Naples in 1750. Their personal encounter inspired his decision to write Pignatelli into his *Della forza de' corpi che chiamano viva* (1752), a dialogue on the question of *vis viva* addressed to his friend, the celebrated physician Giambattista Morgagni.

This episode, however, says more about the preoccupations of the Bologna Academy of Sciences, without revealing much of anything about Pignatelli. Why was she next in line after Bassi as a candidate for admission? Why didn't the short-lived but ambitious Neapolitan Academy of Sciences make her a celebrated female academician? This essay explores Pignatelli's reputation as a student and patron of mathematics and physics in eighteenth-century Naples. Building especially on the pioneering work of Aldo Brigaglia and Pietro Natasi, it reconstructs the network between Naples and Bologna, fostered by Galiani and the Di Martino brothers and eventually by Pignatelli, while paying greater attention to traditions of female learning, patronage, and aristocratic sociability in the Regno.[12]

1 How the Princess Became a Philosopher

Faustina Pignatelli was the only child of the duke of Tolve, Michele Pignatelli (1652–1720), and Faustina Caracciolo of Villamaina (1686–1705) who died giving birth to her daughter. We know very little of her life before her ill-fated marriage to Francesco Domenico Carafa (1695–1746) in February 1724, other than to note that she inherited her father's considerable estate at age fifteen.[13] Related by birth and marriage to three great noble houses of the Kingdom of Naples and born into a family whose various branches produced viceroys, archbishops, and pope Innocent XII (1691–1700), she grew up with enormous wealth and privilege. Upon marriage, her paternal uncle bequeathed the duchy of Tolve in Basilicata to her, and her father-in-law Domenico bestowed the principality of Colubrano upon his son, who already had been declared the sole heir to the family's considerable feudal estate by the Vicaria, including the fabled Renaissance palace embellished by the first count of Maddaloni,

12 Brigaglia and Natasi, "Bologna e il Regno."

13 Archivio di Stato, Naples (hereafter ASN), *Caraffa di Maddaloni*, b. I-D/6a (Ristretto del testamento del Duca di Tolve, padre di Faustina, 9 November 1720).

FIGURE 6.2 The Palazzo Carafa dei Maddaloni in Naples. Pompeo Sarnelli and Antonio Bulifon, *Guida de' forastieri: curiosi di vedere, e d'intendere le cose più notabili della regal città di Napoli, e del suo amenissimo distretto* (Naples, 1697 ed.), engraving by Bulifon between pp. 44–45. Courtesy of the Bowes Art & Architecture Library, Stanford University

Diomede Carafa (ca. 1406–87), with his renowned collection of antiquities.[14] For the rest of her life, Pignatelli would be known as the princess of Colubrano. She enjoyed a prominent place among the baronial elite of the Kingdom of Naples, rich in vast feudal landholdings, palaces filled with books, art, and antiquities, and country villas to escape the summer heat of the city.[15]

Pignatelli's Naples was a world undergoing great social, political, and intellectual transformations, with a lively cultural life in which women played a prominent role.[16] There was a growing desire to understand Cartesian philosophy and other modern systems of knowledge. Cartesian logic, physics, and metaphysics became part of the new curriculum of the University of Naples, with a growing emphasis on teaching many subjects in the vernacular to make them more accessible and meaningful to a local audience.[17] Agostino Ariani (1672–1748), the distinguished professor of mathematics, offered dynamic, well-informed lessons, filled with experimental demonstrations and up-to-date information about the most exciting developments in French and English mathematics and physics. In the city's improvised lecture hall in the cloister of San Domenico Maggiore, he "nourished the best young minds of Naples to cultivate the mathematical sciences, and called here the most prominent young nobles of the country, who did not disdain sitting on the bench with other scholars of inferior condition to listen to him."[18] Scholars observing

14 Elena Papagna, "Faustina Pignatelli," DBI, vol. 83; Maria Antonietta Visceglia, *Il bisogno di eternità. I comportamenti aristocratici a Napoli in età moderna* (Naples: Guida, 1988), pp. 61 62. For the history of the palace, see Guido Ceci, "Il palazzo dei Carafa dei Maddaloni poi di Colubrano," *Napoli nobilissima*, 2 (1893): 148–152, 168–170; and Bianca de Divitiis, "New Evidence for Sculptures from Diomede Carafa's Collection of Antiquities," *Journal of the Warburg and Courtauld Institutes*, 70 (2007): 99–117.

15 On this world, a good starting point is Tommaso Astarita, *The Continuity of Feudal Power: The Carraciolo di Brienza* (Cambridge, U.K.: Cambridge University Press, 1992); Visceglia, *Il bisogno di eternità*; and Gérard Labrot, *Baroni in città. Residenze e comportamenti dell'aristocrazia napoletana 1530–1743* (Naples: SEN, 1979).

16 Elisa Novi Chavarria, "The Space of Women," in Tommaso Astarita, ed., *A Companion to Early Modern Naples* (Leiden: Brill, 2013), pp. 177–196. On the Neapolitan Enlightenment, see John Robertson, *The Case for Enlightenment: Scotland and Naples, 1680–1750* (Cambridge, U.K.: Cambridge University Press, 2005); Girolamo Imbruglia, ed., *Naples in the Eighteenth Century: The Birth and Death of a Nation State* (Cambridge, U.K.: Cambridge University Press, 2000); and Melissa Calaresu, Helen Hills, ed., *New Approaches to Naples c.1500–c.1800: The Power of Place* (Farnham, Surrey: Ashgate, 2013).

17 Raffaele Ajello, "Cartesianismo e la cultura oltremontana al temp dell' 'Istoria civile,'" in Raffaele Ajello, ed., *Pietro Giannone e il suo tempo* (Naples: Jovene, 1980), vol. 1, pp. 102–144, 138.

18 Vincenzo Ariani, ed., *Memoria della vita, e degli scritti di Agostino Ariani* (Naples, 1778), p. 99. For a general overview, see Riccardo De Sanctis, *La nuova scienza a Napoli tra '700 e '800* (Rome: Laterza, 1986). The work of Giuseppe Galasso, Vincenzo Ferrone, Paolo

these developments predicted that now that the Neapolitans had decided to tackle analytic geometry, the Italians would soon have no reason "to envy the Germans, French, and English who seem to claim the cultivate of such studies as theirs alone."[19] The Neapolitan publishing industry was boldly printing controversial works, including Galileo's prohibited *Dialogue* in 1710 as well as the writings of Neapolitan philosophers who gave the inquisition reason to believe that freedom of thought was indeed a dangerous liberty. It seemed possible to publish just about anything in Naples under a false imprint.[20]

Eighteenth-century Naples was increasingly a city of aristocratic and philosophical women. The passion for Cartesian thought, and more generally philosophy and mathematics, migrated between the lecture halls and the city's salons, where it attracted different audiences and took on new meaning. When the vicereine invited Ariani to tutor a young Spanish lady who was a court favorite, he willingly obliged, adding her to his long list of aristocratic pupils, including the viceroy's sons.[21] Female literacy was on the rise though never high; diocesan records reveal that 9.5% of women were able to sign their marriage contracts in 1701 and 8.5% in 1731 in contrast to 3.5% in 1685.[22] Debates about women's learning were part of the fabric of the world in which Pignatelli grew up and there were models for her to emulate.[23]

During the 1710s the duchess of Limatola, Aurelia d'Este, (1683–1719) emerged as the first prominent Cartesian woman of the Regno. She counted

Casini, Massimo Mazzotti, Paola Bertucci, Lucia Dacome, and most recently Salvatore Serrapico offers a deeper understanding of specific episodes.

19 Domenico Guglielmini to Antonio Magliabechi, Padua, 10, October 1700, as quoted in Romano Gatto, *Libri di matematica a Napoli nel Settecento. Editoria, fortuna e diffusione delle opera* (Rome: Edizioni di storia e letteratura, 2010), p. 4.

20 Harold Samuel Stone, *Vico's Cultural History: The Production and Transmission of Ideas in Naples 1685–1750* (Leiden: Brill, 1997), p. 65. For a comprehensive overview of Neapolitan publishing, see Anna Maria Rao, ed., *Editoria e cultura a Napoli nel XVIII secolo* (Naples: Liguori, 1998).

21 Ariani, *Memoria*, pp. 100–101.

22 Antonio Illibato, *La donna a Napoli nel Settecento. Aspetti della condizione e dell'istruzione femminile* (Naples: M. D'Auria, 1985), pp. 9–10. On the status and power of aristocratic women, see Visceglia, *Il bisogno di eternità*, pp. 141–174; and Elisa Novi Chavarria, "Forme e spazi della sociabilità aristocratica nel Settecento," in Roberto Bizzocchi, Arturo Pacini, ed., *Sociabilità aristocratica in età moderna. Il caso genovese: paradigmi, interpretazioni e confronti* (Pisa: Pisa University Press, 2008), pp. 73–86.

23 Luciano Guerci, *La discussione sulla donna nell'Italia del Settecento* (Turin: Tirrenia, 1987); Rebecca Messbarger, *The Century of Women: Representations of Women in Eighteenth-Century Italian Public Discourse* (Toronto: University of Toronto Press, 2002); and Maria Gaetana Agnesi et al., *The Contest for Knowledge: Debates over Women's Learning in Eighteenth-Century Italy*, ed. Messbarger and Findlen (Chicago: University of Chicago Press, 2005).

Ariani among her friends and admirers; upon her death the Venetian *Giornale de' letterati d'Italia* recalled her sonnets inspired by Descartes' *Meditations*.[24] Her mentor Paolo Mattia Doria (1667–1746), with whom she studied Cartesian physics and metaphysics, described Este as a woman who recognized the value of studying modern philosophy. "She has applied her mind so seriously to the science of metaphysics that she knows them as well as every man who has been educated in that subject," he declared in 1716, "and she knows how to deduce truly useful conclusions from it."[25] Doria demonstrated the power of Este's intellect by responding to her criticisms of how he treated women in his earlier writing without entirely ceding ground to her Cartesian understanding of women's intellectual equality.

At the beginning of his *Dialogues*, Doria declared that his work was neither a history of learned and valorous women of the past nor a lighthearted courtly debate comparing the virtues of the perfect male and female courtier, in the spirit of Baldassarre Castiglione's *Book of the Courtier* (1528), but a serious philosophical inquiry into women's role in civil society presented as a conversation with Este. Responding to her accusation that he had unfairly depicted women as "lowly servants" unfit for intellectual pursuits in his book on *Civil Life* (1709), Doria sought to clarify his position in relation to her own. He acknowledged that few men educated their women well, especially in the sciences. Praising Este as a perfect model of women's ability to rise to the challenges of learning difficult subjects, he nonetheless expressed reservations about the Cartesian universality of the mind that she professed. "I do not share the sentiment that many attribute to René Descartes, that is, that God has given everyone an equal ability to understand the sciences," he responded.[26] Doria acknowledged that both men and women could know, yet he nonetheless concluded his conversation with a learned Cartesian woman by producing a lengthy appendix detailing why the natural weakness of women's bodies limited their strength of mind. Exceptions did not alter the general rule. "In truth, if you turn your most judicious mind to consider all the wise, virtuous, and even courageous

24 *Giornale de' letterati d'Italia* 32 (1719): 508; Ariani, *Memoria*, p. 166.

25 Paolo Mattia Doria, *Ragionamenti di Paolo Mattia Doria indirizzati alla Signora d'Aurelia d'Este Duchessa di Limatola ne' quali si dimostra la donna, in quasi che tutte le virtù più grandi, non essere all'uomo inferiore* (Frankfurt, 1716), sig. a4v; Gabriel Maugain, *Étude sur l'évolution intellectuelle de l'Italie de 1657 à 1750* (Paris: Librairie Hachette, 1909), p. 198. On Doria, see especially Vittori Conti, *Paolo Mattia Doria. Dalla repubblica dei togati alla repubblica dei notabili* (Florence: Olschki, 1978).

26 Doria, *Ragionamenti*, p. 344.

women whom I described here, you will find very few legislators and few who have formed their own system of the sciences," he observed.[27]

Doria left no doubts regarding his position. He greatly admired Este's learning and enjoyed public accolades "for having so famously instructed her in philosophical things," but he did not feel that women could create knowledge.[28] Neapolitan nobles and scholars who attended Este's funeral in 1719, in the church adjoining the convent of Regina Coeli in Naples where so many daughters of the Neapolitan nobility were educated before marriage or taking religious vows, observed that the illustrious marriage that brought her to Naples from Modena did not prevent her from pursuing her studies. Her intellectual passions only added to her renown. Nonetheless, Este's eulogy reinforced the passive limits to her knowledge by describing her as a woman who "enjoyed nothing more than the conversation of learned men."[29] Reading and understanding Cartesian philosophy was an accomplishment, but Doria constantly reminded her that it was not the same as being Descartes. This was the world in which Pignatelli came of age under her father's care and tutelage.

As an adolescent, Pignatelli witnessed a bold response to Doria's explanation of the differences between men and women. The next generation of Neapolitan women engaged with escalating polemics about Cartesian thought in their city by making his writings more widely available and using them as a pretext to argue the case for women's education with greater authority. In 1722, Giuseppa Eleonora Barbapiccola translated Descartes' *Principles of Philosophy* into Italian. The niece of the Dominican preacher Tommaso Maria Alfani from Salerno, Barbapiccola traveled in the same circles as Doria since she was a friend of Giambattista Vico's daughter Luisa.[30] In her preface to the first Italian version of Descartes' *Principles of Philosophy*, she declared her intellectual kinship with Princess Elisabeth of Bohemia, to whom Descartes admiringly dedicated his book, and also with his final patron, Queen Christina of Sweden. Barbapiccola resumed the conversation initiated by Doria with Este about whether "there is a need for something other than simply a woman's mind" in discussions of philosophy rather than poetry.[31] Her translation of Descartes,

---

27 Doria, *Ragionamenti*, p. 363.

28 *Giornale de' letterati d'Italia*, 32 (1719): p. 511.

29 *Giornale de' letterati d'Italia*, 32 (1719): p. 510 (description of Este), p. 512 (funeral).

30 Manuela Sanna, "Un'amicizia alla luce del Cartesianesimo: Giuseppa Eleonora Barbapiccola e Luisa Vico," in Pina Totaro, ed., *Donne, filosofia e cultura nel Seicento*, ed. (Rome: Consiglio Nazionale delle Ricerche, 1999), pp. 173–178; and Findlen, "Translating the New Science: Women and the Circulation of Knowledge," *Configurations*, 2 (1995): 167–206.

31 Giuseppa Eleonora Barbapiccola, "The Translator to the Reader," Agnesi et.al., *The Contest for Knowledge*, p. 50.

comparing the French version to the Latin original, rendered the question moot. Descartes had declared Princess Elisabeth his best reader. Barbapiccola transformed this comment into a more general observation that women "are more apt at philosophy than men."[32]

Barbapiccola's ironic rejoinder to the questions that animated Doria's discussions with Este in the previous decade helps us to understand how rapidly Pignatelli's world was changing. At the time of her marriage, less than two years after Barbapiccola published her Cartesian manifesto on behalf of women's learning, Naples had become a city embroiled in public debates about the modern shape of knowledge and the role of women in these philosophical conversations. When the physician Giovanni Bianchi visited Naples in 1724–25, he noted the triumph of Cartesianism in this city.[33] A young Antonio Genovesi (1712–69) spent the late 1720s debating whether his newfound love of novels made his Cartesian education at the Jesuit college seem like a pointless exercise of the mind.[34] Ultimately, his love of mathematics and pursuit of new ideas renewed his passion for philosophy and brought him into Pignatelli's orbit.

At first glance, Pignatelli's marriage to Carafa seemed like the ideal union to facilitate her intellectual ambitions while further cementing her social standing. Carafa had his own literary aspirations that complemented his wife's learned interests. Four children – a daughter Caterina and three sons, Giuseppe, Michele, and Diomede Casimiro – were born starting in 1725. During the first years of their marriage, they initially resided in Naples in the historic Palazzo Carafa di Colubrano, still considered among "the most beautiful in Naples" even though its impressive collection of antiquities was gradually being sold to pay the bills.[35]

In 1728, the young couple migrated to Carafa's castle at Formicola, north of Capua and Caserta. Carafa and Pignatelli's shared love of Arcadian poetry led

32 Barbapiccola, "Translator,", p. 55.

33 Ajello, "Cartesianismo," p. 151. For a broad portrait of women and science in Naples, see Raffaella Simili, "In punta di penna. Donne di scienza e di cultura fra cosmopolitanismo e intimità meridionale," in *La scienza del Mezzogiorno dopo l'Unità d'Italia* (Soveria Manelli: Rubbettino, 2008), pp. 27–89.

34 Antonio Genovesi, *Autobiografia, lettere e altri scritti*, ed. Gennaro Savarese (Milan: Feltrinelli, 1962), pp. 9–10.

35 Carlo Celano, *Delle notizie del bello, dell'antico e del curioso della città di Napoli* (Naples, 1692), giornata III, p. 188; Eloisa Dodero, "Le antichità di Palazzo Carafa-Colubrano: prodromi alla storia della collezione," *Napoli nobilissima*, 5 (2007): 119–140; and Dodero, *Ancient Marbles in Naples in the Eighteenth Century: Findings, Collections, Dispersals* (Leiden: Brill, 2019), pp. 67–68.

FIGURE 6.3 The Academy of Monte Caprario founded by Francesco Carafa at Formicola. Francesco Carafa, *Il Caprario. Accademia di diversi rimatori, che nel medesimo monte si radunarano* (Naples, 1729). Courtesy of Special Collections, University of Leeds

them to found the short-lived Accademia di Monte Caprario.[36] The principal outcome of this literary academy were two volumes of verses dedicated to the prince, "promoter of the learned gathering that has illuminated everyone through the great light of his sublime genius," in 1729 and 1732.[37] Carafa published his own book of poems in 1730, filled with Petrarchan verse lamenting lost and cruel loves, the departure of his brother Giovanni for military service, the return of the skull of his ancestor Don Giuseppe Carafa brutally murdered during the 1647 revolt in Naples, and his own efforts to manage his often resentful vassals eking out a living on family lands. Carafa conspicuously celebrated his wedding, exalting the beauty and virtues of his spouse Faustina. His sonnets meticulously documented this early period of their married life, at least from his perspective. He poetically expressed his hopes for heirs, his joy at her first and second pregnancies, and supreme paternal satisfaction when their eldest son Giuseppe was born.[38]

All seemed to be well in the Carafa-Pignatelli household, though there were hints of discord. Francesco's disliked the social pressures of the city, preferring life on his baronial estates in Terra di Lavoro.[39] Nonetheless, by December 1728 they had returned to the city.[40] Initially, Carafa supported his wife's desire to cultivate a learned salon in Palazzo Carafa di Colubrano on via San Biagio dei Librai, a street lined with bookshops, printers, and publishers. Pignatelli thrived in this environment, planting the seeds of her future reputation as a mathematical philosopher and patron of science and poetry. Carafa presented himself as a doting husband, casting Faustina as the "heroine" of his romance in print. Declaring her a "divine gift from heaven," he described his wife as a "perfect beauty" and "mirror of honor" for everyone to admire.[41]

By February 1731, the honeymoon was over. Carafa insisted on returning to Formicola, Pignatelli preferred to remain in Naples.[42] The fault lines in their deteriorating relationship ran deep enough to force a difficult decision. At the

36 Camillo Minieri Riccio, "Notizie delle Accademie institute nelle province napoletane," *Archivio per le province napoletane*, 3, 1 (1878): 145–146.

37 Damone Tirrenio [Orazio Pacifico], "Ai lettori," *Il Caprario: accademia di diversi rimatori, che nel medesimo monte si radurano. Parte seconda* (Florence, 1732), n.p. See also *Il Caprario: accademia di diversi rimatori, che nel medesimo monte si radurano* (Naples, 1729).

38 Francesco Domenico Carafa, *Rime varie di Francesco Carafa principe di Colobrano* (Florence, 1730), e.g. pp. 61, 69–70, 78–79, 84, 163, 196.

39 Carafa, *Rime varie*, p. 38; see Labrot, *Baroni in città*.

40 ASN, *Carafa di Maddaloni*, busta I-D/6d, c. 1r (Memoriale di Faustina Pignatelli intorno alle sue discordie col marito, 1734).

41 Carafa, *Rime varie*, pp. 61, 354.

42 ASN, *Carafa di Maddaloni*, busta I-D/6d, c. 1r.

advice of her mother-in-law Caterina, Pignatelli abandoned the family estate at Formicola, leaving her three young sons in their father's care while she joined her daughter Caterina who was being educated at the convent of Regina Coeli where Este lay buried. The marriage had become a scandal and a torment for her, indeed she feared that "greater harm" would come to her if she remained in "that feudatory of her husband," as she explained to the viceroy in a lengthy testimonial of 1734.[43] In a world without divorce, the only solution was to petition for a formal separation in the hope of receiving a sympathetic hearing.[44] Retreating to a convent was the standard way to protect oneself from an angry husband while maintaining honor and reputation.[45] Carafa fumed and threatened, and exercised his power over her considerable dowry by withholding payments due to his wife for her maintenance to see if he could extract their daughter and the family jewels from the convent; then all their children and the family estate would be under his control. With the sympathetic support of her mother-in-law, Faustina refused to budge. Regina Coeli became her temporary refuge.

In her *Discourse on Happiness*, written during the 1730–40s, the French Newtonian philosopher Châtelet wisely observed that "the love of study is much less necessary to the happiness of men than it is to that of women." She considered learning a consolation for the intelligent curious woman, "for all the exclusions and all the dependencies to which she finds herself condemned by her place in society."[46] The princess of Colubrano surely agreed with this sentiment. Knowledge, and the company it allowed her to keep, became her other refuge.

Pignatelli seems to have encountered Pietro's older brother, Niccolò Di Martino (1701–69), in the period just before her marital troubles began in earnest. In the second edition of Gianvincenzo Gravina's *Della tragedia* (1731), Stefano Elia advertised Pignatelli's growing fame for her intellect. Praising the rigors of her early education and her fluency in Tuscan, French and Latin, he declared her learning the envy of all men of science, but it was the way she used her mind – "a copious and stupendous union of clear and distinct ideas" – that especially earned his admiration. Pignatelli did not see learning as ornamental but viewed the mind as "an instrument to reach halfway the

43 Ibid., c. 1^{v}.

44 Silvana Seidel, Menchi and Diego Quaglioni, eds., *Coniugi nemici. La separazione in Italia dal XII al XVIII secolo* (Bologna: Il Mulino, 2000).

45 Joanne M. Ferraro, *Marriage Wars in Late Renaissance Venice* (Oxford: Oxford University Press, 2001), p. 127.

46 Émilie du Châtelet, *Selected Philosophical and Scientific Writings*, ed. Judith P. Zinsser, trans. Isabelle Bour and Zinsser (Chicago: University of Chicago Press, 2009), p. 357.

difficult and arduous disciplines that occupy you every day." Discussing her ability to use calculus to solve problems in mathematics and mechanics, Elia celebrated her prudence, wisdom, and love of "the most abstruse mathematical speculations."[47] She was a well-formed, post-Cartesian mind thriving under Di Martino's tutelage.

Like many Neapolitan scholars, Di Martino studied law and theology, but mathematics became his great passion and vocation. His mentor Ariani considered him so talented that he asked his young pupil to supplement his lectures and ultimately substitute for him, starting in 1721, as Ariani's services were in great demand at court. Between 1725 and 1729, Di Martino rapidly proceeded to publish informative and well-designed textbooks on algebra, statics, logic, and planar geometry, cementing his reputation as the leading mathematician in the city. He refused offers to go elsewhere. On 21 November 1732, one day after Pignatelli was admitted to the Bologna Academy of Sciences, Di Martino finally succeeded Ariani as professor of mathematics at the University of Naples, with the influential backing of Galiani who had previously tried and failed to lure him to Rome.[48]

Nurtured in Cartesian Naples, Di Martino was among the first to publicize the arrival of Newtonian mathematical philosophy in this city. He imbibed Newton's discoveries in Ariani's lectures, connected them to the flourishing Galilean tradition of mechanics, and brought these lessons from the classroom into print. Contemporaries described him as a gifted speaker with a seriousness of purpose and wisdom charmingly offset by his ready wit and ease in conversation.[49] In the late 1720s, Di Martino specifically challenged the defense of Cartesian physics, especially its vortices, by demonstrating how recent scientific developments rendered them insufficient to explain the movements of celestial bodies. A careful reader of Newton's *Principia*, he incorporated many elements of Newton's work – his laws of motion, concepts of force and equilibrium, and mathematical understanding of celestial dynamics – into his mechanics. He embraced differential calculus as an essential part of the mathematical philosopher's toolkit. At the same time, Di Martino demonstrated an ability to critically appraise the Newtonian synthesis, indicating a willingness to entertain ideas proposed by critics such as Leibniz.[50]

---

47 Gian Vincenzo Gravina, *Della tragedia* (Naples, 1731), 2nd ed., sig. a3r.

48 Pietro Nastasi, "Nicola Antonio De Martino (Di Martino)," DBI, vol. 38, pp. 600–603.

49 Nicola De Martino, *Nuovi elementi della teoria delle mine*, ed. Giuseppe De Martino (Naples, 1780), p. 16. De Martino's nephew Giuseppe cites physics professor Tommaso Fasano as his source.

50 Paolo Casini, *Newton e la coscienza europea* (Bologna: Il Mulino, 1983), pp. 191, 207–211; Ferrone, *Intellectual Roots*, pp. 199–200.

Widely read and dedicated to improving mathematical instruction in his native city, Di Martino was determined to bring Neapolitan science and mathematics into conversation with current work done in cosmopolitan centers of learning. His ambitions transcended the city where he made his reputation and ultimately encompassed his relationship with his most promising female pupil, Pignatelli, and his younger brother Pietro. Perhaps this was why it was later said, many times over, that *la donna Faustina* corresponded with Sir Isaac Newton.[51] The fact that Newton had died in 1727 had no bearing on the matter. Pignatelli was a woman who should have corresponded with the greatest scientific mind of their age.

2 The Living Force of a Woman

In 1732, Pignatelli was litigating with her husband from the convent, continuing her studies with Di Martino, and beginning to embark on scientific correspondence with colleagues abroad. Despite Elia's fulsome praise, she was virtual unknown to the republic of letters. Galiani had formulated his plan for the new Neapolitan Academy of Sciences, appointing the Cartesian physician Niccolò Cirillo (1671–1735) its president and Francesco Serao (1702–83) as secretary. The Di Martino brothers were invited to become members.[52] There was a carefully orchestrated synergy between the academy and plans to reform university instruction in the sciences. Galiani saw Niccolò and Pietro Di Martino as essential to his plans, which Doria critiqued for being too radically modern.[53] The academy's decision to participate in the European-wide controversy about living force (*vis viva*) in the first year of their existence surely fulfilled Doria's worst fears regarding the ambitions of the Neapolitan academicians.

The controversy over *vis viva* began in 1686, when Gottfried Wilhelm Leibniz pointed out the inadequacies of Cartesian natural philosophy, arguing that matter was not simply inert but conserved energy.[54] Was force purely a product of the momentum that set matter in motion, weight multiplied by velocity (mv) and therefore a direct factor of time, or a dynamic product of mass moving at the square of its velocity through space (mv^2)? How should its impact

51 Ceci, "Il Palazzo di Carafa," p. 169.

52 Ferrone, *Intellectual Roots*, pp. 201, 211, 213.

53 Ferrone, *Intellectual Roots*, p. 213.

54 Thomas L. Hankins, "Eighteenth-Century Attempts to Resolve the Vis Viva Controversy," *Isis*, 56 (1965): 281–297; Larry Laudan, "The *Vis Viva* Controversy, a Post-Mortem," *Isis*, 59 (1968): 130–143; and Carolyn Iltis [Merchant], "Leibniz and the *Vis Viva* Controversy," *Isis*, 62 (1971): 21–35.

be measured? By 1733, new work by English, Italian, Dutch, French, and Swiss mathematical philosophers and physicists reignited the debate over living force. Padua's distinguished physics professor Giovanni Poleni reaffirmed the Leibnizian position in 1718 by dropping balls of equal size but differing weights in tallow to see how force might be empirically measured.[55] The Dutch experimenter Willem Jacob 's Gravesende further developed the idea of measuring impact in 1722 by testing how it might empirically and geometrically demonstrate key precepts of Newtonian physics. Cartesians objected strongly to the idea of a force inherent to substance; the English Newtonians strongly resisted efforts to allow a Leibnizian principle to supplant Newtonian explanations of force and resistance.[56] By the mid-1720s, the Swiss mathematician Johann Bernoulli further complicated the debate with his Leibnizian arguments for the elasticity of bodies that required a sophisticated understanding of integral calculus. The springiness of force and its release by moving balls was now in contention but the Paris academicians were having none of it. In his academy lecture of April 1728, Jean-Jacques Dortous de Mairan roundly critiqued Bernouilli's conclusions by arguing that neither calculus nor metaphysics were required to establish the principle of inertia. He publicly upheld the constancy of Cartesian motion.[57] This position invited further response.

By the early 1730s, most of Europe seemed embroiled in the debate. Except for Poleni and his colleagues in Padua, however, the Italian community of mathematicians and philosophers was conspicuously silent. The Bologna Academy of Sciences had not taken a position on the matter. In Naples, the academicians realized that this was a golden opportunity for their fledgling academy. On 27 October 1733, Bartolomeo Intieri, the Tuscan minister of state in the Kingdom of Naples and a mathematician, informed the Corsini librarian Giovanni Gaetano Bottari in Rome: "Tomorrow we are doing some experiments in our Academy, made by 's Gravesende and the marchese Poleni in Padua, and I believe that they will come out diametrically opposed to the result that

---

55 Silvio Bergia, Paolo Fantazzini, "Dalla regolamentazione delle acque alle leggi dell'urto centrale elastic: una rivalutazione del ruolo di Giovanni Poleni nella disputa delle forze vive," *Giornale di fisica*, 21 (January–March 1980): 46–60; more generally, Maria Laura Soppelsa, *Leibniz e Newton in Italia: Il dibattito padovano* (*1687–1750*) (Padua: Edizioni Lint, 1989).

56 See Henry Pemberton, "A Letter to Dr. Mead ... concerning an Experiment, whereby it has been attempted to shew the falsity of the common Opinion in relation to the force of Bodies in Motion," *Philosophical Transactions of the Royal Society*, 32, 371 (1722–23): 57–68, with its critique of Poleni.

57 Pierre Costabel, *La question des forces vives. La signification d'un débat sur trente ans* (*1728–1758*) (Paris: CNRS, 1983).

these gentlemen affirm."[58] The Neapolitan Academy of Sciences had decided to intervene in the *vis viva* debate, and not in favor of the Leibnizians; they were predisposed to a Galilean-Newtonian worldview even if the president, Cirillo, was a Cartesian. Intieri subsequently reported that their experiments had confirmed the "exactness of the Newtonian theories."[59]

Where was Pignatelli amid all this dropping of balls and measurement of results? Despite her status in Bologna, she had no official role in the Neapolitan Academy. Nonetheless, she found a way to make her own distinctive contribution – her first and only publication. In January 1734, an article containing four mathematical problems appeared in the *Nova Acta Eruditorum* published in Leipzig.[60] They reflected a range of subjects of interest to mathematicians, astronomers, and physicists, including how to determine that the proportion of a perpendicular line bisecting a cone was exactly half the surface length, and how to measure the elasticity of bodies in light of the English royal astronomer James Bradley's efforts to explain stellar aberration by comparing the speed of the earth's motion to the speed of light. The author of the "Mathematical Problems Transmitted from Naples to the Editors of the Acts of the Learned" was unnamed; the journal preferred to publish its contributions without attribution. Commenting on the advantages of anonymity several years later, Châtelet observed that it would have allowed her to "enjoy the pleasure of seeing myself judged without running any risk if the judgment were unfavorable."[61] Discerning readers who consulted the index quickly became aware of the gender of the "Anonymous Neapolitan" (*Anonyma Neapolitana*).[62] Anonymous, as Virginia Woolf famously declared, was a woman. Pignatelli was the only woman in Naples capable of this kind of sophisticated mathematical analysis.

58 Ferrone, *Intellectual Roots*, p. 249n; Serrapico, *Scienza e filosofia*, p. 128 [N.B. the printed book is unpaginated but I am using the electronic pagination].

59 Serrapico, *Scienza e filosofia*, p. 128. On the role of Naples in these debates, see Soppelsa, *Leibniz e Newton*, pp. 146–147.

60 [Pignatelli], "Problemata mathematica Neapoli ad Collectores Actorum Eruditorum transmissa," *Nova Acta Eruditorum*, 1 (January 1734): 28–34.

61 Châtelet to Johann Bernouilli the elder, 30 June 1740, as quoted in Judith Zinsser, *La Dame d'Esprit: A Biography of the Marquise Du Châtelet* (New York: Viking, 2006), p. 190.

62 The index of the *Nova Acta Eruditorum* 1 (January 1734) on p. 572 described the article as *Anonymae Neapolitanae problemata mathematica*; Brigaglia and Nastasi, "Bologna e il Regno," p. 159, n. 27; Augustinus Hubertus Laeven, Lucia Johanna Maria Laeven-Aretz, *The Authors and Reviewers of the Acta Eruditorum 1682–1735* (Göttingen: Publickationsserver der Georg-August-Universität Göttingen, 2014). On the transparent nature of anonymous authorship, see Mary Terrall, "The Uses of Anonymity in the Age of Reason," in Mario Biagioli, Peter Galison, ed., *Scientific Authorship: Credit and Intellectual Property in Science* (New York: Routledge, 2003), pp. 91–112.

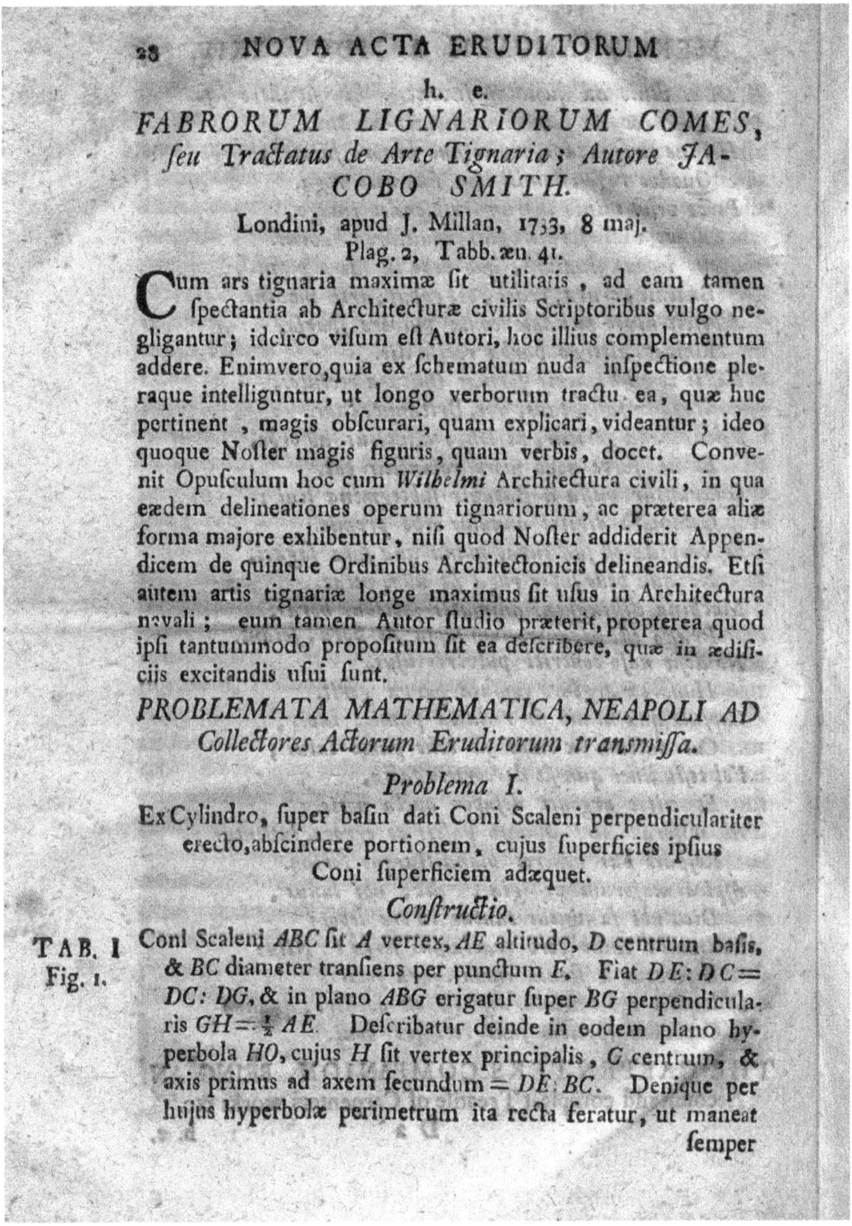

28 NOVA ACTA ERUDITORUM

h. e.

FABRORUM LIGNARIORUM COMES, ſeu Tractatus de Arte Tignaria; Autore JACOBO SMITH.

Londini, apud J. Millan, 1733, 8 maj.
Plag. 2, Tabb. æn. 41.

Cum ars tignaria maximæ ſit utilitatis, ad eam tamen ſpectantia ab Architecturæ civilis Scriptoribus vulgo negligantur; idcirco viſum eſt Autori, hoc illius complementum addere. Enimvero, quia ex ſchematum nuda inſpectione pleraque intelliguntur, ut longo verborum tractu ea, quæ huc pertinent, magis obſcurari, quam explicari, videantur; ideo quoque Noſter magis figuris, quam verbis, docet. Convenit Opuſculum hoc cum *Wilhelmi* Architectura civili, in qua eædem delineationes operum tignariorum, ac præterea aliæ forma majore exhibentur, niſi quod Noſter addiderit Appendicem de quinque Ordinibus Architectonicis delineandis. Etſi autem artis tignariæ longe maximus ſit uſus in Architectura navali; eum tamen Autor ſtudio præterit, propterea quod ipſi tantummodo propoſitum ſit ea deſcribere, quæ in ædificiis excitandis uſui ſunt.

PROBLEMATA MATHEMATICA, NEAPOLI AD Collectores Actorum Eruditorum transmiſſa.

Problema I.

Ex Cylindro, ſuper baſin dati Coni Scaleni perpendiculariter erecto, abſcindere portionem, cujus ſuperficies ipſius Coni ſuperficiem adæquet.

Conſtructio.

TAB. I Fig. 1. Coni Scaleni *ABC* ſit *A* vertex, *AE* altitudo, *D* centrum baſis, & *BC* diameter tranſiens per punctum *E*. Fiat $DE:DC = DC:DG$, & in plano *ABG* erigatur ſuper *BG* perpendicularis $GH = \frac{1}{2} AE$. Deſcribatur deinde in eodem plano hyperbola *HO*, cujus *H* ſit vertex principalis, *C* centrum, & axis primus ad axem ſecundum $= DE:BC$. Denique per hujus hyperbolæ perimetrum ita recta feratur, ut maneat ſemper

FIGURE 6.4A–B Faustina Pignatelli's anonymous solutions to four mathematical problems, 1734. [Faustina Pignatelli], "Problemata mathematica Neapoli ad Collectores Actorum Eruditorum transmissa," *Nova Acta Eruditorum* 1 (January 1734): 28–34. By Courtesy of the Department of Special Collections, Memorial Library, University of Wisconsin, Madison

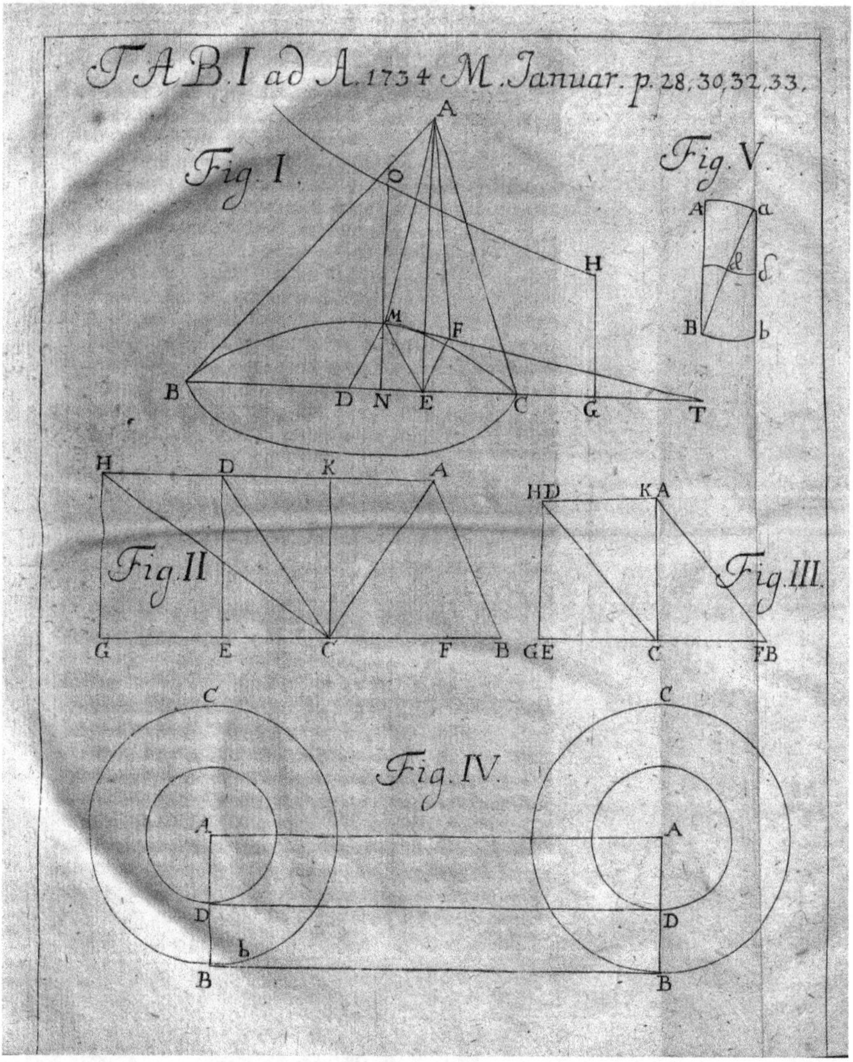

The fourth and final problem presented a "Paralogism to expose the Leibnizian demonstration of moving force, which is, in fact, defined as the square of velocity by mass."[63] Reminding readers that Leibniz first raised his concerns about "a notable error of Descartes and others" and subsequently defined living force in different issues of the *Acta Eruditorum*, making it appropriate to respond in the same journal, Pignatelli clearly and succinctly laid out the Cartesian and Leibnizian definitions of force, referencing Galileo's discussion of the uniform

63 [Pignatelli], "Problemata mathematica Neapoli," p. 33.

acceleration of falling bodies.[64] She introduced Bernoulli's more recent demonstration of how "two weights ascend to different heights at equal times," arriving there "by a cycloidal arch."[65] This was the mathematics of springiness. What, then, was the illogic behind Leibnizian reasoning? Pignatelli provided the answer by discussing how Leibniz handled resistance through an analysis of the logic of rectilinear motion. Using integral calculus, she recalculated Leibniz's results to be $\int F \cdot ds = \frac{1}{2} mv^2$. She was making her own intervention in this debate and demonstrating how well she had absorbed Niccolò Di Martino's lessons about the utility of mathematical physics in resolving thorny philosophical problems.

1734 was truly the *annus mirabilis* for Pignatelli's emergence as Di Martino's celebrity student, capable of making judicious and well-informed interventions in crucial debates. Decades later, Pietro Napoli-Signorelli recalled Pignatelli as "his pupil in mathematics."[66] Di Martino eloquently presented mathematics as the key to all knowledge, including experimental philosophy.[67] He continued to publish his well-regarded instructional manuals. His mentor Ariani inaugurated the teaching of Newtonian mathematics and natural philosophy in Naples, but it was Di Martino who gave the city a reputation.[68] That same year, Pietro returned from Bologna to take up the chair in astronomy, having already published his observations of a lunar eclipse in May 1733 in collaboration with Manfredi.[69] A tightknit community was forming between the Neapolitan Academy, the reformed university, and Pignatelli's salon under the patronage of Galiani and Intieri.

This fortunate conjuncture inspired the elder Di Martino to dedicate his latest book, *Elementa sectionum conicarum* (1734), to Pignatelli. "For many reasons, Most Excellent Princess, I am impelled to bring to light those new Elements of Conic Sections under your name."[70] De Martino described the origins of his work in their private lessons, praising his aristocratic pupil for encouraging

64 Gottfried Wilhelm Leibniz, "A Brief Demonstration of a Notable Error of Descartes and Others Concerning a Natural Law" (1686) and "Specimen dynamicum" (1695), both translated in Leibniz, *Philosophical Papers and Letters*, ed. Leroy. E. Loemker (Chicago: University of Chicago Press, 1956), vol. 1, pp. 455–463, vol. 2, pp. 711–738.

65 [Pignatelli], "Problemata mathematica Neapoli," p. 34; Brigaglia and Nastasi, "Bologna e il Regno," p. 159; Raffaella Simili, "In punta di penna. Donne di scienza e di cultura fra cosmopolitanismo e intimità meridionale," in *La scienza del Mezzogiorno dopo l'Unità d'Italia* (Soveria Manelli: Rubbettino, 2008), Antonio di Meo ed. p. 40. Gatto, *Libri*, pp. 68–70.

66 Pietro Napoli-Signorelli, *Vicende della coltura nelle Due Sicilie* (Naples, 1786), vol. 5, p. 477.

67 Casini, *Newton*, p. 211.

68 Amadeo, "Dai fratelli," pp. 60–61.

69 Serrapico, *Scienza e filosofia*, p. 128; Addabbo, "I *Philosophiae naturalis institutionum libri tres*," pp. 79–80.

70 Niccolò De Martino, *Elementa sectionum conicarum* (Naples, 1734), sig. a3^{r}.

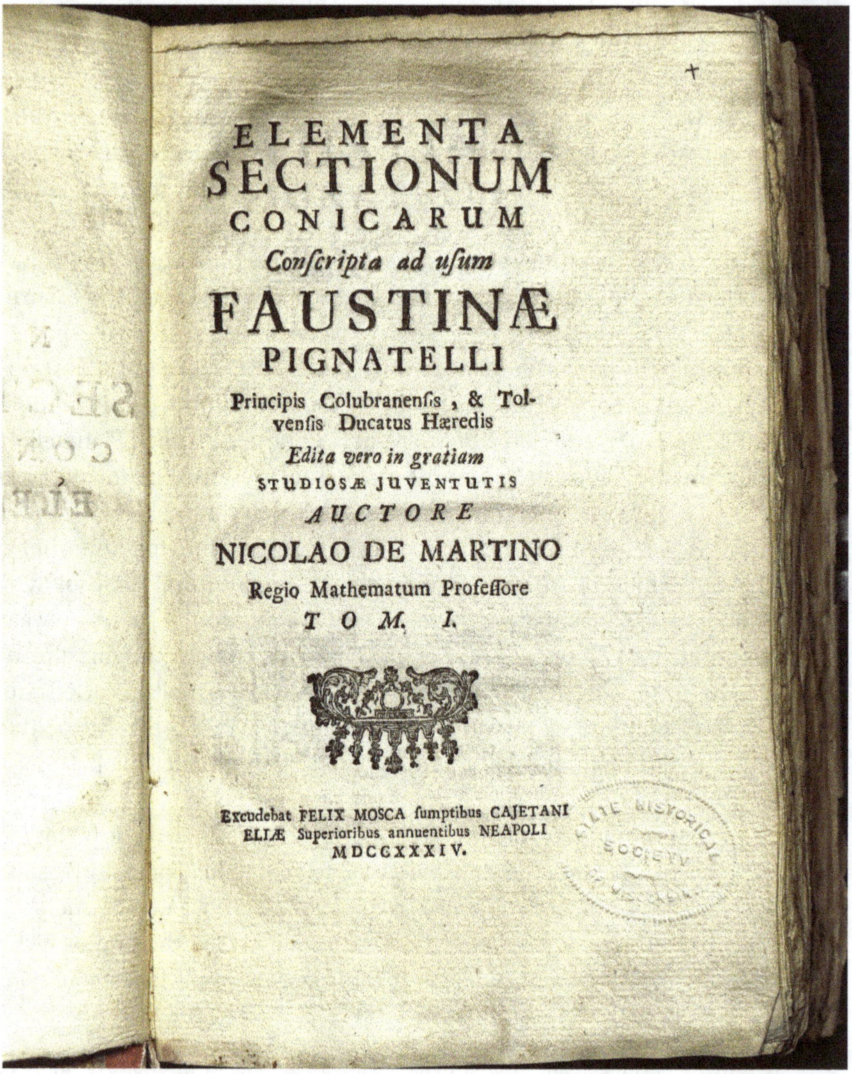

ELEMENTA
SECTIONUM
CONICARUM
Conſcripta ad uſum
FAUSTINÆ
PIGNATELLI
Principis Colubranenſis, & Tolvenſis Ducatus Hæredis
Edita vero in gratiam
STUDIOSÆ JUVENTUTIS
AUCTORE
NICOLAO DE MARTINO
Regio Mathematum Profeſſore
TOM. I.

Excudebat FELIX MOSCA ſumptibus CAJETANI ELIÆ Superioribus annuentibus NEAPOLI
MDCCXXXIV.

FIGURE 6.5 Niccolò di Martino's mathematics textbook written for Faustina Pignatelli, 1734. Niccolò De Martino, *Elementa sectionum conicarum conscripta ad usam Faustinae Pignatelli* (Naples, 1734), 2 vols. Courtesy of the Department of Special Collections, Memorial Library, University of Wisconsin, Madison

him to make such instruction publicly available. His dedication was not simply an act of patronage, but the presentation of a manual written "for the use of Faustina Pignatelli, Princess of Colubrano and heiress to the Duke of Tolve, edited with gratitude for studious youth." She actively helped give shape to the lessons it contained by being his willing pupil. The publication of an important mathematics textbook *ad usum Faustinae Pignatelli* further solidified her place as a singularly talented mathematician, a woman to inspire young Neapolitan men to embrace calculus through her example. She was no paper philosopher in the spirit of Fontenelle, intrigued by science but uninterested in the technical details as he presented his fictional lady philosopher in his *Conversations on the Plurality of Worlds* (1686), but a model of what it meant to apply one's learning to solving problems.

Di Martino especially wanted his readers to appreciate the marvelous quality of Pignatelli's intellect: "you have an exactness of mind by nature, so that you simultaneously seem to learn the lessons by heart and to teach them well."[71] Who was the master and who was the pupil? He informed Pignatelli that words inadequately expressed "the heroic qualities of your intellect." Her impeccable judgment, her restraint and moderation, and her ability to argument her position well had earned his admiration. The moral qualities of Pignatelli's mind, a fruitful "union of all virtues, stained by no proximity to vice," equaled her intellectual talent for precision.[72] On the very first page of his textbook, Di Martino explained his role in initiating Pignatelli into the intellectual pleasures and signification of mathematics as a foundation for every kind of knowledge. "When first you drove your mind away from humanistic learning towards the difficult study of philosophy, I myself was your authority, Princess Faustina, so that you fortified that subject with the elements of planar and solid geometry."[73] In other words, she was an avid reader of his earlier textbooks who came to him for further instruction. She inspired him to write a more advanced manual about conic sections, which explored the curves created by the intersection of a plane with a cone's surface from Apollonius of Perga to Descartes, Fermat, and their early modern readers. Thus, in 1734, De Martino publicly presented Pignatelli as his model student while she put her abilities to use in her published mathematical solutions. Reviewing De Martino's book on conic sections two years later, the editor of *Nova Acta Eruditorum* praised Pignatelli for possessing "every kind of erudition and science, no less philosophical than mathematical."[74] The "Anonymous Neapolitan" had parlayed the admiration of the local community of mathematicians and natural philosophers into the

---

71 De Martino, *Elementa sectionum conicarum*, sig. a4 r. Special thanks to Giovanna Ceserani for translating the Greek word for exactness or precision.

72 De Martino, *Elementa sectionum conicarum*, sig. a4v.

73 De Martino, *Elementa sectionum conicarum*, p. 1.

74 *Nova Acta Eruditorum* (1736): 44.

respect of scientific academies and journals abroad. Pignatelli was now developing an international reputation.

3 Science Interrupted

Di Martino's portrait of his disciple as a woman of talent, judgment, and learning with an impeccable reputation was more than simply an act of courtly praise of an aristocratic patron. In 1734, Pignatelli was embroiled in the latest struggle to assert her rights and defend herself and her children against her husband's patriarchal excesses. Her three young sons were denied access to their mother. Since her arrival in the convent of Regina Coeli, where she remained with her young daughter, Francesco Carafa "not only did not permit them to see her but not even to mention her name." Pignatelli feared even "greater evils" if she dared return to Formicola.[75] Carafa cut off all financial support, forcing Pignatelli to petition the recently departed viceroy, Count Aloys von Harrach. Seven months passed with no response, undoubtedly due his recall to Vienna and the chaotic struggle that ended Austrian Habsburg rule. Rather than pursuing legal options in court, she hoped for "an amicable agreement" that would restore the 2000 ducat annuity specified in their marriage contract.[76]

Carafa accused the viceroy of favoring his wife and mother, who supported Pignatelli's petition. He wanted his daughter removed from the convent and restored to his custody at Formicola. Neapolitan judicial authorities remained undecided – "some to give her to the father, others to the mother."[77] In desperation, Pignatelli wrote to Vienna requesting that the emperor Charles VI intervene. Her husband was threatening her, abusing her considerable dowry, and selling off the family jewels – her property not his, in law if not in fact.

Pignatelli sought justice from the new Austrian viceroy, Count Giulio Borromeo Visconti, who arrived in Naples in 1733. She expressed concern over her sons' future. From Regina Coeli, she heard disturbing reports that none of the 3000 ducats allotted annually for their education was being spent for this purpose. "This is being totally neglected and those boys abandoned by the Prince Father in the said feudatory of Formicola without teachers, without servants, in the midst of four frivolous women, of whom only one was assigned

75 ASN, *Carafa di Maddaloni*, busta I-D/6d, c. 1^{v}.

76 ASN, *Carafa di Maddaloni*, busta I-D/6d, c. 1^{v}.

77 ASN, *Carafa di Maddaloni*, busta I-D/6d, c. 3^{r}.

by the Princess Mother as being capable of serving them at their young age."[78] The eldest Giuseppe was now eight. Pignatelli asked that her sons be brought to Naples where they could be properly educated, learn to take their place in society, "and not be allowed to grow up so estranged from the mother." She hoped that she would finally get a sympathetic hearing regarding "this sorrowful story of her unhappy circumstances."[79]

The struggles between husband and wife became a measure of character as each party aired their grievances. Carafa insisted that he could not afford the expense of bringing his sons regularly to Naples to see their mother. Pignatelli did her own accounting, "with her usual great prudence and zeal for what is right and just." She and her advisors concluded that Carafa was using "the pretext of the expense of a good education" to withhold her annual stipend from her considerable dowry while utterly evading his paternal responsibilities. Pignatelli then calculated the cost of maintaining her sons in Formicola. Food, clothing, and transportation were all more expensive, not to mention attracting and housing tutors willing to live in a feudal backwater until her sons came of age. "That indispensable education ordained by divine and human law that good and worthy teachers could give them … would cost much more in Formicola than in Naples."[80] She was indeed accounting for affection.[81]

Thus, at the very moment that Pignatelli was being publicly praised as the most learned woman in Naples, she was fighting for her economic survival and the right to educate and care for her children. As Carafa left for Sicily with the Austrian army – during their retreat from Naples in March 1733 before the arrival of the Bourbon army – he reiterated orders not to let his estranged wife into the castle. Then he spitefully fired the only servant Pignatelli trusted who had cared for their sons since birth. There was still no tutor and Pignatelli considered the replacement for her own servant unqualified. Fundamentally, Pignatelli felt that Carafa cared little for "the life of his sons." When their middle son Michele fell ill, she sent a physician accompanied by the former servant, but no one would let them into the castle. Those petitioning on her behalf sympathetically described her as "the forlorn princess" (*la sconsolata Prencipessa*).[82] Pignatelli concluded her lengthy memorial with three requests: that her daughter Caterina remain in the convent so that "in this way and with this care, she may be educated by the mother," that her sons be brought to

78 ASN, *Carafa di Maddaloni*, busta I-D/6d, c. 3^r.

79 ASN, *Carafa di Maddaloni*, busta I-D/6d, cc. 3^r, 4^r.

80 ASN, *Carafa di Maddaloni*, busta I-D/6d, cc. 4^v–5^r.

81 Caroline Castiglione, *Accounting for Affection: Motherhood and Politics in Early Modern Rome* (Houndmills, Basingstoke, Hampshire: Palgrave, 2015).

82 ASN, *Carafa di Maddaloni*, busta I-D/6d, c. 5^v.

Naples "to be well educated and instructed by a public lecturer, or another wise and learned person to be chosen by a zealous and righteous Minister," and that her annuity be paid in full.[83] She had brought a dowry of 140,000 ducats into the marriage and wanted Carafa to fulfill his contractual obligations, while allowing her to raise their children. The arrival on May 10 of Charles III (r. 1734–59), the first Spanish Bourbon ruler of Naples, ultimately made this disastrously failed marriage a Bourbon rather than a Habsburg concern.

How exactly did Pignatelli find the time, energy, and resources to pursue her scientific interests while litigating with her husband? She was not in Palazzo Carafa di Colubrano, after all. Did Di Martino conduct their lessons in the convent *parlatorio*? Books and visitors surely kept the princess in touch with scientific conversations going on in the city and their connections to a broader world, since the convent library was unlikely to provide the materials to research solutions to four mathematical problems. She was indeed a woman caught between different worlds.

In summer 1736, Pignatelli resumed her research and correspondence. On 11 September 1736, she wrote to Zanotti in Bologna, describing the circumstances of her publication in the *Nova Acta Eruditorum*. She described sending "a small writing where I made apparent the sophism of Leibniz's first demonstration to determine the quantity of moving force" to Baron Blance in Kassel who published it in the German journal. He subsequently wrote to invite her to demonstrate at greater length the physics of *vis viva*. This request inspired Pignatelli's second paper, "the writing annexed here, which I send you as a gesture of my attention to your merit as much as to this entire noble Institute, which I pray will instruct me to make me sufficiently worthy of its company."[84] Pignatelli aspired to see her paper published by the Bologna Academy. She modified her position in her first paper to reflect the evolution of her understanding of the debate, responding especially to the Bernoulli's recent paper published in the *Nova Acta Eruditorum* in 1735 where he further demonstrated his concept of elasticity in confirmation of Leibniz's position.[85] Pignatelli was no longer certain that Leibniz was entirely wrong. She was searching for a middle ground while hoping to expand her audience in print and play a more active role in the Bologna Academy as a distinguished foreign member.

83 ASN, *Carafa di Maddaloni*, busta I-D/6d, c. 6v.

84 AASB, *Antica Accademia*, Titolo III, sez. I, fasc. 4, *Lettere ricevute* (Pignatelli to Zanotti, Naples, 11 September 1736).

85 Terrall, "Vis Viva Revisited." *History of Science*, 42 (2004): 192–199.

The Bologna Academy recorded the receipt of Pignatelli's "Letter with a Dissertation on Living Forces" on 15 November 1736.[86] They left no account of any discussion of this paper, leaving us in the dark about their collective opinion of Pignatelli's revised conclusions. Zanotti archived the manuscript and corresponded with the princess, undecided what to do with her contribution. Publishing the first volume of the *Commentarii* (1731) had been a complex and vexed undertaking with numerous delays.[87] The academy secretary did not want to risk further alienating any of its members by taking too strong a position on *vis viva*. Subsequent evidence suggests that he continued to correspond with Pignatelli about the position she was now taking as a mediator of this debate, even though the letters do not survive. He did not rush to publish her paper, nor indeed could he since the second volume of the *Commentarii* did not appear until 1745.

The languid reception of Pignatelli's work in Bologna stood in marked contrast to her role in the lively intellectual life taking shape in the Regno. In 1736, Naples was in the early phase of sweeping Bourbon reforms. In November of that year the university finally vacated San Domenico Maggiore to occupy the new Palazzo degli Studi. Galiani's dream of a more modern curriculum, with professorships in subjects such as astronomy and experimental physics, was finally taking shape.[88] Pietro Di Martino hoped to establish an observatory to rival *La Specola* in Bologna. The Neapolitan Academy continued its experimental activities. There was great optimism that Naples had become a cosmopolitan city of modern minds.

Reports of Pignatelli's latest scientific paper reached Paris by fall. The Pignatelli circle had close ties with the community of French mathematicians. Writing to Jean-Jacques Dortous de Mairan on 9 November 1736, Voltaire observed, "I do not know by what trick of fate the ladies have declared themselves for Leibniz. Madame the Princess of Colubrano has also written in favor of living forces. I am no longer astonished that this party is so significant. Neither you, nor I are hardly gallant. But you are like Hercules who fought with restraint against the Amazons, and I am nothing more than a weak volunteer

86 Angelini, ed., *Anatomie accademiche*, p. 331.

87 Mauro De Zan, "L'Accademia delle Scienze di Bologna: l'edizione del primo tomo dei *Commentarii* (1731)," in Maria Vittoria Predaval Magrini, ed., *Scienza, filosofia e religione tra '600 e '700 in Italia* (Milan: Franco Angeli, 1990), pp. 203–259.

88 Addabbo, "I *Philosophiae naturalis institutionum libri tres*," p. 79; Annamaria Rao, *Il Regno di Napoli nel Settecento* (Naples: Guida, 1983), pp. 62–65.

in your army."[89] Voltaire's awareness that Pignatelli shared Châtelet's favorable opinion of Leibniz, which was not his own, is a sign of how rapidly news traveled through correspondence networks. There is indirect evidence that Pignatelli may have been writing to Dortous de Mairan. From Cirey, Voltaire witnessed the convergence of opinion between two leading women of science. It intrigued him and made Pignatelli and Châtelet aware of each other. "Not only France had its Châtelets," remarked Napoli-Signorelli many years after this encounter.[90] Châtelet had yet to publish anything, though she had been studying intensively with Maupertuis and Alexis Clairaut before their departure for Lapland in 1736.[91] At this moment, Pignatelli had more of a reputation.

In celebration of this new age of learning, Niccolò Di Martino decided to expand his geometry manual into a three-volume blockbuster of a book with the help of his mentor Ariani who reviewed drafts. It now included a complete course in algebraic geometry "written for the use of Faustina Pignatelli."[92] He offered his 1737 expanded edition of "these Elements of Algebra that I undertook to write seven years ago for the use of the Most Excellent Princess Faustina Pignatelli" to the Paris Academy of Sciences.[93] Presenting himself as an avid reader of the French mathematical tradition inaugurated by the Marquis de l'Hôpital, who published the first extensive textbook on differential calculus in 1696, Di Martino invited them to look favorably upon his book, which skillfully synthesized Cartesian and Newtonian mathematics. There was no other community whose judgment he respected more than his colleagues in Paris. When Clairaut returned from Lapland, he subsequently published a bestselling French geometry textbook in 1741 followed by an algebra manual in 1746, based on the lessons he gave the Marquise though he did not dedicate them to Châtelet.[94] The Neapolitans were indeed inspiring the French to improve the teaching of mathematics.

"Nothing is more useful than the study of algebra," Di Martino told his readers. He proclaimed calculus a sublime instrument of reason, first among

89 Theodore Besterman, ed., *Correspondance de Voltaire* (Geneva, 1968–2001), vol. 91, p. 459 (Voltaire to Dortous de Mairan, 9 November 1736); Terrall, "Vis Viva Revisited," p. 202. On the connections between Naples and Paris, see Brigaglia and Nastasi, "Un carteggio inedito tra il matematico palermitano Girolamo Settimo and Gabriello Manfredi," *Bollettino di storia delle scienze matematiche*, 3, 1 (1983): 32.

90 Napoli-Signorelli, *Vicende*, vol. 5, p. 497.

91 Zinsser, *La Dame d'Esprit*, pp. 72–74.

92 De Martino, *Algebrae geometria promotae elementa conscripta ad usum Faustinae Pignatelli* (Naples, 1737), 3 vols. On Ariani's role, see Ariani, ed., *Memoria*, p. 212.

93 De Martino, *Algebrae geometria promotae elementa*, sig. a3ʳ.

94 Alexis-Claude Clairaut, *Elémens de géométrie* (Paris, 1741) and *Elémens de algèbra* (Paris, 1746); Judith Zinsser, "Mentors, the Marquise du Châtelet and Historical Memory," *Notes and Records of the Royal Society*, 61 (2007): 94–97.

disciplines and "the origins of all sciences."[95] Mathematics was a clarifying exercise of the mind and a powerful weapon in the struggle against "that universal skepticism." Mathematical certitude was the foundation of philosophical wisdom and political prudence. Pignatelli's passion for the subject, carefully cultivated under his instruction, was presented as inspiration for "young adolescent minds."[96] Di Martino's program advertised the moral and intellectual virtues of analysis while emphasizing its practical benefits for civil society. Little wonder that Antonio Genovesi, newly returned to Naples in 1737, discovered a passion for mathematical philosophy inspired by his studies with Di Martino. This was a mathematical revolution with social implications.[97]

Di Martino's lengthy oration on the virtues and utility of algebra contained a small but telling detail. The prologue to the reader indicates that it came from him and "F. P." What are we to make of this? Should we consider Pignatelli the coauthor of his address to the Paris academicians? It would be entirely in keeping with the resumption of her scientific correspondence the preceding year. Whatever her exact role, Di Martino presented her as a full participant in his public conversation about how mathematics could transform society. Pignatelli's network now expanded to include Paris as well as Bologna and Leipzig. Surely this was why Appiano Buonafede (1716–93), an ambitious young theologian studying in Rome eager to earn Galiani's patronage, chose to dedicate *Delle malignità istorica discorsi tre* (1738) to Pignatelli. Buonafede protested that he did not wish to play the courtier, flattering a prospective patron with empty words, when so much had been written in praise of her, indeed "all of Europe is talking of your glory, both old and new." He offered the following apotheosis of her scientific acumen:

> I restrict myself solely to saying about you that the beauty of your soul and the elevation of your mind applied to mathematical and philosophical studies, and in every kind of learning, makes you the marvel and fortune of learned men who know you. You are among the major ornaments of your sex and our age.[98]

95 De Martino, *Algebrae geometria promotae elementa*, "Ad Lectorem," n.p.

96 De Martino, *Algebrae geometria promotae elementa*, "Ad Lectorem," n.p.

97 Casini, *Newton*, p. 212; Genovesi, *Opuscoli*, pp. 7–8; idem, *Autobiografia*, p.14. For the long-term consequences, see Massimo Mazzotti, "Geometers of God: Mathematics and Reaction in the Kingdom of Naples," *Isis*, 89 (1998): 674–701.

98 Agatopisto Cromaziano [Appiano Buonafede], *Della malignità istorica discorsi tre* (Basel, 1738), p. 6. For the second edition, see *Della malignità istorica discorsi tre di A*** B**** (Bologna, 1757).

When the Bologna Academy of Sciences' printer Lelio dalla Volpe republished this work in 1757, it gave an official imprimatur to this assessment of their second female member.

Grand Tourists now began to seek out Pignatelli when they visited Naples. In November 1739, writing from Rome, Charles de Brosses described the many pleasures of his experience of Naples to friends back home in Dijon. He would later write an entire book on the archeological wonders of Herculaneum, emerging from under centuries of volcanic ash under the watchful eyes of the Bourbon monarch and his antiquarians, but at the time he especially delighted in those evenings when he discussed physics after dinner with minister Intieri "or with the Princess of Palombrano [sic] who also excels in geometry."[99] Pignatelli's reputation also traveled in Roman circles, in part, because of her correspondence with the Corsini librarian, Giovanni Gaetano Bottari, who was part of the Tuscan network that facilitated ties between Rome and Naples and a member of the Bologna Academy.[100] "I know that in Naples the Princess of Colubrano, Barbapiccola, and other women are at the cutting edge of the study of mathematics and philosophy," exclaimed Giovanni Niccolò Bandiera, chaplain of the Oratory of San Girolamo della Carità in Rome, in his 1740 treatise on women's education.[101] Perhaps Bandiera met her when he visited Naples but he was certainly well-informed about the women who mattered in the Regno. Pignatelli was now being written into a new generation of books arguing in favor of women's learning.

In his commentary on Châtelet's recently published *Institutions de physique* (1740), Voltaire reflected on why she "and Madame the Princess of Colubrano follow Leibniz's banner, not like women who commonly side with the theologians out of weakness, taste, and with a stubbornness based on ignorance and often that of their masters. The one and the other have written as mathematicians and the two of them both with some novel views."[102] Pignatelli's philosophical acumen now earned the praise of this celebrated French philosophe, himself a committed Newtonian and therefore deeply skeptical of Leibnizian

99 Charles de Brosses, *Lettres d'Italie*, ed. Frédéric d'Agay (Paris: Mercure d France, 1986), vol. 1, p. 425. (Rome, 24 November 1739).

100 Biblioteca Corsiniana, Rome, ms. 33, c. 13 (=cod. 1633), cc. 47^r–56^r (Pignatelli to Bottari, Naples, 19 November 1735, 7 January, 31 January, 25 February, and 20 March 1736).

101 [Giovanni Niccolò Bandiera], *Trattato degli studi delle donne* (Venice, 1740), vol. 1, p. 148. Bandiera published this work without his name under the guise of the "Accademico Intronato."

102 Voltaire, *Exposition du livre des Institutions physiques* (1740), in *Oeuvres complètes de Voltaire* (Paris: Garnier, 1877–85), vol. 23, p. 140.

physics. Voltaire, however, expressed nothing but admiration for these two women of science, even though he never met or corresponded with Pignatelli.

De Brosses' praise of an evening with Pignatelli begs the question – where was she in the fall of 1739? No longer in the convent. "The Princess of Colubrano has left the monastery of Regina Coeli because of illness and gone to her house on the island, notwithstanding the strong protests of the husband who has put up most unwillingly with her exit," reported Niccolò Fraggianni to Bartolomeo Corsini, viceroy of the Kingdom of Naples since 1737.[103] The following month Francesco Carafa visited Pignatelli as she recovered from her "breast ailment," encouraging her, "with mutual signs of tenderness and tears," to return home with him.[104] Not yet well enough to travel, she nonetheless agreed to a reconciliation. Rewards and recognition followed swiftly on the heels of this announcement. On Christmas Day 1739, Pignatelli was appointed *dama di corte* to Queen Maria Amalia of Saxony who had arrived in Naples after her marriage to Charles in 1738. Pignatelli's mother-in-law Caterina Carafa, who wielded a great deal of influence as principal lady-in-waiting (*cameriere maggiore*), was surely the deus ex machina behind this court position. The Bourbon king described Carafa admiringly as "the best among all the others" and she was known to enjoy the favor of the King and Queen of Spain.[105] Everyone hoped that Pignatelli would return to her husband. There were great expectations that she could now resume her rightful place in Neapolitan society.

As so often happens in tumultuous relations, the honeymoon was brief. One year later, King Charles informed the Spanish monarchs that "the new disagreements between the Prince of Colubrano and his wife are all too true." They incurred royal displeasure by requesting another separation, leading the king to appoint one of his most trusted ministers to attempt a resolution. "It is important that they are made to live together, as God commands between husband and wife," the young Bourbon king piously informed his parents.[106] Charles discovered connubial bliss while fulfilling his duty to crown and state; he hoped to inspire similarly responsible and loving acts of obedience among his courtiers.

Even the interventions of the most experienced diplomats failed to save the Carafa-Pignatelli union. Bernardo Tanucci, another Tuscan at the Bourbon

103 Niccolò Fraggianni, *Lettere a B. Corsini (1739–1746)*, ed. Elia del Curatolo (Naples: Jovene, 1991), p. 42 (Naples, 7 November 1739).

104 Fraggianni, *Lettere*, p. 55 (12 December 1739).

105 Carlo di Borbone, *Lettere ai sovrani di Spagna*. Vol. 2. *1735–1739* (Rome: Ministero per I beni e le attività culturali, Direzione generale per gli archive, 2002), p. 256 (Procida, 19 November 1737); Elena Papagna, *La corte di Carlo di Borbone il re "proprio e nazionale"* (Naples: Guida, 2011), pp. 48–49, 86–87, 142–143.

106 Borbone, *Lettere*, pp. 139–140 (Naples, 27 December 1740).

court who later became prime minister and president of the Council of Regency, felt that it was entirely the fault of *la vecchia Colubrano*. Tanucci intimated that her dislike of the appointed mediators doomed negotiations by entangling them in court politics. Francesco Carafa's perception that the secretary of state, marquis José Joaquin de Montealegre, cared more for his wife than he should only added to climate of rumor and suspicion.[107] Pignatelli accused her husband of being violent, unfaithful, and giving her syphilis. Carafa responded by presenting her as an arrogant and disobedient wife who loved her autonomy far more than any woman should. He chastised her for taking liberties with her position, "under the guise of learning she wanted to deal and converse for many hours of the day" [with] "nobles, foreigners, dull-witted peasants, and some scholars, especially young monks."[108] Pignatelli's salon and her intellectual fame became an additional source of disagreement between husband and wife. Her mentor Di Martino was unavailable for support since he was in Madrid serving as secretary to the Neapolitan ambassador in 1740–44. The Neapolitan Academy of Sciences had faltered due to lack of institutional support and surely the fact that key members were now involved in politics and diplomacy.[109] Pignatelli was acutely aware of the precarity of her situation and her health continued to suffer.

Pignatelli requested permission "to live in a separate house." Her husband insisted that the only honorable way for her to live apart was in a convent; he suggested the Retreat of Mondragone in the Spanish quarter of the city, where unmarried noblewomen, widows, and women separated from their husbands lived honorably. By April 1741, Pignatelli became part of this community "with the ability to leave it every time she must perform her functions at court."[110] The marriage was officially over, just in time for Carafa to be called into action as Neapolitans troops headed north to join the Spanish army gathering to take Milan back from the Austrian Habsburgs. He was part of the first wave to

107 Bernardo Tanucci, *Epistolario*. Vol. 1. *1723–1746*, ed. Romano Paolo Coppini, Lamberto Del Bianco, Rolando Nieri, and Mario D'Addio (Rome: Edizione di Storia e Letteratura, 1980), p. 409 (Tanucci to Corsini, Naples, 24 January 1741).

108 ASN, *Segretaria di Casa Reale*, 777, as discussed in Papagna, "Faustina Pignatelli." Thanks to Elena Papagna for providing me with the longer unpublished version of her essay from which this quote is taken. On these larger problems, see Guerci, *La sposa obeddiente*; and Seidel Menchi, Quaglioni, ed., *Coniugi nemici*.

109 Giuseppe Galasso, "Scienze, istituzioni e attrezzature scientifiche nella Napoli del Settecento," in Sodalitas, ed., *L'Età dei lumi: studi storici sul Settecento europeo in onore di Franco Venturi* (Naples, 1985), vol. 1, p. 208.

110 Fraggianni, *Lettere*, p. 214 (8 April 1739).

disembark and among the first to return to Naples in September 1742, despite malicious rumors of his death in battle.[111]

Pignatelli was surely relieved to see him go and less than thrilled at his return. The Italian phase of the War of the Austrian Succession (1740–48), however, had many victims, starting with those living at the center of this theater of war and devastation. Carafa became one of its political sacrifices. Soon after his return, he was imprisoned under suspicion of colluding with the Austrians.[112] Supporters came to his defense but could not fully cleanse his reputation. As Spanish forces faltered in northern and central Italy, retreating all the way to Velletri by summer 1744 where they finally held their ground, there were growing concerns about his loyalty to the new regime. Carafa was promptly rearrested and accused of conspiring with the Austrians. Spring 1745 found him incarcerated in Castel Nuovo, outraging officials by bribing guards to allow him to leave his cell "under the deceitful dress of an abbot" and go about the city.[113] He was hardly the only Neapolitan noble in prison taking liberties that came with power and privilege, but secretary of state Montealegre was determined to put an end to it.

Carafa's suspect politics put Pignatelli in an awkward position. The arrest of their eldest son Giuseppe, caught fleeing the seminary in August 1745 and sternly remitted to resume his studies, was equally concerning.[114] Would the next generation survive unscathed? Salvaging her family's reputation and position became Pignatelli's most important task; her position at court made it possible. Years later, Tanucci remarked that the king had done for the Carafa sons what "the father didn't do."[115] That would not have happened without his admiration for the women in the family.

At the very moment when things seemed like they couldn't get any worse, providence intervened. Pignatelli's situation began to improve in the late 1740s and different elements of her life started to come together. In April 1743, Châtelet's daughter Françoise-Gabrielle-Pauline (1726–54) married Alfonso Carafa, duke of Montenero. The renowned French Minim François Jacquier, who had just completed his three-volume commentary on Newton's *Principia* with Thomas LeSeur, accompanied Gabrielle-Pauline from Rome to Naples

111 Borbone, *Lettere*, vol. 2, p.244n388 (19 December 1741); Fraggianni, *Lettere*, pp. 337, 361 (16 June and 8 September 1742).

112 Tanucci, *Epistolario*, vol. 1, p. 624 (20 September 1742).

113 Fraggianni, *Lettere*, pp. 516, 520- (9 and 24 April 1745).

114 Fraggianni, *Lettere*, pp. 551–552 (7 and 21 August 1745).

115 Tanucci, *Epistolario*, vol. 12, p. 83.

at her mother's request.[116] It seems likely that Jacquier introduced her to Pignatelli, raising the question of what role Pignatelli played in the successful integration of this young French aristocrat into the Neapolitan court. In 1747, Gabrielle-Pauline received a coveted appointment as principal lady-in-waiting to Maria Amalia.[117] Pignatelli surely knew Châtelet's daughter.

4 Imagined Conversations

In Bologna, secretary Zanotti was preparing the second volume of the academy's *Commentarii* for publication. He had finally decided what he wanted to do with Pignatelli's unpublished paper. In August 1741, Zanotti told the papal physician Antonio Leprotti that he was reading a new "printed dissertation … on living forces" but remained unpersuaded by "Leibniz's opinion."[118] The book had been sent from Naples by Pietro Di Martino. Di Martino published his own account of the history of the *vis viva* controversy, advocating a Newtonian-Cartesian position on the nature of force while expressing a willingness to change his mind if better proofs were offered that would make Poleni's experimental proof of Leibniz's concept more persuasive.[119] He addressed the treatise to the Bologna Academy but also indicated that he was writing on the heels of the sharply worded exchange between Mairan and Châtelet early in 1741, after she critiqued Mairan's Cartesian position in her *Institutions de physique* (1740) indicating her Leibnizian sympathies. Mairan had just replaced Fontenelle as perpetual secretary of the Paris Academy of Sciences. He defended his science with a strongly worded letter to "Madame ***," making the gender and perceived deficiencies of the author of this new French physics textbook apparent. Previously Châtelet fantasized about remaining anonymous.[120] Nonetheless, she was fully prepared to engage in a war of words. The

116 Federica Favino, "Minimi in 'Sapienza.' François Jacquier, Thomas Le Seur e il rinnovamento dell'insegnamento scientifico in Studium Urbis," *Mélanges de l'École française de Rome. Italie et Méditeranée*, 117, 1 (2005): 159–187.

117 Zinsser, *La Dame d'esprit*, p. 221; Châtelet, *Selected Philosophical and Scientific Writings*, p. 255 (Châtelet to Father François Jacquier, Paris, 13 April 1747).

118 AASB, *Antica Accademia, Titolo III, Parte II*, Lettere missive Zanotti, fasc. 5 (30 August 1741); see Brigaglia and Nastasi, "Bologna e il Regno," p. 160.

119 Gatto, *Libri di matematica*, pp. 65–67.

120 Châtelet to Bernoulli, 30 June 1740, as quoted in Zinsser, *La Dame d'Esprit*, p. 190. On this debate, see Keiko Kawashima, "La querelle des 'forces vives.' Le débat entre Mme Du Châtelet et Dortous de Mairan," in *Madame Du Châtelet. La femme des lumières*, ed. Elisabeth Badinter, Danielle Muzerelle (Paris: Bibliothèque Nationale de France, 2006),

second edition of Châtelet's textbook, translated into Italian in 1743, included her exchange with Mairan.[121]

Di Martino had access to Mairan's response, which he cited in his own analysis of the *vis viva* controversy, and he also knew that Châtelet was the author of the critique of the French Cartesian position.[122] Pignatelli also received copies of these publications from her French colleagues. No correspondence between her, Mairan, and Châtelet survives but Di Martino published a passage from a letter Châtelet addressed to her. The marquise described her pleasure at hearing Pignatelli's name invoked while reading her and wrote, hoping to enlist her Neapolitan colleague in making new arguments to confute their adversaries. Once "intelligent men get to know it, they easily understand," she declared.[123] In some fundamental sense, Châtelet perceived Pignatelli as an interesting and potentially influential ally. She was inviting her to participate in the debate and take a stand. Châtelet had certainly read her publication in the *Nova Acta Eruditorum*.[124]

Zanotti joked with Leprotti that everyone in southern Italy was "so decisively Cartesian in this controversy that I can't stand it here."[125] But he knew at least one person in Naples who, as Voltaire had already observed, did not tow the party line. Zanotti reread Pignatelli's paper. It inspired him to write a Latin dialogue set in Bologna in 1738, which he published in the second volume of the *Commentarii* in 1745.[126] He had the Bologna academicians debate the state of the question over several months, with some taking a strongly Cartesian

pp. 99–100. See also Elizabeth Goldsmith and Dena Goodman, ed., *Going Public: Women and Publishing in Early Modern France* (Ithaca: Cornell University Press, 1995).

121 Émilie du Châtelet, *Instituzioni di fisica ... accresciuta con la Dissertazione sopra le forze motrici di M. de Mairan* (Venice, 1743).

122 Pietro Di Martino, *De corporum quae moventur viribus earumque aestimandarum ratione. Dissertatio philosophica ad Instituti Bononiensis Socios inscripta* (Naples, 1741), pp. 93, 105.

123 Addabbo, "I *Philosophiae naturalis institutionum libri tres*," p. 82; Gatto, *Libri di matematica*, p. 71n25. Quote from Federico Amodeo, *Vita matematica napolitana. Parte prima* (Naples: Giannini & Figli, 1905), p. 69: "Interim ego quum legerem quotidie honoris caussa [sic] iterum nomino, quantam in eius judicio praesidio mihi accesserit sive ad confutandos adversaries, sive ad nova argumenta excogitanda, qui eius ingenii Viri norunt, facile intelligent." Due to COVID-19, I have been unable to find the original passage in Di Martino.

124 Châtelet, *Selected Philosophical and Scientific Writings*, p. 109 (30 April [1738]).

125 AASB, *Antica Accademia, Titolo III, Parte II*, Lettere missive Zanotti, fasc. 5 (30 August 1741).

126 Francesco Maria Zanotti, "De vi corporum viva," *De Bononiensi scientiarum et atrium Instituto atque Accademia Commentarii*, vol. 2, part 1 (Bologna, 1745), pp. 377–413; Walter Tega, ed., *Anatomie accademiche*. Vol. 1. *Commentari dell'Accademia delle Scienze di Bologna* (Bologna: Il Mulino, 1986), p. 138; Walter Tega, ed., *Anatomie accademiche*. Vol. 2. *L'Enciclopedia scientifica dell'Accademia delle Scienze di Bologna* (Bologna: Il Mulino, 1987), p. 173.

position and others more open to the Leibnizian viewpoint. Was there any way out of this impasse, they wondered? Zanotti responded:

> If I had here the letters that a most noble and ingenious woman, Faustina Pignatelli, Princess of Colubrano, wrote me a little while ago, and you read them, you would clearly understand how such a controversy can easily be seen as nominal. In fact, she interweaves the most elegant demonstrations and hypotheses, with which she shows that if the Cartesians and Leibnizians agreed on the meaning of a few names, the rest of the matter would be reduced to nothing.[127]

Zanotti praised the elegant style, charm, and brilliance of Pignatelli's letters, allegorizing her as one of the Greek muses. "You would say Thalia had written them." How had Pignatelli explained the relationship between quantity of motion and energy? Zanotti explained that she considered the force that the body has to have acted" to be equivalent to the "greatest of all forces, which prevents the body from exercising its individual actions." Zanotti's Cartesians offered to challenge "the reluctant Pignatelli" but ultimately disputed the matter among themselves, comparing her position to Pietro Di Martino's who considered the differences between living and inert force to be real rather than apparent.[128] Nonetheless, it became clear that both Neapolitan mathematical philosophers considered *vis viva* a methodological conundrum. Increasingly, they felt that the problem lay not in nature but in the limits of human understanding.

Without ever endorsing "Pignatelli's opinion," Zanotti respectfully outlined its merits.[129] He felt that it was an idea with possibilities. Zanotti pointedly suggested that Di Martino was aware of this alternative and should have engaged these ideas in his dissertation, where neither Pignatelli nor Châtelet was given a voice. He lamented that his colleagues in Naples had not taken greater advantage of her rationalist efforts to effect a reconciliation between the two positions. More to the point, Zanotti made his admiration for Pignatelli the highlight of this imagined conversation, creating his own dialogue with her letters. Pignatelli's paper remained unpublished, but its contents had now been cited authoritatively and her scientific correspondence praised publicly. Things were finally looking up for the most visible woman of science in Naples.

---

127 Zanotti, "De vi corporum viva," p. 384.

128 Zanotti, "De vi corporum viva," p. 385; Simili, "In punta di penna," pp. 40–42.

129 Zanotti, "De vi corporum viva," p. 393.

One year later, Carafa died in the military prison at Castel Sant'Elmo, finally relieving Pignatelli of the burden of an unhappy, scandalous, and financially ruinous marriage. She remained at court and fully resumed her place in society. Her eldest son Giuseppe joined the Knights of Malta and renounced his right to inherit in favor of his younger brother Michele in return for a sum equivalent to his mother's annuity and some benefices.[130] The Bologna Academy made Châtelet the third female member.[131] Tragically, Pietro Di Martino died of tuberculosis in 1746, so he did not become part of Pignatelli's renewed salon, unlike his elder brother Niccolò who returned from the Spanish court. Genovesi, who considered Pietro "one of his dearest friends" also participated; he later recalled meeting Count Karl Joseph von Firmian, Minister Plenipotentary of Lombardy, in her home.[132]

Foreign visitors frequently mentioned meeting Niccolò Di Martino *chez Madame la Duchesse de Colubrano* as the abbé Nollet wrote in 1749.[133] Indeed, he was an equal attraction, given his authority as a courtier, diplomat, and mathematician; recalled from Madrid to assist in the development of the royal military academy, in the next decade Di Martino developed a vernacular curriculum in practical mathematics for naval science and engineering, repurposing a great deal of the material he initially presented to *la donna Faustina* for a new generation to the benefit society.[134] He no longer dedicated his books to her but their relationship continued to be important to both of them. Di Martino himself described to mutual friends the pleasure of evenings "at the home of the Signora Princess of Colubrano," awaiting the arrival of English books, reading the latest account of the eruption of Vesuvius, and new mathematics publications. At times, he was a broker of her correspondence with other men of science, including the Sicilian mathematician Guido Settimo (1706–62). On

130 Fraggianni, *Lettere*, pp. 552n2; Gaetano Fusco, "I Carafa della Stradera nella relazioni storiche tra Terra di Lavoro e la Regione Lucana," in Pietro Borraro, ed., *Studi Lucani. Atti del II Convegno Nazionale di Storiografica Lucana* (Galatina: Congedo, 1976), Part II, p. 170.

131 Mazzotti, "Mme Du Châtelet."

132 Paola Zambelli, *La formazione filosofica di Antonio Genovesi* (Naples: A. Morano, 1972), p. 849.

133 Bibliothèque municipale, Soissons, ms. 150, Jean-Antoine Nollet, *Journal du voyage de Piémont et d'Italie en 1749*, p. 191A [in John Heilbron's typescript transcription]; Bertucci, *Viaggio nel paese delle meraviglie. Scienza e curiosita nell'Italia del Settecento* (Turin: Bollato Boringhieri, 2007), pp. 231–232.

134 Nastasi, "Nicola Antonio De Martino (Di Martino)," DBI, vol. 38, pp. 602–603; Amodeo, "Dai fratelli Di Martino a Vito Caravalli, "*Atti della Accademia Pontaniana*, vol. 32, n. 11/ ser. 2, vol. 7 (1902): 6–7, 23–28. On the many reprints of Di Martino's textbooks, see Gatto, "La matematica nell'editoria napolitana del Settecento," in Rao, ed., *Editoria e cultura a Napoli*, pp. 774–776.

one occasion, Di Martino assured his friend in Palermo that he received a "folio for the Signora Princess of Colubrano, whose response is joined with my own," suggesting how intimately their networks were bound together.[135] As his fame as a mathematics professor grew throughout the Kingdom, contemporaries remembered that she was his most interesting disciple.[136]

Now anointed by the respectability that came with widowhood, Pignatelli finally realized her personal ambition of becoming a promoter of science and the arts in her native city. Palazzo Carafa di Colubrano was once again a place to see and be seen. She was no ordinary salonnière but a woman of great erudition who commanded the respect of all who knew her.[137] Nonetheless, when Ferdinando Spinelli, the prince of Tarsia, opened his library to the public with great fanfare on 22 July 1747, inaugurating his short-lived experimental academy, she was not part of the public celebration of this effort to revive Galiani's idea of a modern scientific academy in Naples. Barbapiccola composed a poem in celebration of this auspicious occasion. Mariangela Ardinghelli (1730–1825) made her first public appearance at age seventeen, inaugurating her career as the most important scientific translator in the city.[138] Pignatelli's absence was conspicuous. It has been suggested that rivalries or even philosophical differences kept her way but, in fact, we do not know if she attended but chose not to perform which would have put her in an untenable subordinate role to a fellow aristocrat. Possibly the health issues that had so recently plagued her made it impossible to attend. In so many other ways, she had become more visible than ever.

Early in the summer of 1750, Zanotti finally reached Naples where he met one of the scientific women he greatly admired at a distance. Châtelet was no longer alive, having died from complications giving birth the year before. Perhaps he was feeling nostalgic for the period of his life when he corresponded about science with a French marquise and a Neapolitan princess. Zanotti had already availed himself of his contacts in Naples to publish a small anonymous pamphlet, *Della forza attrattiva delle idee* (1747) without a license, like so many

135 Nicla Palladino, Anna Maria Mercurio and Franco Palladino, ed., *La corrispondenza epistolare tra Niccolò Di Martino e Girolamo Settimo* (Florence: Olschki, 2008), p. 88 (n.d. [end of November 1751]), p. 110 (n.d. [March 1752]), p. 153 (3 February 1753). Quote is from the 1752 letter.

136 Giangiuseppe Origlia Paolino, *Istoria dello Studio di Napoli* (Naples, 1754), vol. 2, p. 394.

137 Cavazza, "Dalle biblioteche dei dotti alle tolette delle dame. La conversazione filosofica e scientifica nell'Italia dei lumi," *I Castelli di Yale*, 12 (2012): 87–102, esp. 93–94.

138 Bertucci, "The Architecture of Knowledge: Science, Collecting and Display in Eighteenth-Century Naples,"in Calaresu and Hills, ed., *New Approaches to Naples*, pp. 149–174; Bertucci, "The In/visible Woman"; and Findlen, "Translating the New Science."

books circulating in the illicit market of ideas. Presented as a translation of the burnt remnants of an unpublished French treatise by a one-eyed, one-armed, one-legged English colonel who retreated after the war to a village to cultivate his passion for natural philosophy and converse with a French marquise about Newton, it combined familiar ingredients from Fontenelle's *Conversations* and his own student Francesco Algarotti's *Newtonianism for Ladies* (1737), with a ludic spirit that anticipated Voltaire's *Candide* (1759). This was a lighthearted thought piece for a world just emerging from seven years of devastating foreign wars and in desperate need of philosophical distraction. Zanotti's encounters with the aristocratic women he knew and tutored in Bologna, his admiration for Châtelet, and his epistolary friendship with Pignatelli further stimulated his imagination.[139]

In this pamphlet, Zanotti proposed an extension of Newton's principle of universal attraction to encompass questions of human cognition. He asked a simple question – what makes ideas attractive? Defining attraction as "the union of soul and body," he argued that how we form our opinions was a subject equally worthy of study.[140] Why do we prefer one idea over another? Which ideas stick in the mind? These were questions that brought the writings of John Locke to mind, but Zanotti carefully did not mention an author whose *Essay Concerning Human Understanding* was placed on the Index of Prohibited Books in 1734. Instead, Zanotti made the attraction of ideas both a Leibnizian and Newtonian principle by arguing that if Leibniz's monadology were true, the attraction of bodies was a direct result of how the concept of attraction took shape in the mind, indeed how ideas themselves formed and adhered. This was a clever homage to both Châtelet and Pignatelli, who felt that it was better to synthesize the best elements of different natural philosophies rather than choose between them.

Several years later, Zanotti finally had the opportunity to converse in person with Pignatelli. He was enthralled by Naples, contrasting its natural beauty to the artifice of Rome. But what he was most eager to tell friends and family in Bologna were his impressions of *la signora donna Faustina*. In June 1750, he told Gabriele Manfredi, "Among all the learned women, I have never seen one who speaks so naturally and so well as she does; and among learned men, no one speaks better. I won't even speak of her courteousness because in this she

139 On his Bolognese pupils, see Findlen, "Women on the Verge of Science: Aristocratic Women and Science in Early Eighteenth-Century Italy," ed., *Women, Gender and Enlightenment*, ed. Sarah Knott and Barbara Taylor (London: Palgrave Press, 2005), pp. 265–287.

140 Zanotti, *Della forza attrattiva delle idee* (Naples, 1747), p. 36.

seems born deliberately to serve as an example to queens."[141] Pignatelli gave him a personal tour of the most interesting sites in the city and introduced him to Di Martino, who promised to send "his very elegant geometric theorems" they discussed to the Bologna Academy. He heard their plans to erect an observatory with her support and joked that if they stopped excavating "so many of the monuments and statues of Herculaneum," which he planned to see the following day, they would certainly have the resources to build it. Zanotti felt certain that "our friends at the Institute" would be pleased with how well his was received.[142] His admiration for Pignatelli only grew with this encounter.

5 A Philosophical Tour of Naples

Pignatelli's Naples captivated Zanotti's mind and heart. On the trip home, he began to sketch an outline for a scientific dialogue about living forces, written in the vernacular and set in Naples. As he told the physician and anatomist Giambattista Morgagni, to whom Zanotti addressed *Della forza de' corpi che chiamano viva* (1752), it was a remake of "that little Latin dialogue you already read in our Commentaries," written in a spirited and playful style that explicitly evoked that great Renaissance classic of courtly conversation, Castiglione's *The Book of the Courtier*. Numerous editions of Castiglione were still being published, including several during the eighteenth century.[143] Fresh from his experience of the Bourbon court and Pignatelli's salon, Zanotti could not resist capturing the sense and spirit of this world to immortalize it just as Castiglione did for Urbino.

"What a wonderful thing are your dialogues!" exclaimed Morgagni upon receiving them in February 1752.[144] He found reading them so pleasurable that he skimmed the entire book after a long day dissecting in Padua's famed public anatomy theater. Zanotti skillfully set the scene by painting a beautiful portrait that suggests he was a great reader of travel literature. "To tell the truth, even in the short time that I stayed there, the city of Naples seemed to me exceedingly noble, magnificent, alluring and delightful, beyond any other city in the world." Naples was the perfect fusion of nature and art populated with such pleasing and learned inhabitants that "they alone made that marvelous city truly

141 Zanotti, *Discorsi e orazioni* (Venice, 1830), p. 232 (Naples, 9 June 1750).

142 Zanotti, *Discorsi e orazioni*, p. 234.

143 Zanotti, *Discorsi e orazioni*, p. 234 (Bologna, 7 September 1751); Peter Burke, *The Fortunes of the Courtier: The European Reception of Castiglione's Cortegiano* (University Park, PA: The Pennsylvania State University Press, 1996), pp. 41, 133, 140, 162.

144 Zanotti, *Discorsi e orazioni*, p. 237 (Padua, 18 February 1752).

DELLA FORZA DE' CORPI
CHE CHIAMANO VIVA
LIBRI TRE
DEL SIGNOR
FRANCESCO MARIA ZANOTTI
AL SIGNORE
GIAMBATISTA MORGAGNI.

Ne quali libri ha proccurato l' Autore, quanto ha potuto, di promovere la quiſtione col ſolo diſcorſo metafiſico, ſenza aſſumere dalla geometria, ne dalla meccanica altro, che le propoſizioni più note, e più comuni.

IN BOLOGNA MDCCLII.

Per gli Eredi di Conſtantino Piſarri, e Giacomo Filippo Primodì, Impreſſori del S. Officio. *Con lic. de' Sup.*

FIGURE 6.6 Francesco Maria Zanotti's dialogue on living force inspired by conversations with Pignatelli and her philosophical circle in Naples. Francesco Maria Zanotti, *Della forza de' corpi che chiamano viva* (Bologna, 1752). Courtesy of Special Collections, Stanford University Library

beautiful."[145] Like many Grand Tourists, he discovered the allure of a capital city that had become the most glorious court society of the Italian peninsula. The other Naples – a kingdom of baronial estates and privileges populated with impoverished and illiterate peasants with few rights, limited possibilities, and teeming urban poverty – did not capture his attention.[146]

Zanotti's destination was not the Bourbon court but Pignatelli's salon in motion. He joined her group of intimates – all learned men of science including Di Martino whom he described as "the brightest light of Italy" – during a singularly memorable day that he brought to life on paper. The court had decided to see the ruins of Baia. Pignatelli proposed a carriage ride to Pozzuoli, followed by a boat excursion to join the court on its archeological explorations, and ultimately an evening return to Pozzuoli spent in Pignatelli's palatial and well-furnished rooms overlooking a beautiful garden on the Bay of Naples. This was a moveable feast of ideas orchestrated by Pignatelli, whom Zanotti now described as "that famous and refined collector of all the most noble and elegant minds."[147] She was indeed holding court in and around Naples.

On this occasion, Zanotti offered his most complete portrait of Pignatelli for those not fortunate enough to have met her, remarking that it would be hard to imagine the effect of her living presence. Admiring her grace and beauty, he especially praised her for having "a most sublime mind and a rare intelligence in all things, including the most subtle and obscure, accompanied by the greatest clarity and by a most gracious way of saying and explaining them."[148] These virtues combined so marvelously to create a singularly bedazzling effect. Zanotti counted himself lucky to be invited to her salon with friends who were her intimates and leading men of science. "I visited her as often as I was able."[149] He admired the quality of conversation in her salon and observed that the "most wonderful issues of every kind" emerged from her probing questions and thoughtful responses, always presented "in few words, because she loved

---

145 Zanotti, *Della forza de' corpi che chiamano viva* (Bologna, 1752), p. 5. The participants in this dialogue were Zanotti, Niccolò Di Martino, Faustina Pignatelli, the physician and *protomedico* Francesco Serao, the astronomer Felice Sabatelli who succeed Pietro as the university professor of this subject after also spending time in Bologna, and the Marchese di Campo Hermoso, an aristocratic mathematician recently arrive at the Neapolitan court from Palermo. I am indebted to Brigaglia and Nastasi, "Bologna e il Regno," pp. 163–165; and Serrapica, *Scienza e filosofia*, among the few publications to discuss this book.

146 On visitors' impressions of Naples, see Rosemary Sweet, *Cities and the Grand Tour: The British in Italy, c.1690–1820* (Cambridge, U.K.: Cambridge University Press, 2012), pp. 164–198; and Calaresu, "From the Street to the Stereotype: Urban Space, Travel, and the Picturesque in Eighteenth-Century Naples, *Italian Studies*, 62 (2007): 189–203.

147 Zanotti, *Della forza*, pp. 5, 7.

148 Zanotti, *Della forza*, p. 7.

149 Zanotti, *Della forza*, p. 8.

better to listen than to be heard. On this issue, we entirely disagreed because everyone of us would willingly have remained silent to hear her."[150] Pignatelli's ability to be playful while exuding gravitas, to bring grace to philosophy and substance to conversation, and to always leave her audience wanting more, made her his perfect courtier. She wore her learning lightly but took it very seriously.

There was a certain *sprezzatura* to Zanotti's lengthy paean to Pignatelli and her world. He made it beautiful and alluring when it was often small and petty, he made it expansive when many elements were undergoing contraction. The Bourbon court was a wonderful place to visit but not to stay. A year after the publication of Zanotti's dialogue, Pignatelli was embroiled in an inheritance dispute with relatives. A male cousin made a will in favor of "the little prince," presumably her son Michele, and the cousin's brother was irate. Di Martino commended the "Signora Princess's efforts to make the peace between the two brothers" but recognized that she would be accused of self-interest.[151] When the testament was read, the terms were so financially ruinous that her son could not accept the inheritance. It seemed obvious to everyone but the aggrieved brother that they had done nothing wrong; their cousin deluded himself into thinking it would benefit the family to consolidate their estates when the reality was that they could barely afford the lands and palaces already in their possession. The affair dragged on for months, wasting Di Martino's time and everyone's good will. Such were the realities of eighteenth-century Naples.

Zanotti's book was a Neapolitan conversation offering a detailed meditation on the most recent Italian publication concerning *vis viva* – the Jesuit mathematician Vincenzo Riccati's *Dialogo ... dove ne' congressi di più giornate delle forze vive e dell'azioni delle forze morte si tien discorso* (1749). Published in Bologna, this lengthy dialogue strongly defended the Leibnizian position held by the scientific community in the Venetian Republic.[152] The Bologna Academy counted the prolific Riccati among its members, making a local debate complicated and unappealing. Naples provided the perfect counterfoil to permit Zanotti to explore the subject since he visited the city the year after its publication. It was indeed scientific news that merited a discussion in Pignatelli's salon. Everyone wanted to know his opinion of a book that only some of them had read. Di Martino knew Riccati's book but Pignatelli had yet to read it.

150 Zanotti, *Della forza*, p. 8.

151 Palladino, Mercurio, Palladino, ed., *La corrispondenza epistolare*, p. 151 (Naples, 27 January 1753); see also pp. 152, 157.

152 Gregorio Piaia, Maria Laura Soppelsa, ed., *I Riccati e la cultura della Marca nel Settecento Europe* (Florence: Olschki, 1992).

The conversation began by establishing Zanotti's accommodationist position since he was no longer purely a Cartesian after years of advocating Newtonian natural philosophy. Pignatelli asks the essential question – if force doesn't produce velocity directly, then what does it do? "That force doesn't produce velocity," he responds, "but a living force that therefore restrains velocity."[153] Zanotti argued that the elasticity Bernoulli attributed to the moving balls produced a velocity proportional to itself but was not infinite. Soon Pignatelli's elegant furniture is covered with sheets with copies of the same diagram – the experiment with two globes of differing weights – for everyone to consider. Those who have read Riccati's book inform the others about its principal arguments, while wryly observing that it is a difficult read – a marked contrast to Zanotti's accessible prose. "In any case, I would like to see such a book," Pignatelli replied, reminding everyone that she liked to think about complex and abstruse matters.[154] Zanotti also makes clear that she is a knowledgeable reader, unlike the imaginary marquises who populate Fontenelle's and Algarotti's bestselling books. The search to retrieve a copy in Naples begins while Zanotti continues to explain its contents, implying that you do not have to actually read Riccati if you prefer to hear his account of *vis viva*.

Soon, however, the conversation becomes more broadly epistemological. What is the best way to understand and explain nature scientifically? Pignatelli invokes "the principle of simplicity" as the foundation of all good philosophical systems, leading them to measure the inchoate nature of the *vis viva* debate against the relative clarity and elegance of the Copernican, Cartesian, and Newtonian systems. There is a history of knowledge at the center of Zanotti's dialogue that demonstrated his agreement with the Di Martino brothers' efforts to make history an instrument to gain perspective on science's evolution.[155] Maupertuis' principle of least action is the foundation for their mechanics. They joke that one need not invoke the divine to arrive at this conclusion – Maupertuis is no God but president of the Berlin Academy of Sciences! "It was enough that he knew a bit of differential calculus."[156] Such comments echoed Neapolitan arguments for a modern epistemology based on mathematics rather than metaphysics. Niccolò Di Martino now introduces his own reading of Riccati's dialogue by recalling a specific passage from memory, inspiring Zanotti to rise to the challenge of identifying other key passages that he remembers. Reading the book becomes a philosophical game of memory

153 Zanotti, *Della forza*, p. 79.

154 Zanotti, *Della forza*, p. 84.

155 Casini, *Newton*, pp. 208–210.

156 Zanotti, *Della forza*, p. 108.

in Pignatelli's salon.[157] When the time for a midday meal arrives, they pause to dine with the governor of Pozzuoli and hear his daughter sings Sicilian songs.

That afternoon, still awaiting the arrival of Riccati's book, they revisit Castiglione as the model for their conversation. In the early sixteenth century, Castiglione famously asked what made a perfect courtier, male and female. Zanotti's eighteenth-century Neapolitans instead debate the ideal characteristics of "that most perfect philosopher."[158] They declare the best philosophy to be a seamless fusion of poetry and science. Modern philosophers need to be able to assess arguments, "knowing how to measure their probability and to be satisfied in case it cannot be joined with evidence." They must also be eloquent to persuade their listeners; Cicero's perfect orator as much as Castiglione's courtier should be their inspiration. Knowing all the major philosophical systems is required. How else can anyone decide which to reject and which to accept as the most "probable"?[159] What to know remains an open question; there are no certain conclusions in an open dialogue. All sciences and arts contribute to this edifice of knowledge. Zanotti ultimately concludes that all his Neapolitan friends taken together make a perfect philosopher. He singled out Pignatelli for coming closer than any of them in achieving perfection on her own, paraphrasing what he wrote in his letter to Manfredi during that languid summer when he came to Naples.[160]

Finally, Di Martino rejoins the conversation bearing a copy of Riccati's book. Pignatelli skims it rapidly while the group continues to query Zanotti and Di Martino to glean greater insight. Since it is now time to join the court at Baia, they take the book and their diagrams on a boat across the Bay of Naples. When Zanotti falters in his explanation, Pignatelli takes charge of the conversation. She adroitly explains the diagram, patiently pointing out why one globe has four times the elasticity of the other, while also questioning the premises of Bernoulli's argument – like many other readers of his work who felt that his experiment described ideal rather than real circumstances. As Pignatelli concludes her explanation, she turns to Zanotti and says: "Well, I'm waiting for you to show your ingenuity!"[161] Zanotti wanted his readers to appreciate Pignatelli's mental agility and wit in conversations with men of science. He made it clear

157 Di Martino points to Riccati, *Dialogo*, p. 234, inspiring Zanotti to recall pp. 175–176, 118, and 22 in that order; ibid., pp. 115–117.

158 Zanotti, *Della forza*, p. 130.

159 Zanotti, *Della forza*, pp. 128, 131.

160 Zanotti, *Della forza*, p. 137: "La Signora Principessa condiva tutte queste scienze, che ottimamente intendeva, di tanta soavità e grazia, e così fattamente le abbelliva, che non parean quasi belle se non per lei sola."

161 Zanotti, *Della forza*, p. 181.

that she was not only in charge of the conversation but fully commanded the subject.

The discussion of how to arrive at the true nature of force revolved around three variables – the role of calculation, the consistency of numerical and experimental results, and the logical premises of the argument. Zanotti expressed a certain exasperation with testing everything by equation since he was, without a doubt, a far more traditional natural philosopher than Pignatelli and Di Martino. When Pignatelli pushed him to explain why "so many demonstrations and so many calculations" resulted in "so much constancy," Zanotti seized the opportunity to remind everyone that she was the real expert. "I see all too well that you, Signora, tempt me by questioning me on those things that you know better than me."[162] He was not going to take the bait and deferred to her opinion.

At the heart of Zanotti's leisurely and at times light-hearted critique of Riccati was a serious subject. It was not who had the right answer to the question of *vis viva,* a subject on which so many words had been wasted by 1752, leading Châtelet and later Jean d'Alembert to famously describe it as *un disput des mots,* but how to approach knowledge.[163] In conversation with his Neapolitan colleagues, Zanotti understood that Riccati's more traditional and less open approach to his subject revolved around false certainties. The Neapolitans joked that Riccati's greatest fear was that the universe would devolve into chaos if nature had one less law. They cite their own reading of Zanotti's Latin dialogue as a pleasing contrast to this kind of philosophical dogmatism. Ultimately, Pignatelli emerges as the boldest mind of all: "I am not as fearful … as Father Riccati, who will never be able to decide if anything outside the laws of nature might nonetheless be possible unto itself. Therefore, not risking to conjecture it, he cannot examine it."[164] A philosophy without fear and indecision is the only path forward through the tangled thicket of knowledge. Modern philosophy can only thrive if its practitioners are willing to ask hard questions and defer final answers. This is the conclusion of the Neapolitans as their boat returns from a day of viewing ruins to Pozzuoli in the twilight of a warm summer's evening.

There was still much hard philosophical work to be done since little had been resolved. The group agrees to a working dinner in Pignatelli's rooms

162 Zanotti, *Della forza,* pp. 182–183.

163 Laudan, "The *Vis Viva* Controversy," pp. 136–137; Carolyn Iltis [Merchant], "D'Alembert and the *Vis* Viva Controversy," *Studies in History and Philosophy of Science* 1(1970): 139–140; and Terrall, "Vis Viva Revisited," p. 199.

164 Zanotti, *Della forza,* p. 208.

whose windows reveal a bright emergent moon as the sun begins to set on the bay. New diagrams are drawn and copied, transforming one of the abstract globes in the previous diagram into an image of the earth. The final measure of any working hypothesis is its applicability to understanding nature and the cosmos. Pignetalli jokes that they are now about to engage in "a metaphysical debate." They tease Zanotti about his love of metaphysics, leading Pignatelli to declare that metaphysicians are "so obscurantist that they lose themselves in useless questions."[165] This opinion had long been the position of the Neapolitan community of mathematical philosophers educated by Di Martino who emphasized the utility of knowledge. Zanotti's defense of metaphysics – his personal intellectual contribution to the entire dialogue – concludes this philosophical day in Naples. As the sun sets, he and the other male philosophers bid farewell, leaving Pignatelli alone in her rooms filled with books and diagrams, as they stroll through the garden before returning to Naples. Zanotti and Pignatelli continued to correspond for the next two decades, but they never saw each other again.

What did Pignatelli think of Zanotti's book? No discussion survives between Zanotti and Pignatelli about this book, or indeed with any of her colleagues. One surviving letter from 1762, a decade after this publication, offers an interesting answer. On March 2, Pignatelli wrote to Zanotti calling him "my most refined fellow disciple" (*mio Gentiliss*[*i*]*mo condiscepolo*). Zanotti also referred to her as *condiscepola*, making it clear that they both considered themselves disciples of Di Martino.[166] This strong sense of intellectual kinship joined them together, even at a distance, and became part of how Pignatelli received his publications. She thanked him for sending several copies of a book – probably his *De viribus centralibus* (1762), a short primer on Newtonian physics, especially the idea of attraction – and indicated that she was distributing them among their friends. She found "the said little work learned, clear, and appealing" but added that she awaited their *maestro*'s opinion.[167] This judicious response seems entirely in keeping with how Zanotti depicted her. In correspondence, she could be the mathematical princess that Zanotti imagined, obscuring to some degree the woman she actually was.

165 Zanotti, *Della forza*, pp. 281, 307.

166 Biblioteca Comunale dell'Archiginnasio, Bolonga (hereafter BCAB), B.160, lettera 94 (2 March 1762); Biblioteca Universitaria, Bologna (hereafter BUB), ms. 4174, caps. XLVIII, n. 12 (draft of Zanotti's letter to Pignatelli, 8 July 1769). Briefly discussed in Mazzei, "Relazioni culturali," p. 296 n. 45.

167 BCAB, B.160, lettera 94.

6 Fleeting Fame

In addition to meeting all the Neapolitan men of science who belonged to Pignatelli's circle, Zanotti encountered a young female poet, Maria Maddalena Morelli (1727–1800), daughter of the first violinist in the Pistoia cathedral who, in 1776, became the only woman to be crowned poet laureate on the steps of the Campidoglio.[168] More commonly known by her Arcadian name Corilla Olimpica which she received in honor of her celebrated poetic improvisations, Morelli also arrived in Naples in 1750. Pignatelli became her most important patron and protector at the Bourbon court, appreciating the unique talents of this young woman of humble origins. Perhaps Zanotti met her during his brief visit. Certainly, he began to hear more about her talent after returning to Bologna. In 1753, he composed a poem for Corilla, having heard that she planned to study of geometry:

> Thus, is it true that you seek new fame elsewhere,
> and the curling tendril of a new garland adorns you,
> that you race through the deep paths of geometry
> and are surrounded by circles and squares?[169]

From Naples, Corilla responded in verse, matching his precise meter and using his own words. She enclosed this sheet in a letter dated 3 July 1753, making it clear that she had no passion for science.[170] Despite Zanotti's wishful thinking that Pignatelli's salon would give birth to a new generation of women versed in science and poetry, Corilla made it clear that she did not wish to emulate her patron. Pignatelli arranged her marriage to a Spanish colonel, Francisco Fernandez, supported their eventual separation, and saw her young protegée leave Naples behind in 1760 to become the toast of Europe, a controversial and talented *improvvasatrice* whose life inspired Germaine de Staël to write *Corinne or Italy* (1807). It was Corilla who called her *la gran Fausta*, pronouncing her a "diva whose name is the sound of fame."[171]

---

168 Ademollo, *Corilla Olimpica*; Marco Catucci, "Maria Maddalena Morelli," DBI vol. 76 (2012).

169 As quoted in Michele Feo, "Il 'canzoniere' di Corilla," in *Corilla Olimpica e la poesia del Settecento*, ed. Moreno Fabbri (Pistoia: Maschietto, 2002), p. 92.

170 BCAB, B. 159 (3 July 1753).

171 Ademollo, *Corilla Olimpica*, p. 62; Paola Giuli, "Tracing a Sisterhood: Corilla Olimpica as Corinne's Unacknowledged Alter Ego," in Karyna Szmurlo, ed., *The Novel's Seductions: Staël's Corinne in Critical Inquiry* (Lewiston, PA: Bucknell University Press, 1999), pp. 165–184.

For the remainder of her life, Pignatelli continued to take her place at the Bourbon court and maintained her salon, as much as her health and dwindling family resources permitted. Grand Tourists continued to pay their respects, mostly to see the antiquities. In May 1755, Frederick the Great's beloved older sister, Wilhelmine of Bayreuth, admired the monumental Carafa bronze horse's head in the courtyard which she claimed locals venerated like a relic. Little else captured her attention. "There is nothing especially curious in the house other than a great quantity of Etruscan vases," she wrote in her travel diary. The great art critic Johann Joachim Winckelmann also briefly described the antiquities in 1758.[172] Neither of them mentioned Pignatelli – and yet she was surely there with her son Michele and his family, her daughter Caterina (who married after her mother's death), perhaps even her youngest son Diomede who became a cleric and antiquarian. The family's financial difficulties forced them to begin to sell off their assets. Tanucci attempted to prevent foreigners from exporting the choicest pieces. In the early 1760s, he commissioned an inventory of the antiquities from a member of the Accademia Ercolanese, the archeological academy he founded.[173] Ultimately, this did not thwart greedy dealers and eager collectors from emptying out the palace.

Zanotti's admiring, lighthearted portrayal of the intellectual life of the salon masked a brutal reality. Pignatelli and her sons could no longer afford the lifestyle to which they had become accustomed. The eldest brothers quarreled as Michele found himself in difficulty paying the annuity given to his oldest brother Giuseppe. Pignatelli contemplated selling some of her own lands to resolve the matter but nothing came of it. Perhaps this is why Caterina did not marry until 1772.[174] The inability of the Carafa di Colubrano to generate adequate income from their vast landholdings, let alone understand and empathize with the plight of their desperate and resentful peasants, was an object lesson in why philosophers such as Genovesi, installed in the newly created chair of political economy (commerce and mechanics) in 1754, argued that it was imperative to rethink the entire premise of their society.

Genovesi loved mathematics, physics, and modern philosophy but he disliked the court and wondered why Di Martino spent so much time there. He was critical of Galiani's failed academy because it offered no tangible support

---

172 Helke Kammerer-Grothaus, "'Voyage d'Italie' (1755). Markgrafin Wilhelmine von Bayreuth im Königreich Neapel," in Helke Kammer-Grothaus and Detlve Kreikenbom, ed., *Wilhelmine und Friedrich II und die Antiken* (Tübingen: Winckelmann-Gesellschaft Stendal, 1998), p. 13; Dodero, *Ancient Marbles in Naples in the Eighteenth Century: Findings, Collections, Dispersals* (Leiden: Brill, 2019), p. 69.

173 Dodero, *Ancient Marbles*, pp. 75–77.

174 Fusco, "I Carafa della Stradera," pp. 170–172.

for science the way that Paris and, to a lesser degree, Bologna did.[175] No revolutionary himself, he nonetheless imagined reform from the middle. A shoemaker's son from a small town near Salerno, he valued the ascent that came through work, talent, and above all education.[176] We should read Zanotti's portrait of Neapolitan science in light of Genovesi's *Discorso sopra il vero fine delle lettere e delle scienze* (1753), a manifesto of Neapolitan Enlightenment emerging from this same moment yet branching in another direction. Genovesi argued that human elasticity was what distinguished us from inanimate bodies because the mind was much more than a product of brute force. He felt that "the most useful sciences" should be made accessible to the common people (*il basso popolo*) for the good of the Regno and all its inhabitants. He envisioned a pedagogy of "infinite utility" educating and enlightening all rather than some.[177] Genovesi considered it imperative that those who already had this education share it with others who could not arrive there on their own.

It is tempting to think that Genovesi made these arguments in Pignatelli's salon which he certainly frequented. The final words of his second autobiography, written in 1755–56, recalled "the princess of Colubrano, a most intelligent lady in mathematics and philosophy who was very kind to me."[178] Other aristocratic women at court invited Genovesi to tutor them, the way Di Martino had done with her. Despite his avowed dislike of courtiers, he did not always refuse their requests. Genovesi dedicated his anonymous *Lettere accademiche su la questione se sieno più felici gl'ignoranti che gli scienziati* (1764) to Giustiniana Pignatelli, from another branch of this noble family, complimenting her on her excellent education. "I know a Frenchman who made amorous love to peasant women after reading Fontenelle's *Plurality of Worlds*," he joked, observing that the greatest passion in life was not necessarily a love of learning.[179]

Genovesi certainly knew and appreciated the wit and erudition of Pignatelli's world. Yet he also understood how to put it into perspective. In September 1764, the year of a great famine that devasted the Regno, he argued that the Neapolitan barons and the educated urban classes had a duty to look

175 Genovesi, *Autobiografia*, pp. 15, 38.

176 Zambelli, *La formazione filosofica*; Giuseppe Cacciatore, Salvatore Cicenia, ed., *Antonio Genovesi a trecento anni della nascita* (Battipaglia: Laveglia & Carlone, 2016); Ferrone, *Intellectual Roots*, pp. 248–262; and Galasso, *La filosofia in soccorso de' governi: la cultura napolitana del '700* (Naples: Guida, 1989); also Robertson, "Antonio Genovesi: The Neapolitan Enlightenment and Political Economy," *History of Political Thought*, 8 (1987): 335–344.

177 Genovesi, *Opuscoli*, pp. 74–77 (quotes on p. 77)

178 Genovesi, *Autobiografia*, p. 42.

179 Genovesi, *Autobiografia*, p. 393.

after the "poor country people." Contrasting Naples unfavorably to Tuscany and England, he observed that those in power had a moral obligation "to help and enlighten the peasants in all that they can."[180] Whether he offered his critique of feudal privilege in Pignatelli's salon is anyone's guess. In the end, she remained a woman of the Ancien Regime.

In the last year or two of her life, Pignatelli engaged in an intense correspondence with the Bologna Academy through her friend Zanotti, who was now the president. She shared news of herbal remedies from Surinam said to alleviate "Muscovite fever." In 1768, she sent a copy of Sebastiano Bartoli's *Thermologia aragonia sive historia naturalis thermarum* (1679), which Zanotti donated to the library of the Institute for Sciences. When the talented naturalist Domenico Cirillo (1739–99) came to Bologna, "she gave him a very valuable piece of sulfur found in the Solfatura of Naples, entiredly speckled with stones that seemed like rubies, to donate to the Institute."[181] Sebastiano Canterzani, who replaced Zanotti as the academy secretary, wrote Pignatelli in May 1769 to thank her for the gift of this rare Neapolitan book on the thermal springs of Campania which he officially presented to the academicians. "Not content to honor the academy catalogue with your most distinguished and respectable name, you also deign to make space for it among your noblest thoughts and signal this by also making the academy the object of one of your gifts."[182] Even in Bologna, no one ever forgot who she was.

Pignatelli was evidently pleased with this letter, responding almost immediately. Mentioning that Zanotti had spoken of him in their correspondence, she thanked Canterzani – indeed the entire Bologna Academy of Sciences – for the gift they had sent in return. It was surely the history of the Institute for Sciences, containing the list of its academicians. On this occasion, Pignatelli reflected on the arc of her long career, including what she had not been able to do given her circumstances:

> The honor of finding my name inscribed in your catalogue is entirely mine. I only regret that the sequence of events in my life have not allowed me to be able to travel and enjoy it personally. Although, in that case, I should have had to rely upon all the kindness of the professors to excuse my inadequacy.[183]

180 Genovesi, *Autobiografia*, p. 174; Rosario Villari, "Antonio Genovesi e la ricerca delle forze motrici dello sviluppo sociale," *Studi storici*, 11 (1970): 26–52, esp. 41–42.

181 BUB, ms. 4150, fasc. 25.

182 AASB, *Antica Accademia*, Missive Canterzani, n. 1 (10 May 1769).

183 AAAS, *Antica Accademia*, Titolo III, Sez. I, fasc. 4, Lettere ricevute (30 May 1769). For an example of this kind of printed list, see BCAB, *Bassi-Veratti*, series 6, box 7, folder 7,

In an ironic twist of fate, her long dead husband Francesco had been to Bologna, traveling from one battlefield to another. She instead had never had the opportunity to see this city of learning, yet she nonetheless belonged to its community of women scientists.[184] The list had reminded her of this important fact which she recalled with great pleasure.

Pignatelli's story should end here but there was a final coda. In July 1769, responding to a letter from Zanotti requesting her advice on how to approach Tanucci, Pignatelli indicated that she was fully in command of court protocol as well as the politics of science in Bourbon Naples. She mentioned philosophical discussions at court among the young princes, now the offspring of Ferdinand IV, about the nature of mortality and divinity. Pignatelli was having none of it. "I, who am accustomed to consider objects in their geometric simplicity stop there, and I do not embellish them with other ideas, save for the obligation to listen and thus work for the straight path of justice."[185] Whatever elicited this comment, it is apparent that her mind was as sharp and clear as ever.

In fall, Pignatelli's health took a turn for the worse. She went to Portici to seek a cure. One observer remarked disapprovingly, "The princess of Colubrano is here to recuperate and those who court her are the people of this town, uncultivated young men who have nothing to say. She takes them about in her carriage and debates with them and plays cards in the evening. What an unhappy bunch!"[186] Perhaps there was a bit of Genovesi in her view of the world, after all. Both of them died in 1769, Pignatelli on December 30. Genovesi would be remembered, his sometime patron Pignatelli soon forgotten.

Matteo Barbieri's 1778 history of mathematicians and philosophers in the Kingdom of Naples does not mention Pignatelli by name, or refer to any women of science.[187] Instead, it was a foreigner, the French astronomer Jérôme de Lalande who composed Pignatelli's epitaph in his bestselling travel book. After seeing Naples in 1765, he wrote: "The princess of Colubrano is another lady distinguished as much for her learning as her birth, and she corresponds with

Catalogus academicorum Bononiensis Scientiariam Instituti (1747). Pignatelli is described in it as "the most illustrious Princess of Colubrano."

184 Fraggianni, *Lettere*, p. 337 (16 June 1742): "Egli è ora in Bologna."

185 BUB, ms. 4174, caps. XLVIII, n. 12 (17 July 1749). The final section of this letter is slightly unclear and might need a word or two corrected from this transcription, but I cannot consult the original during COVID.

186 Franco Strazzullo, *Carteggi eruditi del Settecento* (Naples: Edizioni Fondazione Pasquale Corsicato, 1993), p. 339 (Giacomo Martorelli to Francesco Vargas Macciucca, [Portici], 24 September [1769?]). The editor dates it 1770, but notes (p. 337) that it must be between 1768 and 1770. Given the date of her death, I have chosen 1769.

187 Matteo Barbieri, *Notizie istoriche del mattematici e filosofi del Regno di Napoli* (Naples, 1778).

many learned men in Europe."[188] During the 1770s and 1780s, her son Michele sold most of the antiquities on the open market, but this could not halt the family's inevitable decline into genteel poverty. With the abolition of feudalism, they lost everything except their titles; around 1809 her grandson finally sold the palace.[189] Pignatelli's library, papers, and her considerable correspondence disappeared in the process. The woman who captivated her admirers in Bologna and beyond, for her ability to have a good scientific conversation, had simply and utterly vanished.

Acknowledgements

Thanks to Paolo Savoia, Marco Beretta, and Antonella Romano for inviting me to participate in this workshop, Paola Govoni for her comments, Paola Bertucci for sharing my love of the Neapolitan scientific women of this era, Marta Cavazza for always being ready to talk about the eighteenth-century women of science, and Mary Terrall for asking me many years ago why Voltaire was interested in Pignatelli's views on the *vis viva* debates. Graylin Harrison photographed the relevant documents from the Caraffa di Maddaloni archive while doing her own research in Naples. Elena Papagna generously shared a longer version of her article on Pignatelli, allowing me to further incorporate her insights and research.

References

Addabbo, Claudia, "The *Philosophiae Naturalis Principia Mathematica* in Naples," in Elizabethanne Born and Mordechai Feingold, ed., *Reading Newton in Early Modern Europe* (Leiden: Brill, 2017), pp. 23–63.

Ademollo, Alessandro, *Corilla Olimpica* (Florence: C. Ademollo, 1887).

Agnesi, Maria Gaetana, et. al., *The Contest for Knowledge: Debates over Women's Learning in Eighteenth-Century Italy* (Chicago: University of Chicago Press, 2005).

Ajello, Raffaele, "Cartesianismo e la cultura oltremontana al temp dell' 'Istoria civile,'" in Raffaele Ajello, ed., *Pietro Giannone e il suo tempo* (Naples: Jovene, 1980), pp. 102–144.

Amodeo, Federico, *Vita matematica napolitana. Parte prima* (Naples: Giannini & Figli, 1905).

188 Jérôme de Lalande, *Voyage d'un François en Italie, fait dans les années 1765 & 1766* (Yverdon, 1769), vol. 6, p. 238.

189 Fusco, "I Carafa della Stradera," pp. 171–172; Ceci, "Il Palazzo Caraffa," p. 170.

Angelini, Annarita, ed., *Anatomie accademiche*. Vol. 3. *L'Istituto delle Scienze e l'Accademia* (Bologna: Il Mulino, 1993).

Ariani, Vincenzo, ed., *Memoria della vita, e degli scritti di Agostino Ariani* (Naples, 1778).

Astarita, Tommaso, *The Continuity of Feudal Power: The Carraciolo di Brienza* (Cambridge, U.K.: Cambridge University Press, 1992).

Bandiera, Giovanni Niccolò, *Trattato degli studi delle donne* (Venice, 1740).

Barbieri, Matteo, *Notizie istoriche del mattematici e filosofi del Regno di Napoli* (Naples, 1778).

Bergia, Silvio, Paolo Fantazzini, "Dalla regolamentazione delle acque alle leggi dell'urto centrale elastica: una rivalutazione del ruolo di Giovanni Poleni nella disputa delle forze vive," *Giornale di fisico*, 21 (January–March 1980): 46–60.

Bertucci, Paola, "The In/visible Woman: Mariangela Ardinghelli and the Circulation of Knowledge between Paris and Naples in the Eighteenth Century," *Isis*, 104 (2013): 226–249.

Bertucci, Paola, *Viaggio nel paese delle meraviglie. Scienza e curiosita nell'Italia del Settecento* (Turin: Bollato Boringhieri, 2007).

Besterman, Theodore, ed., *Correspondance de Voltaire* (Geneva, 1968–2001), vol. 91.

Biagioli, Mario, Peter Galison, ed., *Scientific Authorship: Credit and Intellectual Property in Science* (New York: Routledge, 2003).

Borbone, Carlo di, *Lettere ai sovrani di Spagna*. Vol. 2. *1735–1739* (Rome: Ministero per i beni e le attività culturali, Direzione generale per gli archivi, 2002).

Brigagli, Aldo, Pietro Nastasi, "Bologna e il Regno delle Due Sicilie: aspetti di un dialogo scientifico (1730–1760)," *Giornale critico della filosofia italiana*, 63 (1984): 145–178.

Burke, Peter, *The Fortunes of the Courtier: The European Reception of Castiglione's Cortegiano* (University Park, PA: The Pennsylvania State University Press, 1996).

Cacciatore, Giuseppe, Salvatore Cicenia, ed., *Antonio Genovesi a trecento anni della nascita* (Battipaglia: Laveglia & Carlone, 2016).

Calaresu, Melissa, "From the Street to the Stereotype: Urban Space, Travel, and the Picturesque in Eighteenth-Century Naples," *Italian Studies*, 62 (2007): 189–203.

Calaresu, Melissa, Helen Hills, ed., *New Approaches to Naples c.1500–c.1800: The Power of Place* (Farnham, Surrey: Ashgate, 2013).

Carafa, Francesco Domenico, *Rime varie di Francesco Carafa principe di Colobrano* (Florence, 1730).

Casini, Paolo, *Newton e la coscienza europa* (Bologna: Il Mulino, 1983).

Castiglione, Caroline, *Accounting for Affection: Motherhood and Politics in Early Modern Rome* (Houndmills, Basingstoke, Hampshire: Palgrave, 2015).

Catucci, Marco, "Maria Maddalena Morelli," DBI (Rome: Enciclopedia Italiana, 1960–), vol. 76.

Cavazza, Marta, "Dalle biblioteche dei dotti alle tolette delle dame. La conversazione filosofica e scientifica nell'Italia dei lumi," *I Castelli di Yale*, 12 (2012): 87–102.

Cavazza, Marta, "Les femmes à l'Académie: le cas de Bologne," in Daniel-Odon Hurel and Gérard Laudin, ed., *Académies et sociétés savants en Europe (1650–1800)* (Paris: Honoré Champion, 2000), pp. 161–175.

Cavazza, Marta, *Laura Bassi. Donne, genere e scienza nell'Italia del Settecento* ((Milan: Editrice Bibliografica, 2020).

Ceci, Guido, "Il palazzo dei Carafa dei Maddaloni poi di Colubrano," *Napoli nobilissima*, 2 (1893): 148–152.

Celano, Carlo, *Delle notizie del bello, dell'antico e del curioso della città di Napoli* (Naples, 1692).

Ceranski, Beate, *"Und sie fürchtet sich vor niemandem": Die Physikerin Laura Bassi (1711–1778)* (Frankfurt: Campus Verlag, 1996).

Châtelet, Émilie du, *Instituzioni di fisica ... accresciuta con la Dissertazione sopra le forze motrici di M. de Mairan* (Venice, 1743).

Châtelet, Émilie du, *Selected Philosophical and Scientific Writings*, ed. Judith P. Zinsser, tr. Isabelle Bour and Zinsser (Chicago: University of Chicago Press, 2009).

Clairaut, Alexis-Claude, *Elémens de algèbra* (Paris, 1746).

Clairaut, Alexis-Claude, *Elémens de géométrie* (Paris, 1741).

Conti, Vittori, *Paolo Mattia Doria. Dalla repubblica dei togati alla repubblica dei notabili* (Florence: Olschki, 1978).

Costabel, Pierre, *La question des forces vives. La signification d'un débat sur trente ans (1728–1758)* (Paris: CNRS, 1983).

Cremante, Renzo, Walter Tega, ed., *Scienza e letteratura nella cultura italiana del Settecento*, ed. (Bologna: Il Mulino, 1984).

Cromaziano, Agatopisto [Appiano Buonafede], *Della malignità istorica discorsi tre* (Basel, 1738).

De Brosses, Charles, *Lettres d'Italie*, ed. Frédéric d'Agay (Paris: Mercure d France, 1986), 2 vols.

De Divitiis, Bianca, "New Evidence for Sculptures from Diomede Carafa's Collection of Antiquities," *Journal of the Warburg and Courtauld Institutes*, 70 (2007): 99–117.

De Martino, Niccolò, *Algebrae geometria promotae elementa conscripta ad usum Faustinae Pignatelli* (Naples, 1737).

De Martino, Niccolò, *Elementa sectionum conicarum* (Naples, 1734).

De Martino, Niccolò, *Nuovi elementi della teoria delle mine*, ed. Giuseppe De Martino (Naples, 1780).

De Sanctis, Riccardo, *La nuova scienza a Napoli tra '700 e '800* (Rome: Laterza, 1986).

De Zan, Mauro, "L'Accademia delle scienze di Bologna: l'edizione del primo tomo dei Commentarii (1731)", in Maria Vittoria Predaval Magrini, ed., *Scienza, filosofia e religione tra '600 e '700 in Italia* (Milan: FrancoAngeli, 1990), pp. 203–259.

Di Martino, Pietro, *De corporum quae moventur viribus earumque aestimandarum ratione. Dissertatio philosophica ad Instituti Bononiensis Socios inscripta* (Naples, 1741).

Dodero, Eloisa, "Le antichità di Palazzo Carafa-Colubrano: prodromi alla storia della collezione," *Napoli nobilissima* 5, (2007): 119–140.

Dodero, Eloisa, *Ancient Marbles in Naples in the Eighteenth Century: Findings, Collections, Dispersals* (Leiden: Brill, 2019).

Doria, Paolo Mattia, *Ragionamenti di Paolo Mattia Doria indirizzati alla Signora d'Aurelia d'Este Duchessa di Limatola ne' quali si dimostra la donna, in quasi che tutte le virtù più grandi, non essere all'uomo inferiore* (Frankfurt, 1716).

Favino, Federica, "Minimi in 'Sapienza.' François Jacquier, Thomas Le Seur e il rinnovamento dell'insegnamento scientifico in Studium Urbis," *Mélanges de l'École française de Rome. Italie et Méditeranée*, 117, 1 (2005): 159–187.

Feo, Michele, "Il 'canzoniere' di Corilla," in Moreno Fabbri, ed., *Corilla Olimpica e la poesia del Settecento* (Pistoia: Maschietto, 2002).

Ferraro, Joanne M., *Marriage Wars in Late Renaissance Venice* (Oxford: Oxford University Press, 2001).

Ferrone, Vincenzo, *The Intellectual Roots of the Italian Enlightenment: Newtonian Science, Religion, and Politics in the Early Eighteenth Century*, tr. Sue Bretherton (Atlantic Highlands, NJ: Humanities Press, 1995).

Findlen, Paula, "Tra uomini. Laura Bassi all'Istituto delle Scienze (1732–78)"/"Always among Men: Laura Bassi at the Bologna Academy of Sciences (1732–78)," in Luisa Cifarelli and Raffaella Simili, ed., *Laura Bassi. Emblema e primato nella scienza del Settecento* (Bologna: Società Italiana di Fisica, 2012), pp. 71–88 (English pp. 189–206).

Findlen, Paula, "Translating the New Science: Women and the Circulation of Knowledge," *Configurations*, 2 (1995): 167–206.

Findlen, Paula, "Women on the Verge of Science: Aristocratic Women and Science in Early Eighteenth-Century Italy," in Sarah Knott, Barbara Taylor, ed., *Women, Gender and Enlightenment* (London: Palgrave Press, 2005), 265–287.

Fraggianni, Niccolò, *Lettere a B. Corsini (1739–1746)*, ed. Elia del Curatolo (Naples: Jovene, 1991).

Fusco, Gaetano, "I Carafa della Stradera nella relazioni storiche tra Terra di Lavoro e la Regione Lucana," in Pietro Borraro, ed., *Studi Lucani. Atti del II Convegno Nazionale di Storiografica Lucana* (Galatina: Congedo, 1976).

Galasso, Giuseppe, *L'Età dei lumi: studi storici sul Settecento europeo in onore di Franco Venturi* (Naples, 1985).

Gatto, Romano, *Libri di matematica a Napoli nel Settecento. Editoria, fortuna e diffusione delle opera* (Rome: Edizioni di storia e letteratura, 2010).

Genovesi, Antonio, *Autobiografia, lettere e altri scritti*, ed. Gennaro Savarese (Milan: Feltrinelli, 1962).

Giornale de' letterati d'Italia 32 (1719).

Giuli, Paola, "Tracing a Sisterhood: Corilla Olimpica as Corinne's Unacknowledged Alter Ego," in Karyna Szmurlo, ed., *The Novel's Seductions: Staël's Corinne in Critical Inquiry* (Lewiston, PA: Bucknell University Press, 1999), pp. 165–184.

Goldsmith, Elizabeth, Dena Goodman, eds., *Going Public: Women and Publishing in Early Modern France* (Ithaca: Cornell University Press, 1995).

Gravina, Gian Vincenzo, *Della tragedia* (Naples, 1731).

Guerci, Luciano, *La discussione sulla donna nell'Italia del Settecento* (Turin: Tirrenia, 1987).

Hankins, Thomas L., "Eighteenth-Century Attempts to Resolve the Vis Viva Controversy," *Isis*, 56 (1965): 281–297.

Illibato, Antonio, *La donna a Napoli nel Settecento. Aspetti della condizione e dell'istruzione femminile* (Naples: M. D'Auria, 1985).

Iltis [Merchant], Carolyn, "D'Alembert and the *Vis* Viva Controversy," *Studies in History and Philosophy of Science*, 1 (1970): 139–159.

Iltis [Merchant], Carolyn, "Leibniz and the *Vis Viva* Controversy," *Isis*, 62 (1971): 21–35.

Imbruglia, Girolamo, ed., *Naples in the Eighteenth Century: The Birth and Death of a Nation State* (Cambridge, U.K.: Cambridge University Press, 2000).

Kammerer-Grothaus, Helke, "'Voyage d'Italie' (1755). Markgrafin Wilhelmine von Bayreuth im Königreich Neapel," in Helke Kammer-Grothaus, Detlve Kreikenbom, ed., *Wilhelmine und Friedrich II und die Antiken* (Tübingen: Winckelmann-Gesellschaft Stendal, 1998).

Kawashima, Keiko, "La querelle des 'forces vives.' Le débat entre Mme Du Châtelet et Dortous de Mairan," in Elisabeth Badinter, Danielle Muzerelle, ed., *Madame Du Châtelet. La femme des lumières* (Paris: Bibliothèque Nationale de France, 2006).

Laeven, Augustinus Hubertus, Lucia Johanna Maria Laeven-Aretz, *The Authors and Reviewers of the Acta Eruditorum 1682–1735* (Göttingen: Publickationsserver der Georg-August-Universität Göttingen, 2014).

Lalande, Jérôme de, *Voyage d'un François en Italie, fait dans les années 1765 & 1766* (Yverdon, 1769).

Laudan, Larry, "The *Vis Viva* Controversy, a Post-Mortem," *Isis*, 59 (1968): 130–143.

Leibniz, Gottfried Wilhelm, *Philosophical Papers and Letters*, ed. Leroy. E. Loemker (Chicago: University of Chicago Press, 1956).

Maugain, Gabriel, *Étude sur l'évolution intellectuelle de l'Italie de 1657 à 1750* (Paris: Librairie Hachette, 1909).

Mazzei, Raffaele, "Relazioni culturali fra Bologna e Napoli nel secolo XVIII," *Il Carobbio*, 3 (1977): 285–297.

Mazzotti, Massimo, "Geometers of God: Mathematics and Reaction in the Kingdom of Naples," *Isis*, 89 (1998): 674–701.

Mazzotti, Massimo, "Mme Du Châtelet, académicienne de Boulogne," in Olivier Courcelle, Ulla Kölving, ed., *Émilie Du Châtelet, éclairages et documents nouveaux* (Ferney-Voltaire: Centre international d'étude du XVIIIe siècle, 2008), pp. 121–126.

Messbarger, Rebecca, *The Century of Women: Representations of Women in Eighteenth-Century Italian Public Discourse* (Toronto: University of Toronto Press, 2002).

Minieri Riccio, Camillo, "Notizie delle Accademie institute nelle province napoletane," *Archivio per le province napoletane* 3, 1 (1878): 145–146.

Napoli-Signorelli, Pietro, *Vicende della coltura nelle Due Sicilie* (Naples, 1786).

Nastasi, Pietro, "Pietro De Martino (Di Martino), DBI (Rome: Enciclopedia Italiana, 1960–), vol. 38, 603–606.

Nicolini, Fausto, *Un grande educatore italiano. Celestino Galiani* (Naples: Giannini, 1951).

Novi Chavarria, Elisa, "Forme e spazi della sociabilità aristocratica nel Settecento," in Roberto Bizzocchi, Arturo Pacini, ed., *Sociabilità aristocratica in età moderna. Il caso genovese: paradigmi, interpretazioni e confronti* (Pisa: Pisa University Press, 2008), 73–86.

Novi Chavarria, Elisa, "The Space of Women," in Tommaso Astarita, ed., *A Companion to Early Modern Naples*, (Leiden: Brill, 2013), 177–196.

Origlia, Paolino, Giangiuseppe, *Istoria dello Studio di Napoli* (Naples, 1754).

Palladino, Nicla, Anna Maria Mercurio, Franco Palladino, ed., *La corrispondenza epistolare tra Niccolò Di Martino e Girolamo Settimo* (Florence: Olschki, 2008).

Papagna, Elena, "Faustina Pignatelli," DBI (Rome: Enciclopedia Italiana, 1960–), vol. 83.

Papagna, Elena, *La corte di Carlo di Borbone il re "proprio e nazionale"* (Naples: Guida, 2011).

Pemberton, Henry, "A Letter to Dr. Mead ... concerning an Experiment, whereby it has been attempted to shew the falsity of the common Opinion in relation to the force of Bodies in Motion," *Philosophical Transactions of the Royal Society*, 32, 371 (1722–23): 57–68.

Piaia, Gregorio, Maria Laura Soppelsa, ed., *I Riccati e la cultura della Marca nel Settecento Europe* (Florence: Olschki, 1992).

Pignatelli, Faustina, "Problemata mathematica Neapoli ad Collectores Actorum Eruditorum transmissa," *Nova Acta Eruditorum* 1 (January 1734): 28–34.

Rao, Anna Maria, ed., *Editoria e cultura a Napoli nel XVIII secolo* (Naples: Liguori, 1998).

Rao, Annamaria, *Il Regno di Napoli nel Settecento* (Naples: Guida, 1983).

Robertson, John, "Antonio Genovesi: The Neapolitan Enlightenment and Political Economy," *History of Political Thought*, 8 (1987): 335–344.

Robertson, John, *The Case for Enlightenment: Scotland and Naples, 1680–1750* (Cambridge, U.K.: Cambridge University Press, 2005).

Sanna, Manuela, "Un'amicizia alla luce del Cartesianesimo: Giuseppa Eleonora Barbapiccola e Luisa Vico," in Pina Totaro, ed., *Donne, filosofia e cultura nel Seicento* (Rome: Consiglio Nazionale delle Ricerche, 1999), pp. 173–178.

Seidel Menchi, Silvana, Diego Quaglioni, ed., *Coniugi nemici. La separazione in Italia dal XII al XVIII secolo* (Bologna: Il Mulino, 2000).

Serrapica, Salvatore, *Scienza e filosofia nella Napoli settecentesca. I fratelli Di Martino* (2011).

Simili, Raffaella, "In punta di penna. Donne di scienza e di cultura fra cosmopolitanismo e intimità meridionale," in *La scienza del Mezzogiorno dopo l'Unità d'Italia*, Antonio Di Meo ed. (Soveria Manelli: Rubbettino, 2008), pp. 27–89.

Soppelsa, Maria Laura, *Leibniz e Newton in Italia: Il dibattito padovano (1687–1750)* (Padua: Edizioni Lint, 1989).

Stone, Harold Samuel, *Vico's Cultural History: The Production and Transmission of Ideas in Naples 1685–1750* (Leiden: Brill, 1997).

Strazzullo, Franco, *Carteggi eruditi del Settecento* (Naples: Edizioni Fondazione Pasquale Corsicato, 1993).

Sweet, Rosemary, *Cities and the Grand Tour: The British in Italy, c.1690–1820* (Cambridge, U.K.: Cambridge University Press, 2012).

Tanucci, Bernardo, *Epistolario*. Vol. 1. *1723–1746*, ed. Romano Paolo Coppini, Lamberto Del Bianco, Rolando Nieri, and Mario D'Addio (Rome: Edizione di Storia e Letteratura, 1980).

Tega, Walter, ed., *Anatomie accademiche*. Vol. 2. *L'Enciclopedia scientifica dell'Accademia delle Scienze di Bologna* (Bologna: Il Mulino, 1987).

Tega, Walter, ed., *Anatomie accademiche*. Vol. 1. *Commentari dell'Accademia delle Scienze di Bologna* (Bologna: Il Mulino, 1986).

Terrall, Mary, "Vis Viva Revisited." *History of Science*, 42 (2004): 192–199.

Tirrenio, Damone [Orazio Pacifico], "Ai lettori," *Il Caprario: accademia di diversi rimatori, che nel medesimo monte si radurano. Parte seconda* (Florence, 1732).

Villari, Rosario, "Antonio Genovesi e la ricerca delle forze motrici dello sviluppo sociale," *Studi* storici, 11 (1970): 26–52.

Visceglia, Maria Antonietta, *Il bisogno di eternità. I comportamenti aristocratici a Napoli in età moderna* (Naples: Guida, 1988).

Voltaire, *Exposition du livre des Institutions physiques* (1740), in *Oeuvres complètes de Voltaire* (Paris: Garnier, 1877–85), vol. 23.

Zambelli, Paola, *La formazione filosofica di Antonio Genovesi* (Naples: A. Morano, 1972).

Zanotti, Francesco Maria, *Discorsi e orazioni* (Venice, 1830).

Zanotti, Francesco Maria, *Della forza de' corpi che chiamano viva* (Bologna, 1752).

Zanotti, Francesco Maria, *Della forza attrattiva delle idee* (Naples, 1747).

Zanotti, Francesco Maria, "De vi corporum viva," *De Bononiensi scientiarum et atrium Instituto atque Accademia Commentarii*, vol. 2 (Bologna, 1745).

Zinsser, Judith, "Mentors, the Marquise du Châtelet and Historical Memory," *Notes and Records of the Royal Society*, 61 (2007): 94–97.

Zinsser, Judith, *La Dame d'Esprit: A Biography of the Marquise Du Châtelet* (New York: Viking, 2006).

CHAPTER 7

Note-taking and Self-promotion: Marie-Anne Paulze-Lavoisier as a *Secrétaire* (1772–1792)

Francesca Antonelli

On August 30, 1788 Louis-Bernard Guyton de Morveau wrote to his friend and colleague Antoine-Laurent Lavoisier (1743–1794), asking for some news about his latest projects.[1] The letter began with a mention of Lavoisier's trip to the French provinces, where he had been followed by his wife, Marie-Anne Paulze-Lavoisier (1758–1836). On that occasion, Guyton de Morveau noted, she probably acted as a "secrétaire":

> J'imagine que vous êtes dejà depuis quelques jours de retour du voyage que vous projetiez avec Madame Lavoisier, et où vous n'aurez pas manqué de recueillir à votre ordinaire de bonnes observations redigées sur place par un *secretaire* aussi aimable qu'intelligent.[2]

When Guyton de Morveau was writing this letter, Paulze-Lavoisier was already known as one of Lavoisier's most active associates. That summer, for instance, she had completed the French translation of Richard Kirwan's *Essay on Phlogiston* (1788), a collective and ambitious project involving both Guyton and other *savants*.[3] About at the same time, she begun working on the illustrations for Lavoisier's *Traité élémentaire de chimie* (1789), which resulted in their

1 I warmly thank all the participants to the conference and especially Stéphane Van Damme, who made an insightful commentary of this paper. I owe a great deal also to Marco Beretta, Antonella Romano, and Silvia Sebastiani, with whom I have been having many fruitful discussions in the last few years. I am grateful, in addition, to the friends and colleagues of the seminar *La curiosité des Lumières* (held by Antoine Lilti and Silvia Sebastiani at the EHESS, Paris) who kindly accepted to discuss with me an earlier version of this manuscript. I finally thank Paolo Savoia for his attentive readings and Christine Lehman for helping me explore the Fonds Lavoisier at the Archives de l'Académie des Sciences in Paris.

2 Louis-Bernard Guyton de Morveau to Antoine-Laurent Lavoisier, 30 August 1788, pub. in Antoine-Laurent Lavoisier, *Correspondance*, vol. 5: 1787–1788 (Paris: Académie des Sciences, 1993), pp. 206–207: 206. My emphasis.

3 On this project, see Keiko Kawashima, "Madame Lavoisier et la traduction de l'*Essay on Phlogiston* de Kirwan," *Revue d'histoire des sciences*, 53 (2000): 235–263.

 | DOI:10.1163/9789004512610_009

final version in thirteen engraved plates.[4] These achievements were celebrated among others by Jacques-Louis David who, in 1788, realized the famous a double-portrait in which the Lavoisiers are represented as a collaborating couple.[5]

However, referring to Paulze-Lavoisier as *secrétaire*, he had chosen a term which, in late eighteenth-century French, had a quite specific meaning. According to the *Encyclopédie*, it designated a high-ranking official who wrote letters and formal documents on behalf of kings, princes, and ecclesiastic authorities. Although in some cases it could also describe a scribe and a servant, within the society of the Ancien Régime its main sense was that of a crucial political and cultural agent.[6] Furthermore, a secondary but still important meaning connected the term to a specific occupation within early-modern academies which, since their foundation, used to promote a secretary at the top of their rankings and charge him with the important tasks of collecting the *mémoires* delivered by the members and of representing the instutition by writing its history.[7] As it is well known, no woman in eighteen-century France was admitted to any of these positions.[8] It is therefore obvious that in calling Paulze-Lavoisier a *secrétaire*, Guyton de Morveau was making a somewhat metaphorical use of the term.

There was, however, something true in his words. The link between Paulze-Lavoisier's role as a scientific associate and the *secrétaire* in its most traditional sense lies in the specific act of writing in which they both engaged.

4 On Paulze-Lavoisier's work as an illustrator, see Madeleine Pinault Sørensen, "Madame Lavoisier, dessinatrice et peintre," *La Revue. Musée des Arts et Métiers*, 6 (1994): 23–25.

5 For an insightful analysis of this portrait, see Mary Vidal, "David Among the Moderns: Art, Science and the Lavoisiers", *Journal of the History of Ideas*, 56 (1995): 592–623. A recent and compelling account of the Lavoisiers as a collaborating couple can be found in Meghan Roberts, *Sentimental Savants. Philosophical Families in Enlightenment France* (Chicago: University of Chicago Press, 2016), esp. chap. 2: "Working Together". See also Keiko Kawashima, *Émilie du Châtelet et Marie Anne Lavoisier. Science et genre au XVIIIe siècle* (Paris: Honoré Champion, 2013).

6 Boucher d'Argis, "Secrétaire", in Denis Diderot, Jean Bapiste Le Rond d'Alembert, ed., *Encyclopédie, ou dictionnaire raisonné des sciences, des arts et des métiers, par une société de gens de lettres*, vol. 14 (Paris: 1765), pp. 863–871: http://enccre.academie-sciences.fr/encyclopedie/article/v14-3042-0/. For a social and political history of this charge in early-modern France, see Nicolas Schapira, *Maîtres et secrétaires (XVI–XVIIIe siècles). L'exercice du pouvoir dans la France d'Ancien* Régime (Paris: Albin Michel, 2020); for a "global" perspective, see Paul M. Dover, ed., *Secretaries and Statecraft in the Early Modern World* (Edinburgh: Edinburgh University Press, 2016).

7 On the case of the Académie des Sciences and its *secrétaire perpetuel*, see Roger Hahn, *The Anatomy of a Scientific Institution. The Paris Academy of Science 1666–1803* (Berkley: University of California Press, 1971), esp. pp. 78–79.

8 On this topic, see the classic Londa Schiebinger, *The Mind Has No Sex? Women in the Origins of Modern Science* (Cambridge/London: Harvard University Press, 1991).

Paulze-Lavoisier's trajectory in science certainly do not correspond to that of a "professional" writer and differs radically from other cases of eighteenth-century women authors, such as, just to name one, Marie-Geneviève Thiroux d'Arconville, today known for her prolific scientific and literary production.[9] And yet, throughout his collaboration with her husband, Paulze-Lavoisier wrote hundreds of pages, not so much in view of publishing her own texts, but rather to record informations during experimental sessions, travels, and observations of various kinds, compiling a significant amount of what are today considered as Lavoisier's personal and scientific papers. In fact, soon after her marriage in 1771 until Lavoisier's death in 1794 but also, to a lesser extent, during the first decade of the nineteenth century, she engaged in what recent anglo-american historiography calls "note-taking practices", that is a wide set of scribal practices aiming at storing of information for further use.[10] It is to this particular kind of writing that Guyton de Morveau was referring to in the above mentioned letter; in that specific case, it produced one of Lavoisier's *Carnets de voyage*, which clearly bears Paulze-Lavoisier's handwriting.[11]

It has to be noted that since the early 1760s, when he was travelling throughout France with one of his mentor, the naturalist Jean-Étienne Guettard, Lavoisier himself used to compile diaries, where he mostly recorded observations on the physical (and sometimes social) features of the lands he explored.[12] This

9 Patrice Bret, Brigitte van Tiggelen, ed., *Madame d'Arconville. Une femme de lettres et de sciences au siècle des Lumières* (Paris: Hermann, 2011).

10 The literature on note-taking practices is now quite vast. Here I will just mention the analysis that significantly influenced my view on the topic: José Beltrán, "Ciencia amanuense: cultura manuscrita e historia natural en la Francia moderna (*c.*1660–1830)", *Asclepio. Revista de Historia de la Medicina y de la Ciencia*, 71 1 (2019): http://asclepio.revistas.csic.es/index.php/asclepio/article/view/814; Ann M. Blair, *Too Much to Know. Managing Scholarly Information before the Modern Age* (New Haven and London: Yale University Press, 2010); Marie-Noëlle Bourguet, *Le monde dans un carnet. Alexander von Humboldt en Italie (1805)* (Paris: Éditions du Félin, 2017); Ann M. Blair, Richard Yeo, ed., *Note-taking in Early Modern Europe*, special issue of *Intellectual History Journal*, 20, 3 (2010), esp. the contribution by Bourguet, "A Portable World: The Notebooks of European Travellers (Eighteenth to Nineteenth Centuries)", 377–400; Élizabeth Décultot, ed., *Lire, copier, écrire. Les bibliothèques manuscrites et leurs usages au XVIIIe siècle* (Paris: CNRS Éditions, 2003); Christian Jacob, ed., *Lieux de savoir*, vol. 2: *Les mains de l'intellect* (Paris: Albin Michel, 2011), section "Dynamiques de la notation", pp. 283–353. On Paulze-Lavoisier's trajectories as a note-taker, before and after her husband's death, see my PhD dissertation: Francesca Antonelli, *Scrittura, sociabilità e strategie di persuasione: Marie-Anne Paulze-Lavoisier*, secrétaire *(1758–1836)*, Università di Bologna/Écoles des Hautes Études en Sciences Sociales, 2021.

11 Archives de l'Académie des Sciences, Paris, Fonds Lavoisier, *Carnet de voyage n. 17*.

12 There is no literature on Lavoisier's approach to note-taking while travelling; however, the majority of his travel diaries can be consulted on the database *Panopticon Lavoisier*: http://moro.imss.fi.it/lavoisier/.

practice was actually quite common among eighteenth-century naturalists and travellers, who frequenlty relied on note-taking as an instrument to gather data and compensate the weakness of memory. Often taken *à la volée*, these notes could be consulted in different places and after long periods of time; they thus provided their author with a unique representation of what he or she could no longer observe.[13] In this sense, notebooks and diaries can be considered as scientific instruments in their own rights or, as Marie-Noëlle Bourguet has put it, as "instruments de papier".[14] However, as regards to the Lavoisiers, it has to be noted that the existence of around 30 notebooks, relating to various subjects, has so far been neglected and the few scholars who have worked on them have mostly confined themselved to extrapolate isolate information related to specific chemical theories and experiments.[15]

In what follows I will focus on what the Lavoisiers called the *Registres de laboratoire*, namely the 14 small folio laboratory notebooks where they stored observations and data pertaining to their usual experimental pratice. Many things could be said about these elegant and well-crafted notebooks, carefully preserved in Paulze-Lavoisier's library until 1835, when they were donated to her friend François Arago.[16] Throughout the *Registres* we find evidences suggesting, for instance, that they were designed to be consulted more than once and not just by the note-taker. Hence the need for a clear organization of the topics, the introduction of writing tools that help the reader to find the information, such as titles and tables of contents, along with the reproduction of excerpts taken from other sources, material traces of experiments, and other tiny objets connected to everyday life. In all likelihood, these notebooks were some of the instruments used by the Lavoisiers in their daily experimental activity.

It should also be taken into account that while Lavoisier's research had always been pursued in a social context, there were very few people in charge of the management of the *Registres de laboratoire*. In fact, until the mid-80s, these notebooks bear exclusively his and her wife handwritings. In my analysis I will follow these traces, with a double aim. On the one hand, I will try to show that note-taking practices were crucial in Lavoisiers' working *routine*. Their collaboration cannot in fact be reduced to Paulze-Lavoisier's work as a

---

13 Bourguet, "A Portable World".

14 Bourguet, *Le monde dans un carnet*.

15 See esp. Marcellin Berthelot, *La révolution chimique: Lavoisier. Ouvrage suivi de notices et extraits des registres inédits de laboratoire de Lavoisier* (Paris: Félix Alcan, 1890) and Frederic L. Holmes, *Lavoisier and the Chemistry of Life. An Exloration of Scientific Creativity* (Madison: The University of Wisconsin Press, 1983).

16 See the manuscript note sent by Paulze-Lavoisier to François Arago in 1835: Archives de l'Académie des Sciences, Paris, Fonds Lavoisier, dation Chabrol, carton 2, *Note manuscrite d'un envoie de manuscrit par Mad.e de Rumford à M. Arago*.

translator and illustrator of chemical texts, but rested on her role as a *secrétaire* as well. One the other hand, I will explore the ways in which the art of note-taking was used by the two associates as a mean of self-promotion. As we will see, taking notes was not only essential to the experimental practice, which costantly relied on the information that were stored in the *Registres de laboratoire*; it also intertwined with the personal ambitions of both their compilers. As some of Lavoisier's scholars have already noticed, the French chemist and farmer general was an ambitious man, eager to stand out in the scientific community of his time since he was very young.[17] Especially during the 1780s, this attitude materialized in a "campaign of persuasion" aiming at "converting" his rivals to his chemical theories.[18] In my view, Paulze-Lavoisier's work as a note-taker can be considered as an integral part of her husband's tireless search for public attention. But in the process, as I will try to show, she managed to promote her own image as a scientific associate as well.

1 Burning Diamonds at the Jardin de l'Infante

> Le 22 8bre 1773
> 6 diamans dont 5 transparens ~~pesant~~ provenant de l'experience du 20. ~~pes~~ et un noir charbonneux à la surface pesant 3 grains 1/8 on les exposé au soleil un quart d'heure et ils ont été retirés à 29 grains 1/8 c'est ~~un~~ 1 grains juste qu'ils ont perdu ~~et~~ il y en avoit plusieurs qui etoit charbonneux qui qui noircissoit le papier.[19]

At first glance, this excerpt may appear just as one of the many annotations left by Paulze-Lavoisier in the *Registres de laboratoire* at the beginning of the 1770s.

17 This side of Lavoisier's personality is stressed above all in the short biography by Herly Guerlac, "Lavoisier, Antoine-Laurent", in *Dictionary of Scientific Biography*, edited by Charles C. Gillispie, vol. 7 (New York: Charles Scribner's Sons, 1971), pp. 66–91. Lavoisier's efforts to construct his public image have been examined from a different perspective in Roberts, *Sentimental Savants*, esp. pp. 136–165.

18 I borrow the expression "campaign of persuasion" from Carleton E. Perrin, "The Triumph of the Antiphlogistians", in Harry Woolf, ed., *The Analytic Spirit. Essays in the History of Science in Honor of Henry Guerlac* (Ithaca and London: Cornell University Press, 1981), pp. 40–63. On science and persuasion see the classic Steven Shapin, Simon Schaffer, *Leviathan and the Air-Pump. Hobbes, Boyle, and the Experimental Life* (Princeton: Princeton University Press, 1985), and Simon Schaffer, *La fabrique des sciences moderne* (*XVII*^{e}*–XIX*^{e} *siècle*), tr. Frédérique Aït Touati, Loïc Marcou, and Stéphane Van Damme (Paris: Seuil, 2014).

19 Archives de l'Académie des Sciences, Paris, Fonds Lavoisier, *Registre de laboratoire n.4*, f. 4-Bv.

It deals with the exposure of diamonds to the heat produced by a burning lens, which is today considered as a minor step in Lavoisier's scientific career.[20] By paying more attention to some of the material features of these notes, however, it is possible to delve deeper in the two associates' working practices. First of all, we see that the notes have been taken on a small paper slip, with a rushed and inelegant handwriting [**Fig. 7.1**]. The paper shows also some traces of burns, which damaged the slip right in the middle. This detail suggests that the notes were written while the experiment was performed or, at least, when the instruments were still functioning. It is thus very likely that Paulze-Lavoisier was present on the experimental setting, writing down her observations on the spot. Sometime later, Lavoisier would have copied these notes in the second volume of the *Registres*, reworking and extending Paulze-Lavoisier's version of the text.[21] Finally, the original slip would have been kept in another volume and probably reconsulted a few years later.[22] These practices were not exceptional in themselves, but actually reproduced some standard procedures in the management of the *Registres*. They also recall some of the most common techniques through which eighteenth-century *savants* used to handle their personal notes.[23]

The reason why these slips are so important for us lies in fact that they can be considered as the material traces of Paulze-Lavoisier's first appearance as a *secrétaire* in front of an audience. In order to highlight this point it is necessary, however, to put her notes in a broader context. It has in fact to be to noted that the combustion of diamonds with a burning lens described by Paulze-Lavoisier was part of a more complex research program, which began to attract of the public attention in the summer of 1771. Since 1768, the French chemist Jean Darcet had been reporting on these experiments at the Académie des Sciences, pointing out a "curious" fact: the diamond, the hardest and most precious of stones, when exposed to the heat of a furnace for a few minutes, first blackened and then simply disappeared. This phenomenon had already been observed at the end of the seventeenth century, when the Grand

20 Archives de l'Académie des Sciences, Paris, Fonds Lavoisier, Fonds Lavoisier, *Registre de laboratoire n. 4*, ff. B4ar–B6v.

21 Archives de l'Académie des Sciences, Fonds Lavoisier, *Registre de laboratoire n. 2*, ff. 25–28.

22 The original slips are kept in *Registre n. 4*, whose content is mostly dated 1774 and devoted to the analysis of air; it is thus probable that the Lavoisiers reused the notes on diamonds while working on pneumatic chemistry. On these different and yet related research, see Henry Guerlac, *Lavoisier – The Crucial Year. The Background and Origin of His First Experiments on Combustion in 1772* (Ithaca: Cornell University Press, 1961), pp. 79–107.

23 On this kind of practices see Claire Bustarret, "Couper, coller dans les manuscrits de travail du XVIIIe au XXe siècle", in Jacob, ed., *Lieux de savoir*, vol. 2, pp. 353–375 and Carla Bittel, Elaine Leong, Christine von Oertzen, ed., *Working with Paper. Gendered Practices in the History of Knowledge* (Pittsburgh: University of Pittsburgh Press, 2019).

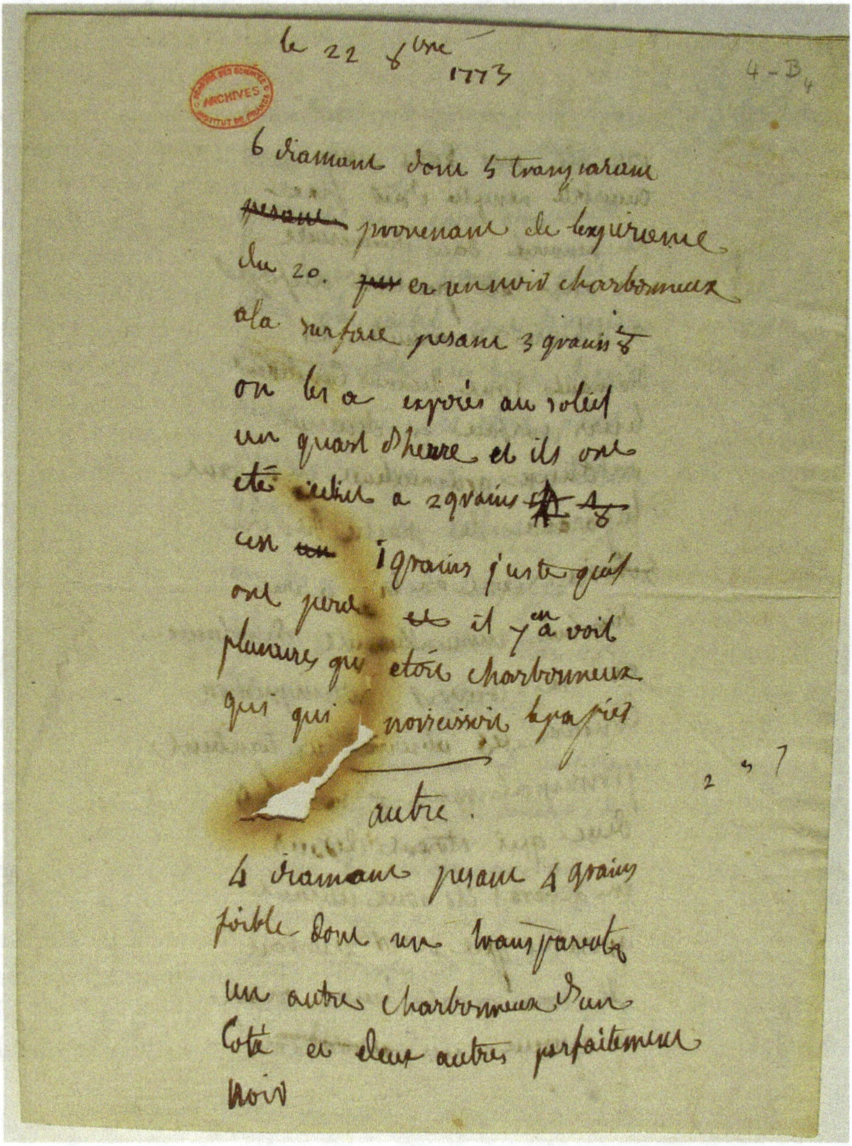

le 22 8bre 1773

4-B4

6 diamant dont 5 transparant
provenant de lexperience
du 20. et un noir charbonneux
a la surface pesant 3 grains 7/8
on les a exposés au soleil
un quart d'heure et ils ont
eté reduit a 2 grains 1/8
un 1 grains juste quil
ont perdu et il y en avoit
plusieurs qui etoit charbonneux
qui qui noircissoit le papier

autre.

4 diamant pesant 4 grains
foible dont un transparent
un autre charbonneux d'un
Coté et deux autres parfaitement
noir

FIGURE 7.1 Experiments on diamonds, 22 October 1773. Paulze-Lavoisier's handwriting. Archives de l'Académie des Sciences, Paris, Fonds Lavoisier, *Registre de laboratoire n. 4*, f. 4-Bv

Duke Cosimo III commissioned a series of experiments on precious stones to Tuscan naturalists.[24] It was however after the perfomance held in his Parisian

24 Guerlac, *Lavoisier*, pp. 79–107; Christine Lehman, "What Is the True Nature of Diamond?", *Nuncius*, 31 (2016): 361–407.

laboratory by Pierre-Joseph Macquer, on 26 July 1771, that the disappearing diamonds raised the interest of the Parisian community. Academicians, artisans, and amateurs were invited to this social event, that was soon reported about in several journals and gazettes.[25] The experiment was then repeated in similar circumnstances by other renown demonstrators. The session organized by Hilaire-Marin Rouelle e Jean Darcet in the *amphiteâtre* of the Jardin du Roi in the following August was apparently highly successful. Attended by aristocrats and renown *gens de lettres*, such as Diderot, Grimm, and Suard, it was described in the press as a public attraction. It is worth quoting from one of those reports, where the social dimension of these experiments was explicity emphasized:

> On a mis quatre diamans sous une moufle, placée dans un fourneau de reverbere, en présence du Margrave & de la Margave de Bade-Dourlach, du Prince héréditaire, & du Prince son frere, leurs fils, de plusieurs seigneurs & dames de la premiere qualité, & de plus de cent cinquante personnes, tant françois qu'étrangers, & choisis de tous les ordres & de différens états de cette capitale. Ces quatre diamans ont d'abord été vus, touchés par tout le monde, & essayés sur le crystal de roche; ensuite on les a numérotés & pesés avec des poids de carat, à la balance d'essai.[26]

The popularity of this investigation was later recalled by Macquer who, in the second edition of his *Dictionnaire de chimie* (1778), discussed at length the "curious" nature of the diamond and did not hesitate to describe his own interest in the subjet as the result of a sense of wonder and astonishement.[27] The

25 Among the journals that reported on these experiments there were the *Journal de sçavans*, the *Journal de médecine, chirugie et pharmacie*, the *Observations sur la physique*, the *Journal Encyclopédique*, the *Avantcourer*, and the *Gazette de France*. Some descriptions appeared also in Louis Petit de Bachaumont, *Mémoire sécrets pour servir à l'histoire de la république des lettres en France*, tome 5 (Londres: chez Adamnson, 1780), p. 306 and in *Correspondance littéraire, philosophique et critique par Grimm, Diderot, Raynal, Meister, etc*, depuis 1753 jusqu'en 1790, tome 7 (Paris: chez Furne, 1829), pp. 334–345.

26 "Expériences nouvellement faites à Paris sur le diamant", *Journal Encyclopédique*, vol. 7, partie 2, 1771, pp. 287–292.

27 The growing importance of this investigative field is evidenced also by the different degree of analysis provided by Macquer in first and the second edition of his *Dictionnaire de chimie contenant la théorie et la pratique de cette science, son application à la phisique, à l'histoire naturelle et à la médecine, et aux arts dépendans de la chimie* (the few lines dedicated to the entry "Diamant" in the 1766 edition are substituted by around 7 pages in the 1778 edition).

same feeling was shared by "the less initiated"[28] who, in his view, could easily reproduce the phenomena:

> [I]l étoit seulement bien naturel que tous ceux qui s'intéressent à l'avancement des sciences, désirassent de jouir eux-mêmes du spectacle de ces espèces de merveilles; & chacun pouvoit satisfaire d'autant plus facilement sa curiosité à cet égard, que M. d'Arcet s'étoit encore assuré que, ni le feu long & violent des grands fours à porcelaine, ni celui du foyer des forts verres ardens, n'étoient nécessaires pour l'entière destruction du diamant; & que ce lui d'un fourneau ordinaire tirant passablement bien, étoit suffisant.[29]

It is in this climate of general curiosity that Lavoisier became interested in the problem. Already present at the public experiment performed by Rouelle in 1771, he asked Macquer and Cadet to join the investigative group appointed by the Académie.[30] The experiments were resumed in the spring of 1772 and, shortly after, Macquer's group observed something unexpected: by preventing the contact of the air the diamonds did not undergo any alteration, but turned out to be perfectly intact at the end of the experiment. New experimental

28 Pierre-Joseph Macquer, "Diamant", in Id., *Dictionnaire de chimie contenat la théorie et la pratique de cette science, son application à la phisique, à l'histoire naturelle et à la médecine, et aux arts dépendans de la chimie*, vol. 1 (Paris: De l'Imprimérie de Monsieur, 1778), pp. 321–388.

29 Ibid., pp. 324–325. For a classic study on early-modern cultures of curiosities, see Lorraine Daston, Katharine Park, *Wonders and the Order of Nature, 1150–1750* (New York: Zone Books, 1998), although the case of the experiments on diamonds and precious stones discussed here differs from their account of eighteenth-century attitudes toward curiosity and wonder. See also J.W. Evans, Alexander Marr, eds., *Curiosity and Wonder from the Renaissance to the Enlightenment* (Aldershot: Ashgate, 2006). For a fundamental study on curiosity and collecting within the Parisian context, see Krzysztof Pomian, *Collectionneurs, amateurs et curieux. Paris, Venise, XVIe–XVIIIe siècle* (Paris, Gallimard, 1987). The seminar *La curiosité des Lumières* held by Antoine Lilti and Silvia Sebastiani at the Ehess during the academic year 2018–2019 also helped me to approach this topic from a more complex perspective.

30 "Je communiquai à Mrs Macquer & Cadet le projet où j'étois de les suivre, je leur demandai leurs conseils, je les priai de vouloir bien permettre qu'elles fussent faites de concert, & nous nous assemblames à cet effet dans le laboratoire de M. Cadet"; Antoine-Laurent Lavoisier, "Premier mémoire sur la destruction du diamant par le feu", *Mémoire de l'Académie des Sciences pour l'année 1772* (Paris: Imprimérie Royale, 1776), pp. 564–591: 574–575.

sessions followed, aimed at solving this new puzzle.[31] In summer 1772, these experiments became even more spectacular. In order increase the heat applied to the diamonds, the Académie des Sciences made available the large burning lens made by Graf Ehrenfried Walther von Tschirnaus at the beginning of the century. The lens was placed in the Jardin de l'Infante at the Louvre, a place well exposed to the sun and adjacent to the Académie. The enthusiasm was such that, between August 1772 and October 1773, every kind of mineral was submitted to the lens, almost forgetting the primary purpose of the investigation.[32] These events shows the degree of popularity attained by chemistry in late eighteenth-century Paris, an effect of the successful courses held by Rouelle and other demonstrators since the 1750s and attended by crowds of Parisians, aristocrats and bourgois alike. From a broader perspective, they higlights the importance of public demonstrations and spectacles in *savants*' working *routine* and sociability which, as number of historians has shown, took place outside institutional venues as well.[33] The experiments on diamonds at the Jardin de l'Infante have thus to be considered within this specific context. As noted by Christine Lehman,

> Like chemistry courses, this great investigation both challenged the learned world and that of the curious. It excited the curiosity not only of chemists but also of geologists and naturalists, not to mention bourgeois, owners of curiosity cabinets, jewellers and noblemen who offered precious metals for the experiments, etc. The whole of the learned world had its eyes fixed on the experiments being performed at the focus of the academy's impressive lens.[34]

31 Ibid. On the public impact of this new observation within the Parisian context, see *Journal des sçavans*, juin 1772, pp. 435–439.

32 Among among 225 exposures to the burning glass conducted in 1772, diamond was exposed only once: see Lehman, "What is the True Nature of Diamond?", p. 381.

33 See esp. Stéphane Van Damme, *Paris, capitale philosophique. De la Fronde à la Revolution* (Paris: Odile Jacob, 2005), part II.

34 Ibid., pp. 380–381. On chemistry courses in eighteenth-century Paris, see Bernadette Bensaude-Vincent, Christine Blondel, ed., *Science and Spectacle in the European Enlightenment* (Aldershot: Ashgate, 2008); Bernadette Bensaude-Vincent, Christine Lehman, "Public Lectures of Chemistry in Mid-eighteenth-century France", in Lawrence M. Principe, ed., *New Narratives in Eighteenth-Century Chemistry* (Dordrecht: Springer, 2007), pp. 77–96 and John Perkins, "Chemistry Courses, the Parisian Chemical World and the Chemical Revolution, 1770–1790", *Ambix*, 57, 2010: 27–47.

Further evidence of the public's interest in these experiments can be found in the frequent donations of precious stones by jewellers and aristocrats, eager to be part of the event. Even Lavoisier, whose role in the research progressively expanded, gave his financial support.[35] Trudaine de Montigy, an honorary member of the Académie, *intendant des finances* and head of the Bureau de commerce, shared the same views and commissioned the construction of a new and larger burning lens, which became operative in October 1774.[36] This donation was also celebrated in an engraving, dedicated to Trudaine by the instrument maker Charpentier, which is interesing as one of the few existing visual representations of those experiments. The image shows a man submitting a substance to the heat produced by the lens in what seems to be an open and spatious venue, probably a public garden. Right behind him we see a gender mixed audience, composed by two women and four men; while some of them are following the performance, others appears to be distracted and caught up in conversations [**Fig. 7.2**]. From this point of view, this engraving recalls at least two important features of the experiments on diamonds, which were often stressed in the written reports as well: the presence of women in the double role of spectactors and witnesses and the sociable exchange that took place during the performance. However, in the autumn of 1774, when this artwork was published, the general attention raised by the spectacular effects of the burning lens declined and also Lavoisier, after proposing his own interpretations, set aside the problem.[37]

In the light of these remarks we can look at Paulze-Lavoisier's manuscript note dating October 1773 from a new perspective. She was by then a 15-year-old girl, very recently married and her scientific apprenticeship, probably led under the supervision of her husband, was just at the beginning.[38] Nevertheless, even at this early stage of her collaboration with Lavoisier, she

35 Macquer, "Diamant", p. 335.

36 "Premier essai du grand verre ardent de M. Trudaine, établi au Jardin de l'Infante, au commencement du mois d'octobre 1774, Par M.rs Trudaine de Montigny, Macquer, Cadet & Brisson", *Mémoires de l'Académie Royale des Sciences pour l'année 1774* (Paris: Imprimérie Royale, 1778), pp. 62–72.

37 Christine Lehman, "What Is the True Nature of Diamond?".

38 On Paulze-Lavoisier's scientific education see Carleton E. Perrin, "The Lavoisier-Bucquet Collaboration: A Conjecture", *Ambix*, 1989, *36*:5–40. On women's education in eighteenth-century Parisian *élites*, see Martine Sonnet, *L'éducation des filles au temps des Lumières* (Paris: Les Éditions du Cerf, 2011, 1st ed. 1987); for a wider and more recent account see Mónica Bolufer Peruga, "Gender and the Reasoning Mind. Introduction" in Barbara Taylor, Sarah Knott, ed., *Women, Gender and Enlightenment* (Basingstoke: Palgrave Macmillan, 2008), pp. 189–194.

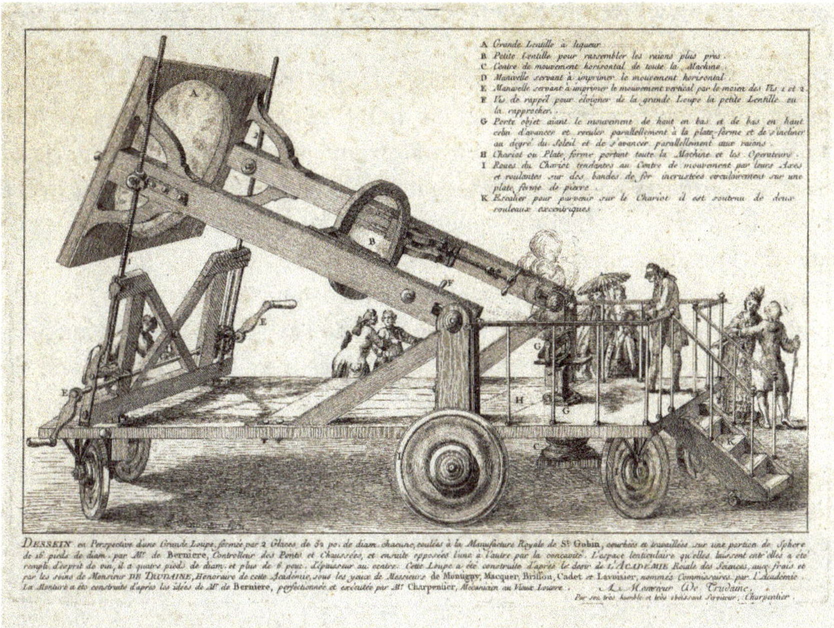

FIGURE 7.2 Trudaine's burning lens, 1774. Engraving. Reproduced in Antoine-Laurent Lavoisier, *Œuvres*, vol. 3 (Paris: Imprémerie Imperiale, 1864), Planche IX

took an active role and, what is more important here, this role was was deliberately *exposed* to the eyes of the audience. Her name never appears in the *mémoires* that were published on the nature of the diamond; Macquer, for instance, while discussing Lavoisier's involvement in this investigation, only mentioned an undefined group of "friends" who helped him in the autumun of 1773.[39] And yet Paulze-Lavoisier's work as a note-taker could be easily noticed by those who were present at the experiments which, as we have seen, were public. She thus moved from being a spectator in a gendered mixed audience, such as the women represented in the 1774 engraving, to being, at least potentially, the object of the audience's attention. As we shall see below, this was by no means an isolated case. It was rather a sort of *debut* for her, which was followed by other public (or semi-public) of her skills as a *secrétaire*. So let us turn to the site were these most of these displays took place in the following years: her Parisian residence.

39 Macquer, "Diamant", p. 336.

2 Burning Precious Stones at the Arsenal

Shortly after their marriage in December 1771, the Lavoisiers settled in an apartment in rue des Bons Enfants, not far from the Académie des Sciences. Here Lavoisier built his first chemical laboratory and, together with his wife, began to organize *dîners* and other social gatherings since at least 1774. It is in this context, for instance, that the famous conversation between Joseph Priestley and the French chemist on the discovery of oxygen took place: according to Priestley's account, on that occasion his hosts had invited for dinner "the most philosophical people of the city".[40] The reception of guests in a domestic space was in fact a common pratice in eighteenth-century Parisian *élites* and the Lavoisiers, both belonging to the *haute bourgeoisie*, were no exception.[41]

However, only by 1776, when the couple moved to the Arsenal, these social gatherings became regular and increasingly intertwined with the experimental activity. In fact, Lavoisier's design and organisation of the chemical laboratory included not only the introduction of high precision instruments, as often stressed by scholars, but also the involvement of different social and professional figures.[42] These people, among whom we find academicians, *amateurs*, instrument makers and artisans, regularly met in his residence. The social dimension of this space was highlighted by Paulze-Lavoisier herself who, after her husband's death in 1794, recalled the *routine* of the Arsenal as follow:

> [U]n jour tout entier dans chaque semaine était consacré aux expériences; c'était disait Lavoisier son jour de bonheur, quelques amis éclairés, quelques jeunes gens fiers d'être admis à l'honneur de coopérer à ses expériences se réunissaient dès le matin, dans le laboratoire; c'était là

40 Joseph Priestley, *The Doctrine of Phlogiston Established and that of the Composition of Water Refuted* (Norththumberland: Bryne, 1803, 1st ed. 1800), p. 116.

41 The fundamental studies on this topic are Dena Goodman, *The Republic of Letters. A Cultural History of the French Enlightenment* (Cornell: Cornell University Press, 1994) and Antoine Lilti, *Le monde des salons. Sociabilité et mondanité à Paris au XVIIIe siècle* (Paris: Fayard, 2005). On domestic sociability within the context of the Ferme générale, to which both Lavoisier and Paulze-Lavoisier's father belonged, see Yves Durand, *Les fermiers généraux au XVIIIe siècle* (Paris: Presse Universitaire de France, 1971), esp. pp. 445–89 and 526–544.

42 Marco Beretta, "Big Chemistry: Lavoisier's Design and Organisation of His Laboratories", in Marta Lourenço, Ana Carneiro, ed., *Spaces and Collections in the History of Science* (Lisbon: Museum of Science of the University of Lisbon, 2009), pp. 65–80; Id., "Between the Workshop and the Laboratory: Lavoisier's Network of Instrument Makers", *Osiris*, 29 (2014): 197–214. For an in-depth reconstruction of Lavoisier's laboratories see also Marco Beretta, Paolo Brenni, *The Arsenal of Eighteenth-century Chemistry. The Laboratories of Antoine Laurent Lavoisier (1743–1794)* (Leiden: Brill, forthcoming).

que l'on déjeunait, que l'on dissertait, que l'on travaillait, que l'on faisait des expériences, que naissait cette belle théorie qui a immortalisé son auteur.[43]

The liveliness of the Arsenal's social life increased over time and seemingly reached a climax in the 1780s, when it became a venue for public experiments and spectacles, aimed at promoting Lavoisier's chemical theories on a international level.[44]

A series of experiments dating to the beginning of 1780s deserves special attention. In 1782 Lavoisier resumed his research on the combustion of precious stones. Dissatisfied with the burning lens, he devised a new apparatus through which the heat was produced and controlled by a flame alimented by oxygen. He was particularly proud of this method which, in his view, had the merit of solving an old problem of chemical practice, namely the need for great amounts of heat and, at the same time, relied on a gas, oxygen, that was then at the core of many scholarly debates.[45] Once again he took advantage of a social event to make his invention known to the public. It is probably not a coincidence, in fact, that several months before the reading of his *mémoires* on the subject, he made a first public demonstration at the *rentrée publique* of the Académie des Sciences, on 5 June 1782.[46] Because of the exceptional presence of Paul I of Russia and his wife Marie Feodorovna, then visiting Paris, that session was announced and reported in various gazettes and brochures. As we read in those reports, on that occasion Lavoisier made an experiment on platinum, a precious metal discovered in 1748 and very hard to melt. Thanks to his apparatus Lavoisier succeded in fusing it, we read, in "very little time".[47] With

43 Marie-Anne Paulze-Lavoisier, cit. in Charles C. Gillispie, "Notice biographique de Lavoisier par Madame Lavoisier," *Revue d'histoire des sciences et de leurs applications,"*, 9, 1 (1956): 52–61, p. 57.

44 Carleton E. Perrin, "The Triumph of the Antiphlogistians", pp. 40–63.

45 Antoine-Laurent Lavoisier, "Mémoire sur un moyen d'augmenter considérablement l'action du Feu & de la chaleur, dans les opérations chimiques", *Mémoires de l'Académie Royale des Sciences pour l'année 1782* (Paris: Imprimérie Royale, 1785), pp. 457–465; Id., "Mémoire sur l'effet que produit sur les pierres précieuses un degré de feu très-violent", ibid., pp. 476–485. The instrument designed for these experiments is described in Jean-Baptiste Meusnier, "Description d'un appareil propre à manœvrer différentes espèces d'airs", ibid., pp. 466–475. The results were later published in Antoine-Laurent Lavoisier "De l'action du feu animé par l'air vital, sur les substances minérales les plus réfractaires", *Mémoires de l'Académie Royale des Sciences pour l'année 1783* (Paris: Imprimérie Royale, 1786), pp. 563–614.

46 Archives de l'Académies des Sciences, Paris, *Procès-verbaux*, tome 101, 1782, ff. 102–103.

47 *Journal de Paris*, vol. 1, 10 avril 1782, p. 397; *Journal Encyclopédique*, vol. 5, partie 2, 1783, pp. 332–333; Louis Petit de Bachaumont, *Mémoire sécrets pour servir à l'histoire de la république des lettres en France*, tome 21 (Londres: chez Adamson, 1783), pp. 18–19; *Le*

this achievement Lavoisier was celebrated by the Académie as the author of a method that, according to the report, allowed many "curious discoveries and observations" as well as "useful" applications in the arts.[48]

Between 22 June 1782 and 6 January 1783 these experiments were repeated at the Arsenal and recorded in the *Registres de laboratoire*.[49] Paulze-Lavoisier managed the organisation of the written protocols, preparing the notebooks, taking notes, and insuring the preservation of the specimens after the experiments. As a first step she listed the substances to be tested in the *Registres*; then she noted down the duration of each experiment and described the effects produced by the lamp on the substances; finally, she wrapped up and catalogued the specimens [**Fig. 7.3**].[50] However, it has to be stressed that these procedures, while stemming almost exclusively from Paulze-Lavoisier's pen, were anything but solitary. Firstly, from the dates reported in the *Registres* we notice that the experiments were often made on Mondays, Saturdays, and Sundays, the days when the Lavoisiers used to receive their guests for dinners and other social events. Secondly, and more importantly, a further confirmation comes from a recently discovered account by Barthélemy Faujas de Saint-Fond, a geologue and traveller who, in the autumn of 1782, visited the Arsenal more than once and personally witnessed those performances. The entry of 28 November of his travel diary is particularly interesting: that day the diner was followed, as it was customary, by a visit of the laboratory, where Faujas de Saint-Fond, who had contributed to the experiments by giving some of his minerals, saw Paulze-Lavoisier acting as a *secrétaire*. Thus he wrote:

> Je suis allé diner chès Lavoisier de l'Académie Française, il y avoit à diner M. le duc de La Rochefoucauld plusieurs savants et madame de Lavoisier qui est jolie et fort aimable, après le diné il est arrivé encore plusieurs savants pour assister à des expériences que M. de Lavoisier devait faire sur l'air déphlogistiqué qui produit sur un charbon un feu d'une force extraordinaire. Madame de Lavoisier à servi de la meilleur grâce du monde de secrétaire pour écrire les résultats de ces belles expériences, avec un montre à seconde à la main pour compter les secondes. [...] [L]'on à fait une multitude d'expérience très curieuses qui ont duré environ cinq

Comte et la Comtesse du Nord; Anecdote Russe, mise au jour par M. le Chevalier du Coudray (Paris: Belin, 1782), pp. 77–78.

48 *Histoire Académie Royales de Sciences pour année 1782* (Paris: Imprimérie Royale 1785), pp. 28–31.

49 "Lampe d'air dephlogistiqué. Pierres soumies à l'action du feu animé par l'air gaz oxygène": Archives de l'Académie des Sciences, Paris, Fonds Lavoisier, *Registre de laboratoire n.6*, ff. 151r–192r.

50 Musée des Arts et Métiers, Paris, item n. 20185-0000.

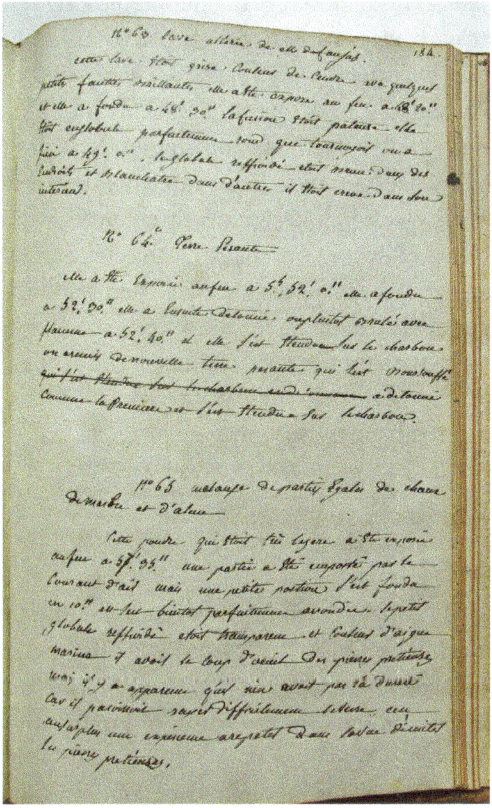

FIGURE 7.3
A page from Lavoisier's *Registres de laboratoire* dealing with the experiments on minerals and precious stones (January 6, 1783). Paulze-Lavoisier's handwriting. Archives de l'Académie des Sciences, Paris, *Registre de laboratoire n. 6*, f. 184r

> quarts d'heures pendant le quel tems on n'a usé pour environ 30# d'air déphlogistiqué tiré du précipité rouge après les expériences chimiques qui ont présenté plusieurs phénomènes remarquables.[51]

From several points of view, these experiments echo the above mentioned debut of Paulze-Lavoisier at the Jardin de l'Infante in 1773. Leaving aside her husband's interest in the effect of fire on minerals, which is in itself an element of continuinity with the investigation on diamonds, it is easy to see that in both

---

51 Barthélemy Faujas de Saint-Fond, cit. in Guillaume Comparato, *Barthélemy Faujas de Saint-Fond, parcours d'un homme de science mondain au tournant des Lumières (1741–1819)*, Ph.D. dissertation, 2018, Université Grenoble Alpes 2 vols., vol. 2, p. 206. It should also be noted that by the time of Faujas de Saint-Fond's visit, Lavoisier's new method for the combustion of minerals was by then already known outside his inner circle of associates: see esp. Jean-Louis Giraud Soulavie to Antoine-Laurent Lavoisier, 12 April 1782, pub. in Antoine-Laurent Lavoisier, *Correspondance*, vol 3: 1776–1783 (Paris: Albin Michel, 1964), pp. 719–720.

circumnstances her role as a *secrétaire* was exposed to the public, althought to different kinds of audiences. In the early 1770s Lavoisier was still a young scientist on the rise and his first book, the *Opuscules de physique et de chimie*, had yet to be published. It is thus easy to imagine him, at that time, seeking for public occasions, as in the case of the spectacular experiments on diamonds, to demonstrate his talents to a wide and varied audience. Ten years later, when he resumed his research on minerals, his position had signicantly changed: while his fame as a scientist was acknowledged by the main European scientific circles, his political influence within the Académie des Sciences had undoubtley increased. At this stage it was probably enough, for his purposes, to invite to the Arsenal a selected audience to witness the experiments. In both cases, however, Paulze-Lavoisier's work as a *secrétaire* was deliberatly displayed, as an object of potential glances and comments by those who were present.

3 Imaging a Scientific Collaboration

As the years went on, the practice of displaying Paulze-Lavoisier's qualities as a scientific associate became somewhat established. As already mentioned, in the 1780s the Arsenal residence was used as a sort of headquarter of Lavoisier's supporters who, after the famous experiments on the composition of water in 1783, formed a sort of sect engaged in the defense of the "new theory" from the attacks coming from the French and European chemical communities.[52] During this eventful period, Paulze-Lavoisier became more and more visibile within her husband's circle, especially through the means of correspondences, social gatherings, spectacles, and semi-public experiments whose purpose was, in her own words, to "convert" the audience. A visit of the Arsenal's laboratory was integral part of this campaign and, in that context, her skills were repeatedly observed and admired. In October 1787, for example, Arthur Young reported on a visit he made to the couple: the exhibition of the instruments, guided by Lavoisier, was first introduced by a conversation with Paulze-Lavoisier who, once again, was appreciated for being a woman "working with her husband in his laboratory". It is interesting to see how Young expressed his surprise for what he had seen at the Arsenal:

> The 16th. To Mons. Lavoisier, by appointment. Madame Lavoisier, a lively, sensible, scientific lady, had prepared a *dejeuné Anglois* of tea and coffee, but her conversation on Mr. Kirwan's *Essay on Phlogiston*, which she

52 Perrin, "The Triumph of the Antiphlogistians", pp. 40–63.

FIGURE 7.4 The experiments on human respiration and transpiration in a drawing by Marie-Anne Paulze-Lavoisier, 1790 circa. Pen and brown ink and wash. The original title of the drawing is unknown. Wellcome Collection, London, item no. 37197i

> is translating from the English, and on other subjects, which a woman of understanding, that works with her husband in his laboratory, knows how to adorn, was the best repast. That apartment, the operations of which have been rendered so interesting to the philosophical world, I had pleasure in viewing. In the apparatus for aerial experiments, nothing makes so great a figure as the machine for burning inflammable and vital air, to make, or deposit water; it is a splendid machine.[53]

From the same perspective it is possible to reconsider the famous drawings in which Paulze-Lavoisier illustrated a moment of laboratory life and, in particular, the experiments on human respiraition carried out by Lavoisier and Armand Séguin in the early 1790s [**Fig. 7.4**].

These images, considered for a long time "fanciful" and of little interest for historians of science,[54] have recently been presented as scientific illustrations

53 Arthur Young, *Travels during the years 1787, 1788 and 1789* (London: 1792), p. 95.

54 Holmes, *Lavoisier and the Chemistry of Life*, p. 546.

in their own rights.[55] In fact, they were probably meant to be published in Lavoisier's *Mémoires de physique et de chimie*, written in collaboration with his youngest associates which, in its author's view, would have offered a comprehensive ovweview of both his past and present research. The publication of the *Mémoires* began around 1792 and was interrupted by the revolutionary events and the subsquent arrest (1793) and execution (1794) of Lavoisier. However, beyond the unfortunate fate of this project, the striking feature of these illustrations lies once again in the visibility of Paulze-Lavoisier, here represented in the act of taking notes during the experimental sessions. In the early 1790s, the representation of a woman in a framework connected to experimental pratice was not, in itself, completely new. A renown example could be found in Jean-Antoine Nollet, Lavoisier's master and author of extremely successful textbooks of experimental physics where he printed several illustrations with men and women engaging in experiments, mostly dealing with electricity.[56] But while it is possible that Paulze-Lavoisier had this model in mind, she certainly introduced an element of novelty: for the first time a woman figured as the author of *both* the written protocols of the experiments and the images that illustrated that action. These drawings can therefore be seen as a further public affirmation of her work as a *secrétaire*: a sort of visual crowing of a role that, as we have seen, had first been displayed in the early 1770s and then progressively reaffirmed in the following years.[57]

4 Conclusions

The historical success of Lavoisier's chemical theories should not distract us from the fact that, throughout his career, he engaged in a clever and accurate

---

55 Marco Beretta, "Imaging the Experiments on Respiration and Transpiration of Lavoisier and Séguin: Two Unknown Drawings by Madame Lavoisier", *Nuncius*, 27 (2012): 163–191 and Sørensen, "Madame Lavoisier".

56 Lewis Pyenson, Jean-François Gauvin, ed., *L'art d'enseigner la physique. Les appareils de démonstration de Jean-Antoine Nollet 1700–1770* (Sillery: Septentrion, 2002), which includes some useful considerations on Nollet's approach to scientific illustration. To consider Nollet's work in a wider cultural context, see Paola Bertucci, *Viaggio nel paese delle meraviglie. Scienza e curiosità nell'Italia del Settecento* (Torino: Bollati Boringhieri, 2007).

57 Although devoted to a different topic, Marie-Josèphe Bonnet, "Femmes peintres à leur travail: de l'autoportrait comme manifeste politique XVIIIe–XIXe siècles)," *Revue d'histoire moderne et contemporaine*, 49, 3 (2002): 140–167 offers an interesting perspective on women and auto-portraiture. Dena Goodman, *Becoming a Woman in the Age of Letters* (Ithaca and London: Cornell University Press, 2009) also influenced my view on this point.

effort of self-promotion. To this purpose, as I have tried to highlight, he relied on what eighteenth-century society and culture made available to him: these ressources included curious phenomena and spectacles, as well as public or semi-public experiments and domestic sociability. There is no real contradiction, I believe, between the episodes examined thus far and the better-known image of the French chemist as the promoter of a new way of doing chemistry, based on objectivity and measurement.[58] After all, as Herly Guerlac has put it, he was "a Parisian through and trough and a son of the Enlightenment".[59]

It is throught this lens that we should look at his collaboration with Paulze-Lavoisier. To a great extent, their work as a collaborating couple was marked by a sexual division of intellectual labor. In any case, for instance, her handwritten annotations resulted in a text published under her name, whereas the male assistants were often presented as co-authors of Lavoisier in several *mémoires*.[60] This choice was somehow consistent with her inclination to appear as a modest person, very careful in avoiding the criticism and social blame often addressed to female authors by French polite society.[61] Modesty was in fact a key attribute for femininity in Enlightenment thinking; it is therefore not surprising that she conformed to this gender norm, chosing to never publish in her own name and not to engage in any "autonoumous" research.[62] But even so, she never really became one of those "invisible" assistants who parallaled eighteenth-century *savants* and natural philosophers in domestic

58 See, for instance, Jean-Luc Chappey, "Enjeux sociaux et politiques de la 'vulgarisation scientifique' en révolution (1780–1810)", *Annales historiques de la Révolution Française*, 338 (2004): 11–51 and Christian Licoppe, *La formation de la pratique scientifique: le discours de l'expérience en France et an Angleterre (1630–1820)* (Paris: La Découverte, 1996).

59 Guerlac, "Lavoisier, Antoine-Laurent", p. 66.

60 See, for instance, the cases of Armand Séguin, studied in Beretta "Imaging the Experiments on Respiration and Transpiration" and Jean-Pierre Poirier, "La contribution d'Armand Seguin (1767–1835) aux programmes de recherche de Lavoisier", in Antoine-Laurent Lavoisier, *Correspondance*, vol. 6 (Paris: Académie des Sciences, 1997), pp. 427–438. For other examples of male collaborations in Lavoisier's laboratory: Beretta, "Big chemistry".

61 Paulze-Lavoisier's approach to publishing has been examined in Keiko Kawashima, "Madame Lavoisier: assistante invisible d'une communauté scientifique", *Bulletin of Nagoya Institute of Technology*, 4 (1995):249–259 and Id., *Émilie du Châtelet et Marie Anne Lavoisier*.

62 On modesty as a "feminine universal" see, among others, Londa Schiebinger, *Nature's Body. Gender in the Making of Modern Science* (New Brunswick: Routledge, 1993), pp. 99–106; Marta Cavazza, "Between Modesty and Spectacle. Women and Science in Eighteenth-Century Italy", in Paula Findlen, Wendy Wassyng Roworth, and Catherine M. Sama, ed., *Italy's Eighteenth Century. Gender and Culture in the Age of the Grand Tour* (Stanford: Stanford University Press, 2009), pp. 275–302, recently re-published in Marta Cavazza, *Laura Bassi: Donne, genere e scienza nell'Italia del Settecento* (Milan: Bibliografica, 2020), pp. 217–254.

laboratories.[63] Instead, as we have seen, her role as a note-taker was repeatedly exposed in front of an audience which, especially in the 1780s, reacted to her performances with a positive sense of wonder and astonishement. Those reactions were probably due to the fact that a woman working as a *secrétaire* was, in many respects, an exception. Her presence in the experimental setting could therefore appear as something to observe and admire.

References

Antonelli, Francesca, *Scrittura, sociabilità e strategie di persuasione: Marie-Anne Paulze-Lavoisier, secrétaire (1758–1836)*, PhD diss., Università di Bologna/Écoles des Hautes Études en Sciences Sociales, 2021.

Bachaumont, Louis Petit de, *Mémoire sécrets pour servir à l'histoire de la république des lettres en France*, tome 21 (Londres: chez Adamson, 1783).

Beltrán, José, "Ciencia amanuense: cultura manuscrita e historia natural en la Francia moderna (c.1660–1830)", *Asclepio. Revista de Historia de la Medicina y de la Ciencia, 71*, 1 (2019): http://asclepio.revistas.csic.es/index.php/asclepio/article/view/814.

Bensaude-Vincent, Bernadette, Christine Blondel, ed., *Science and Spectacle in the European Enlightenment* (Aldershot: Ashgate, 2008).

Bensaude-Vincent, Bernadette, Christine Lehman, "Public Lectures of Chemistry in Mid-eighteenth-century France", in Lawrence M. Principe, ed., *New Narratives in Eighteenth-Century Chemistry* (Dordrecht: Springer, 2007), pp. 77–96.

Beretta, Marco, "Imaging the Experiments on Respiration and Transpiration of Lavoisier and Séguin: Two Unknown Drawings by Madame Lavoisier", *Nuncius*, 27 (2012):163–191.

Beretta, Marco, "Big Chemistry: Lavoisier's Design and Organisation of His Laboratories", in Marta Lourenço, Ana Carneiro, ed., *Spaces and Collections in the History of Science* (Lisbon: Museum of Science of the University of Lisbon, 2009), pp. 65–80.

Beretta, Marco, "Between the Workshop and the Laboratory: Lavoisier's Network of Instrument Makers," *Osiris*, 29 (2014): 197–214.

63 On this topic see, among others, Londa Schiebinger, *The Mind Has No Sex?* and Donald L. Optiz, Staffan Bergwik, and Brigitte Van Tiggelen, ed., *Domesticity in the Making of Modern Science* (Basingstoke: Palgrave Macmillan, 2016). A different and inspiring approach to the problem of women invisibility in eighteenth-century society can be found in Marta Cavazza, "Between Modesty and Spectacle". See also Paola Bertucci, "The In/visible Woman: Mariangela Ardinghelli and the Circulation of Knowledge between Paris and Naples in the Eighteenth Century", *Isis*, 104, 2 (2013): 226–249.

Beretta, Marco, Paolo Brenni, *The Arsenal of Eighteenth-century Chemistry. The Laboratories of Antoine Laurent Lavoisier (1743–1794)* (Leiden: Brill, forthcoming).

Berthelot, Marcellin, *La révolution chimique: Lavoisier. Ouvrage suivi de notices et extraits des registres inédits de laboratoire de Lavoisier* (Paris: Félix Alcan, 1890).

Bertucci, Paola, *Viaggio nel paese delle meraviglie. Scienza e curiosità nell'Italia del Settecento* (Torino: Bollati Boringhieri, 2007).

Bertucci, Paola, "The In/visible Woman: Mariangela Ardinghelli and the Circulation of Knowledge between Paris and Naples in the Eighteenth Century", *Isis*, 104, 2 (2013): 226–249.

Bittel, Carla, Elaine Leong, and Christine von Oertzen, ed., *Working with Paper. Gendered Practices in the History of Knowledge* (Pittsburgh: University of Pittsburgh Press, 2019).

Blair, Ann M., *Too Much to Know. Managing Scholarly Information before the Modern Age* (New Haven and London: Yale University Press, 2010).

Blair, Ann M., Richard Yeo, ed., *Note-taking in Early Modern Europe*, special issue of *Intellectual History Journal*, 20, 3 (2010).

Bret, Patrice, Brigitte van Tiggelen, ed., *Madame d'Arconville. Une femme de lettres et de sciences au siècle des Lumières* (Paris: Hermann, 2011).

Bolufer, Mónica, "Gender and the Reasoning Mind. Introduction" in Barbara Taylor, Sarah Knott, ed., *Women, Gender and Enlightenment* (Basingstoke: Palgrave Macmillan, 2008), pp. 189–194.

Bonnet, Marie-Josèphe, "Femmes peintres à leur travail: de l'autoportrait comme manifeste politique (XVIIIe–XIXe siècles)", *Revue d'histoire moderne et contemporaine*, 49, 3 (2002): 140–167.

Boucher d'Argis, "Secrétaire", in Denis Diderot, Jean Bapiste Le Rond d'Alembert, ed., *Encyclopédie, ou dictionnaire raisonné des sciences, des arts et des métiers, par une société de gens de lettres*, vol. 14 (Paris: 1765): http://enccre.academie-sciences.fr/encyclopedie/article/v14-3042-0/.

Bourguet, Marie-Noëlle, "A Portable World: The Notebooks of European Travellers (Eighteenth to Nineteenth Centuries)", in Ann M. Blair, Richard Yeo, ed., *Note-taking in Early Modern Europe*, special issue of *Intellectual History Journal*, 20, 3, (2010): 377–400.

Bourguet, Marie-Noëlle, *Le monde dans un carnet. Alexander von Humboldt en Italie (1805)* (Paris: Éditions du Félin, 2017).

Bustarret, Claire, "Couper, coller dans les manuscrits de travail du XVIIIe au XXe siècle", in Christian Jacob, ed., *Lieux de savoir*, vol. 2: *Les mains de l'intellect* (Paris: Albin Michel, 2011), pp. 353–375.

Cavazza, Marta, "Between Modesty and Spectacle. Women and Science in Eighteenth-Century Italy", in Paula Findlen, Wendy Wassyng Roworth, and Catherine M. Sama, ed., *Italy's Eighteenth Century. Gender and Culture in the Age of the Grand Tour* (Stanford: Stanford University Press, 2009), pp. 275–302.

Cavazza, Marta, *Laura Bassi: Donne, genere e scienza nell'Italia del Settecento* (Milan: Bibliografica, 2020).

Chappey, Jean-Luc, "Enjeux sociaux et politiques de la 'vulgarisation scientifique' en révolution (1780–1810)", *Annales historiques de la Révolution Française*, 338 (2004): 11–51.

Comparato, Guillaume, *Barthélemy Faujas de Saint-Fond, parcours d'un homme de science mondain au tournant des Lumières (1741–1819)*, Ph.D. Dissertation, 2018, Université Grenoble Alpes.

Correspondance littéraire, philosophique et critique par Grimm, Diderot, Raynal, Meister, etc, depuis 1753 jusqu'en 1790, tome 7 (Paris: chez Furne, 1829).

Daston, Lorraine, Katharine Park, *Wonders and the Order of Nature, 1150–1750* (New York: Zone Books, 1998).

Décultot, Élizabeth, ed., *Lire, copier, écrire. Les bibliothèques manuscrites et leurs usages au XVIIIe siècle* (Paris: CNRS Éditions, 2003).

Dover, Paul M., ed., *Secretaries and Statecraft in the Early Modern World* (Edinburgh: Edinburgh University Press, 2016).

Durand, Yves, *Les fermiers généraux au XVIIIe siècle* (Paris: Presse Universitaire de France, 1971).

Evans, J.W., Alexander Marr, eds., *Curiosity and Wonder from the Renaissance to the Enlightenment* (Aldershot: Ashgate, 2006).

"Expériences nouvellement faites à Paris sur le diamant", *Journal Encyclopédique*, vol. 7, partie 2, 1771, pp. 287–292.

Gillispie, Charles C., "Notice biographique de Lavoisier par Madame Lavoisier," *Revue d'histoire des sciences et de leurs applications*, 9, 1 (1956): 52–61.

Goodman, Dena, *The Republic of Letters. A Cultural History of the French Enlightenment* (Cornell: Cornell University Press, 1994).

Goodman, Dena, *Becoming a Woman in the Age of Letters* (Ithaca and London: Cornell University Press, 2009).

Guerlac, Henry, *Lavoisier – The Crucial Year. The Background and Origin of His First Experiments on Combustion in 1772* (Ithaca, NY: Cornell University Press, 1961).

Guerlac, Henry, "Lavoisier, Antoine-Laurent", in Charles C. Gillispie, ed., *Dictionary of Scientific Biography*, vol. 7 (New York: Charles Scribner's Sons, 1971), pp. 66–91.

Hahn, Roger, *The Anatomy of a Scientific Institution. The Paris Academy of Science 1666–1803* (Berkley: University of California Press, 1971).

Holmes, Frederic L., *Lavoisier and the Chemistry of Life. An Exloration of Scientific Creativity* (Madison: The University of Wisconsin Press, 1983).

Jacob, Christian, ed., *Lieux de savoir*, 2 vols. (Paris: Albin Michel, 2007–2011).

Kawashima, Keiko, "Madame Lavoisier et la traduction de l'*Essay on Phlogiston* de Kirwan", *Revue d'histoire des sciences*, 53 (2000): 235–263.

Kawashima, Keiko, "Madame Lavoisier: assistante invisible d'une communauté scientifique", *Bulletin of Nagoya Institute of Technology*, 4 (1995): 249–259.

Kawashima, Keiko, *Émilie du Châtelet et Marie Anne Lavoisier Émilie du Châtelet et Marie Anne Lavoisier. Science et genre au XVIIIe siècle* (Paris: Honoré Champion, 2013).

Lavoisier, Antoine-Laurent, *Correspondance*, 7 vols, vol 3: 1776–1783 (Paris: Albin Michel, 1964), vol. 5: 1787–1788 (Paris: Académie des Sciences, 1993), and vol. 6 (Paris: Académie des Sciences, 1997).

Lavoisier, Antoine-Laurent, "Premier mémoire sur la destruction du diamant par le feu", *Mémoire de l'Académie des Sciences pour l'année 1772* (Paris: Imprimérie Royale, 1776), pp. 564–591.

Lavoisier, Antoine-Laurent, "Mémoire sur un moyen d'augmenter considérablement l'action du Feu & de la chaleur, dans les opérations chimiques", *Mémoires de l'Académie Royale des Sciences pour l'année 1782* (Paris: Imprimérie Royale, 1785), pp. 457–465.

Lavoisier, Antoine-Laurent, "Mémoire sur l'effet que produit sur les pierres précieuses un degré de feu très-violent", *Mémoires de l'Académie Royale des Sciences pour l'année 1782* (Paris: Imprimérie Royale, 1785), pp. 476–485.

Lehman, Christine, "What Is the True Nature of Diamond?", *Nuncius*, 31 (2016): 361–407.

Le Comte et la Comtesse du Nord; Anecdote Russe, mise au jour par M. le Chevalier du Coudray (Paris: Belin, 1782).

Licoppe, Christian, *La formation de la pratique scientifique: le discours de l'expérience en France et an Angleterre* (1630–1820) (Paris: La Découverte, 1996).

Lilti, Antoine, *Le monde des salons. Sociabilité et mondanité à Paris au XVIIIe siècle* (Paris: Fayard, 2005).

Macquer, Pierre-Joseph, *Dictionnaire de chimie contenat la théorie et la pratique de cette science, son application à la phisique, à l'histoire naturelle et à la médecine, et aux arts dépendans de la chimie*, 3 vols., (Paris: De l'Imprimérie de Monsieur, 1778, 1st ed. 1766).

Meusnier, Jean-Baptiste, "Description d'un appareil propre à manœvrer différentes espèces d'airs", *Mémoires de l'Académie Royale des Sciences pour l'année 1782* (Paris: Imprimérie Royale, 1785), pp. 466–475.

Optiz, Donald L., Staffan Bergwik, and Brigitte Van Tiggelen, ed., *Domesticity in the Making of Modern Science* (Basingstoke: Palgrave Macmillan, 2016).

Panopticon Lavoisier, http://moro.imss.fi.it/lavoisier/.

Perkins, John, "Chemistry Courses, the Parisian Chemical World and the Chemical Revolution, 1770–1790", *Ambix* 57 (2010): 27–47.

Perrin, Carleton E., "The Triumph of the Antiphlogistians", in Harry Woolf, ed., *The Analytic Spirit. Essays in the History of Science in Honor of Henry Guerlac* (Ithaca and London: Cornell University Press, 1981), pp. 40–63.

Pinault Sørensen, Madeleine, "Madame Lavoisier, dessinatrice et peintre", *La Revue. Musée des Arts et Métiers, 6* (1994): 23–25.

Pomian, Krzysztof, *Collectionneurs, amateurs et curieux. Paris, Venise, XVIe–XVIIIe siècle* (Paris: Gallimard, 1987).

Poirier, Jean-Pierre, "La contribution d'Armand Seguin (1767–1835) aux programmes de recherche de Lavoisier", in Antoine-Laurent Lavoisier, *Correspondance*, vol. 6 (Paris: Académie des Sciences, 1997), pp. 427–438.

"Premier essai du grand verre ardent de M. Trudaine, établi au Jardin de l'Infante, au commencement du mois d'octobre 1774, Par M.rs Trudaine de Montigny, Macquer, Cadet & Brisson", *Mémoires de l'Académie Royale des Sciences pour l'année 1774* (Paris: Imprimérie Royale, 1778), pp. 62–72.

Priestley, Joseph, *The Doctrine of Phlogiston Established and that of the Composition of Water Refuted* (Norththumberland: Bryne, 1803, 1st ed. 1800).

Pyenson, Lewis, and Jean-François Gauvin, eds., *L'art d'enseigner la physique. Les appareils de démonstration de Jean-Antoine Nollet 1700–1770* (Sillery: Septentrion, 2002).

Roberts, Meghan, *Sentimental Savants. Philosophical Families in Enlightenment France* (Chicago: University of Chicago Press, 2016).

Schaffer, Simon, *La fabrique des sciences moderne* (*XVIIe–XIXe siècle*), traduit de l'anglais par Frédéreique Aït Touati, Loïc Marcou, Stéphane Van Damme (Paris: Seuil, 2014).

Schapira, Nicolas, *Maîtres et secrétaires* (*XVI–XVIIIe siècles*). *L'exercice du povuoir dans la France d'Ancien Régime* (Paris: Albin Michel, 2020).

Schiebinger, Londa, *The Mind Has No Sex? Women in the Origins of Modern Science* (Cambridge/London: Harvard University Press, 1991).

Schiebinger, Londa, *Nature's Body. Gender in the Making of Modern Science* (New Brunswick: Routledge, 1993).

Shapin, Steven, Simon Schaffer, *Leviathan and the Air-Pump. Hobbes, Boyle, and the Experimental Life* (Princeton: Princeton University Press, 1985).

Van Damme, Stéphane, *Paris, capitale philosophique. De la Fronde à la Revolution* (Paris: Odile Jacob, 2005).

Vidal, Mary, "David Among the Moderns: Art, Science and the Lavoisiers", *Journal of the History of Ideas*, 56 (1995): 592–623.

Young, Arthur, *Travels during the years 1787, 1788 and 1789* (London: 1792).

CHAPTER 8

Musical Bodies: Materiality, Gender, and Knowledge in Musical Performance in 18th-century France

Amparo Fontaine

To the eighteenth-century mind, musical performance was a highly visual and corporeal phenomenon. Historical studies have identified the development of a rich musical culture in France during this period, stressing both the surprisingly varied literature about music and the prominent place of music in social life. In the second half of the century, France saw an unprecedented number of performers on musical instruments, including both musicians who wished to make a living from performing music and musical amateurs coming from the social elites.[1] The increasing fashion for learning to play music coincides with an avalanche in the number and types of musical instruments that were manufactured, invented, and commercialised in this period. The extended interest on playing on musical instruments, this essay shows, was associated with the stress placed upon the cultivation and display of the body in polite culture. This essay's main argument is that performers of musical instruments expressed and mobilised different understandings of the physical body. Musical performers and musical instruments appropriated and negotiated contemporary physiological and medical knowledge, as well as changing moral, social, and political values. Accordingly, the structure of this essay moves from examining the appreciation of musical instruments *as* human bodies, to the consideration of the performer's body as instrumental to notions of physiology, gender, feeling, and musical taste.

Musical learning, this essay argues, involved a process of regulating body language which was transferred to polite sociability from the court. This body language was gendered: the performer's body followed to a great extent a "feminine" ideal in music, which, according to Matthew Head, was overarching in

1 See e.g. Boris Schwarz, *French Instrumental Music Between the Revolutions, 1789–1839* (Boston, MA: Da Capo Press, 1987); Bruno Brévan, *Les changements de la vie musicale parisienne de 1774 à 1799* (Paris: Presses Universitaires de France, 1980).

 DOI:10.1163/9789004512610_010

eighteenth-century German culture.[2] Interrogating how contemporaries crafted this ideal of the musical body, this essay draws heavily on published didactic music manuals designed to teach how to play musical instruments. The number of these published manuals rose considerably over the century, a clear demonstration of the growing demand for these skills.[3] Music manuals are examined in tandem with other prescriptive literature, including deportment manuals and fencing, dancing, and acting methods, which taken together shed new light on the place of the musical body alongside other languages of the body and politeness. Additionally, representations of the body in non-prescriptive literary sources reveal how these notions of the body were appropriated and contested. The visual representation of performers, critiques to musical performances, physiological treatises, and musical instruments all speak to the existence of different constituencies of musical performance and ideals of the body.

The "performative turn" in musicological studies has stressed the crucial role played by musical performers – hitherto relegated to second place in the history of music – and a flurry of studies have focused on a broad notion of "musical gesture" in musical theory and analysis.[4] Nevertheless, strikingly little scholarly attention has been paid to the historical study of what the musical performing body meant for men and women of the eighteenth century, based on considerations of their own cultural, social, and scientific concerns.[5]

This essay establishes three models of the musical body adopted by musical performers, which I label the "courtly body", "mechanical body", and "expressive body". While the courtly body crosses most of the century, the mechanical and expressive bodies emerged around mid-century and co-existed until the end of the century, representing two deviations of the courtly body. Mechanical and expressive bodies involved a specific range of body motions and training, and expressed materially in different types of musical objects, such as the musical automata and the piano-forte. They were respectively associated with different social actors and were grounded in contemporary medical views. On the

---

2 Matthew Head, *Sovereign Feminine: Music and Gender in Eighteenth-Century Germany* (Berkeley: University of California Press, 2013). See also Richard Leppert, *The Sight of Sound. Music, Representation and History of the Body* (Berkeley: University of California Press, 1995).

3 Philippe Lescat, *Méthodes et traités musicaux en France 1660–1800, réflexions sur l'écriture de la pédagogie musicale en France* (Paris: Institut de pédagogie musicale et chorégraphique-la Villette, 1991).

4 See Lydia Goehr, *The Imaginary Museum of Musical Works: An Essay in the Philosophy of Music* (Oxford: Oxford University Press, 1992); Anthony Gritten, Elaine King, ed., *Music and Gesture* (London and New York: Routledge, 2006).

5 Some exceptions include Elisabeth Le Guin, *Boccherini's Body. An Essay in Carnal Musicology* (Berkeley: University of California Press, 2006); Rebecca Cypess, *Curious and Modern Inventions: Instrumental Music as Discovery in Galileo's Italy* (Chicago: University of Chicago Press, 2016).

one hand, the notion of the body as a machine underpinned the association of instruments and performers and justified mechanistic views of practice, which climaxed in the celebrated figure of the "virtuoso" performer. Similarly, the proliferation of android musical automata in this period not only showed that music could result from machines, but also that performers functioned like machines.

On the other hand, the emergence of the expressive body responds to new notions of naturalness, feeling, authenticity, and interiority, for which the second half of the century has often been labelled the "Age of Sensibility".[6] This rubric of the body was eventually embraced by musical amateurs, who contested "excessive" musical training and viewed musical sensibility as both an intimate experience and an effective way for claiming membership in a wider community of "sensible" people.[7] Drawing upon medical views of the body as flux, movement, and vibration, the ideal of the expressive body was paired with the development of new "expressive" musical instruments and technologies at the time, which allowed a broader spectrum of sounds and nuances. Ultimately, this essay's focus on musical bodies sheds new light on wider debates in the period, including the degree of mechanisation of the body versus the expression of feeling, the "artificiality" of courtly life versus the values of naturalness and authenticity in civil society, and the increasing hostility to ideals associated to the "feminine" in late eighteenth-century France.

1 Musical Instruments as Human Bodies

Since the Renaissance, musical instruments had largely been associated with the visual appearance and physiology of human bodies.[8] Musical instruments both embodied and conveyed knowledge about human physiology, and informed about the gendered and social features of their performers and

6 Northrop Frye, 'Towards Defining an Age of Sensibility', *ELH*, 23, 2 (1956): 144–52; Jessica Riskin, *Science in the Age of Sensibility: The Sentimental Empiricists of the French Enlightenment* (Chicago: The University of Chicago Press, 2002), pp.5–8. See also Jean Starobinski, *Jean-Jacques Rousseau: la transparence et l'obstacle* (Paris: Gallimard, 1976); Dorinda Outram, *The Body and the French Revolution. Sex, Class, and Political Culture* (New Haven, CT: Yale University Press, 1989); Robert Chartier, *The Cultural Origins of the French Revolution* (Durham, NC: Duke University Press, 1991).

7 For similar claims in the visual arts and the history of reading, see Charlotte Guichard, *Les Amateurs d'art à Paris au XVIIIe siècle* (Seyssel: Champ Vallon, 2008); Anne Vincent-Buffault, *The History of Tears* (London: Macmillan, 1991); Robert Darnton, *The Great Cat Massacre: And Other Episodes in French Cultural History* (New York: Basic Books, 2009).

8 Penelope Gouk, *Music, Science, and Natural Magic in Seventeenth-Century England* (New Haven, CT: Yale University Press, 1999).

listeners. String instruments were mostly associated to female bodies. This association was entrenched in the instruments' form, materiality and visual decoration. Instruments like viols, lutes, violins, guitars, and hurdy-gurdies often displayed carved scrolls with busts of men and women, angels, demons, and animals. This decorative motif spread throughout Europe.[9] In France, the majority of eighteenth-century instruments with heads represented women.[10] This preference could be associated with the fashion for female singers in this period; yet, significantly, it relates to the view of music and musical instruments as eminently "feminine".

Furthermore, the bodies of musical instruments tied them to specific types of owner, performer, and listener. The fact that the heads of females carved onto the scrolls of instruments were often richly ornamented suggests these instruments were frequently displayed before well-to-do audiences. Ornamented busts conformed to the fashion and taste of the same niche of consumers. Some viols from the first decades of the century, like the *pardessus de viole* made by Jean Baptiste Voboam in 1719 (Figure 8.1), bore heads wearing diadems, signs of sovereignty or distinction which probably alluded to the noble origins of the instruments' patrons, owners, or listeners. Moreover, some carvings displayed "exotic" fashions and contained precious materials, as with many musical instruments deemed as curiosities and objects of collecting. Thus, these visual elements were instructive about the specific performativities and social settings that were bound up with certain musical instruments.

Carved female busts have frequently been associated with musical instruments as an object of seduction. The curved body of the viols was especially suggestive of the shape of women's bodies, upon which men demonstrated their skilful touch. The body of the performer nestled against the body of the viol, in intimate contact with the bust. Yet, when carvings represented men, these heads also participated in intimate settings. The "musical lesson" or "duet" depicted by Jean-Marc Nattier in 1710 (Figure 8.2), represented a viol with a bearded male head. While the viol represented a masculine body, it participated in a scene of gallant courtship, in which a man and a woman are shown singing Nicolas Bernier's air "On ne peut s'empêcher d'aimer" from a music sheet. In this image, the musical instrument acts as a third person, witnessing the complicity between the musicians. Nattier's detailed and reliable depiction of the viol and hands of the performer frames the image as if

9 See e.g. Jeremy Montagu, *The World of Baroque and Classical Musical Instruments* (Woodstock, NY: Overlook Press, 1979).

10 See instruments in Florence Gétreau, *Aux origines du Musée de la Musique: les collections instrumentales du Conservatoire de Paris. 1793–1993* (Paris: Klincksieck, 1996).

FIGURE 8.1 Jean Baptiste Voboam, Pardessus de viole with diadem (1719). Coll. Musée de la Musique, Paris. INV. N°: E.998.11.1. © Cité de la musique - Philharmonie de Paris / Photo: Jean-Philippe Échard

FIGURE 8.2 Jean-Marc Nattier, "La Leçon de Musique" (France, 1710). Oil on canvas, 131 × 99,5 cm. Coll. Musée de la Musique, Paris. INV. N°: E.997.13.1. © Jean-Marc Anglès

to show three equally relevant bodies. The delicate skills demonstrated by the viol player mirrored the manoeuvres he was making towards the woman. The picture thus represents a sequence of touch: from the lady touching the score, to the bodies of the musicians close to the point of almost overlapping, and the violist touching the humanised body of the musical instrument. The polysemy of the word *toucher* in eighteenth-century French language at the time – meaning the sense of touch, emotional impact, as well as "to play" (*joüer*) on a musical instrument – suggests the imbrication of musical performance with the senses, the passions, and materiality.[11]

That musical instruments were built as human bodies is not surprising, given that instruments were endowed with "character". Eighteenth-century writers conceptualised the relationship between music and the passions through a broad notion of character. This notion followed the wide-ranging efforts exerted throughout the century towards conceptualising and classifying individual and national characters. Music was a protagonist in debates of national character in the three musical *querelles* that spanned the century and divided public opinion around national styles in opera.[12] The concern with character in writings about music thus reveal that music was mutually imbricated with morality, national identity, the passions, and language. According to eighteenth-century dictionaries, *caractère* was a writing or printing mark or graphic sign, as well as a marked "disposition" or "inclination" of the soul.[13] The notion of character is found in musical manuals and scores throughout the eighteenth century with the aim of indicating the predominant passion of a piece of music in the score. Characters in musical performance were, thus, a means of binding passions to specific types of sound and instrumental techniques. Furthermore, each instrument was endowed with a specific character or a range of characters determined by its materiality, tonal and volume range, timbre, as well as by performance traditions and musical genres. Referring to a musical instrument's character was a way of categorising its desired sonority, yet it also envisaged the musical instrument as similar to a human whose emotions were embodied, malleable and morally disciplined.

Musical instruments were metaphors for the body, as well as tools to explain how music affected the body. They were thus connected to changing conceptions of human anatomy and the functioning of the body. Since the Renaissance, musical instruments had been models of "resonance" and were identified as metaphors of the "vibrating strings" that composed the body.[14] Moreover,

11 See Luc Charles-Dominique, "'*Jouer*', '*sonner*', '*toucher*': Une taxinomie française historique et dualiste du geste musical", *Cahiers d'ethnomusicologie*, 14 (2001): 111–123.

12 See e.g. Andrea Fabiano, *Histoire de l'opéra italien en France (1752–1815). Héros et héroïnes d'un roman théâtral* (Paris: CNRS Editions, 2006); Mark Darlow, *Dissonance in the Republic of Letters: The Querelle des Gluckistes et de Piccinnistes* (London: Routledge, 2013).

13 "Caractère", in *Dictionnaire de l'Académie française* (Paris: 1694). The ARTFL Project, University of Chicago: http://artfl-project.uchicago.edu.

14 Penelope Gouk, Ingrid Sykes, "Hearing Science in the Eighteenth Century", *Journal of the History of Medicine and Allied Sciences*, 66, 4 (2011): 507–545, pp. 532–533; Veit Erlmann, *Reason and Resonance: A History of Modern Aurality* (New York: Zone Books, 2010), pp. 117–118.

after Descartes instruments epitomised mechanistic understandings of the human body, which was increasingly portrayed as a machine, most notably by Julien Offray de La Mettrie in the middle of the eighteenth century.[15] This view echoed the wider conceptualisation of musical instruments as "machines" in contemporary dictionaries, compendia of musical instruments, and academic reports.[16] Not only were musical instruments described as machines and metaphors of the body, but human bodies were also often imagined as stringed instruments or harpsichords. Through his metaphor of the "homme-clavecin", Denis Diderot associated various anatomic models with bodies and instruments throughout the eighteenth century; he represented the body as a resonant body whose fibres equated to the strings of musical instruments, and viewed it as a machine, like La Mettrie.[17] Furthermore, during the second half of the century, there was a renewed stress on the phenomena of vibration, sympathy, and the nerves, particularly in relation to the new sensibility and vitalist physiology. The musical instrument now came to be conceived as a moving body which triggered powerful emotions and sensations through the action of vibrating.[18] As sensation and feeling became more important, so musical instruments were regarded as embodiments of feeling.

Consequently, these humanised and socialised musical instruments were feeling bodies, which in turn, made bodies feel. The expressive body of the musical instrument – visually ornamented, endowed with gender, social status, character, morality, and anatomy – was translated into the expressive requirements to the performer, which, I shall argue, relied on the performer's physical body.

2 Courtly Bodies

Since the beginning of the century, it is striking to notice the great emphasis that didactic music manuals placed on the visual appearance and body gestures of the musical performer. Manuals placed strong emphasis on the

15 Julien Offray de La Mettrie, *L'Homme-machine* (Leyden: Elie Luzac, 1748).

16 See "Instrumens, (*Musiq. & Luth.*)", in *Encyclopédie* (Paris: 1765), edited by Denis Diderot and Jean le Rond d'Alembert, vol. 8, pp. 803–804; Académie Royale des Sciences, *Machines et inventions approuvées par l'Académie royale des sciences* (Paris: 1735–1777).

17 Philippe Sarrasin Robichaud, *L'Homme-clavecin, une analogie diderotienne* (Paris: Classiques Garnier, 2017).

18 Erlmann, *Reason and Resonance*; James Kennaway, *Bad Vibrations. The History of the Idea of Music as a Cause of Disease* (Aldershot: Ashgate, 2012).

posture, facial gestures, and bodily movements which did not always rely on the ways sound and instruments were to be handled. The publication of music manuals increased considerably throughout the century, and those devoted solely to the teaching of musical instruments surpassed theoretical and singing methods from the 1750s onwards.[19] These music manuals are evidence of the growing practice of and desire for musical performance, as well as of the ongoing circulation of new musical instruments. Most manuals were authored by teachers, which carved out a particular domain of expertise for themselves that they advertised and commercialised through the growing culture of consumption. Despite well-known composers such as François Couperin, Francesco Geminiani, and Jean-Philippe Rameau, the majority of these instructional texts were written by music teachers about whom we know very little. Philippe Lescat's work on music didactic manuals shows that most authors belonged to a small, literate bourgeoisie – often related to small-scale trade and craftsmanship – who self-funded their works, and most of them taught privately, given the scarcity of institutional training for instrumental performance.[20]

Most manuals addressed both amateur and professional types of performer at the same time by offering a range of pieces written at different levels of difficulty. Yet some music manuals addressed a specific type of musical performer, linked to a specific instrument: for instance, didactic manuals for learning the mandolin and guitar often recommended these instruments for amateur performers, whereas the violin or the transverse flute were intended for more technically advanced students.[21] Similarly, a number of manuals were devoted to female performers, especially during the second half of the century. Yet most musical instruments and genres labelled as "easy", were dedicated to both women and male amateur performers.[22]

The taste for musical performance among the elite was greatly influenced by the broader tradition of cultivating the body in courtly life. As Baldassare

---

19 Lescat, *Méthodes et traités musicaux*.

20 Lescat, *Méthodes et traités musicaux*, pp. 17–20.

21 See Joseph Carpentier, *Méthode distribuée par leçons pour apprendre en peu de temps à jouer de l'instrument appellé cytre où guitthare allemande* (Paris : Joseph Carpentier, 1771). See also Ancelet, *Observations sur la musique, les musiciens, et les instrumens* (Amsterdam [i.e. Paris]: 1757), p. 28; Lescat, *Méthodes et traités musicaux*, p. 96.

22 For example, see Giovanni Battista Gervasio, *Méthode très facile pour apprendre à jouer de la mandoline à quatre cordes, instrument fait pour les dames* (Paris: Boüin, ca. 1767); François Boüin, *La Vielleuse habile, ou nouvelle méthode courte, très facile et très sure pour apprendre à jouer de la vielle* (Paris, 1761).

Castiglione had written in 1528, the courtier had to follow strict standards of behaviour, which included paying great attention to bodily movements and appearance.[23] In seventeenth-century France, numerous treatises addressed social performance and manners under the headings of "honnêteté", "civilité", and "politesse".[24] These values were intertwined in the pursue of taste, which, as scholars have documented, was a mark of social privilege and encompassed social, moral, and aesthetic criteria.[25] Louis XIV, with his celebrated posture and sober facial expression, cultivated his physical appearance carefully, and established a protocol that regulated physiognomy and body language throughout the century.[26] The king exemplified the tasteful body, and his physical model was replicated throughout the eighteenth century in different practices and social settings, such as physiognomy, dance, theatre, oratory, salon conversation – and, I contend, musical performance.

It is striking to notice the resemblances between the prescriptive literature on those varied forms of physical training and manuals instructing readers in musical performance. This similarity suggests, on the one hand, that musical performance was considered to belong with other physical activities central to the education of young aristocrats. Indeed, the education of the French nobility integrated music alongside dancing, fencing, and horse-riding. Kate van Orden has documented the uses of music as a crucial element for military and civil disciplining of sixteenth- and seventeenth-century French aristocrats.[27] On the other hand, the similarity between music and physical training reveals the extent to which ideals of courtliness were extended and pursued beyond the court. The cultivation of the body was a means by which musical performers bound themselves to a broader ideal of courtliness. The popularity of both the figure of Louis XIV and the seventeenth-century deportment treatises – as expressed in their many editions throughout the eighteenth century – suggests

23 Baldassare Castiglione, *Il cortegiano* (Venice: 1528).

24 See Jean-Baptiste Morvan, *Réflexions sur la politesse des mœurs* (Paris: 1698); Antoine de Courtin, *Nouveau traité de la civilité qui se pratique en France parmi les honnestes gens* (Amsterdam: 1708).

25 See Rémy Gilbert Saisselin, *Taste in Eighteenth Century France. Critical Reflections on the Origins of Aesthetics; or, an Apology for Amateurs* (Syracuse, NY: Syracuse University Press, 1965); Dena Goodman, *The Republic of Letters: A Cultural History of the French Enlightenment* (Ithaca, NY: Cornell University Press, 1996).

26 Norbert Elias, *The Civilising Process*, trans. Edmunf Jephcott (1939; Oxford: Blackwell Publishing, 2000). See also Peter Burke, *The Fabrication of Louis XIV* (New Heaven, CT: Yale University Press, 1992); Sarah R. Cohen, *Art, Dance, and the Body in French Culture of the Ancien Régime* (Cambridge: Cambridge University Press, 2000).

27 Kate van Orden, *Music, Discipline, and Arms in Early Modern France* (Chicago: The University of Chicago Press, 2005).

that ordinary literate people were eager to cultivate body language as a means for social advancement outside the court.[28]

This motivation was explicitly mentioned by Pierre Rameau, author of a dance treatise published in 1725. Rameau prescribed bodily movements based on court etiquette, something which, he claimed, was not only useful for young pupils to learn, "mais encore aux personnes honnêtes & polies, & qui leur donne des regles pour bien marcher, saluer & faire les reverences convenables dans toutes sortes de compagnies".[29] Therefore, Pierre Rameau's method was simultaneously a dancing method and a deportment treatise for court etiquette. Music manuals and music criticism, in turn, portrayed the ideal musician as a person of taste, with appropriate displays of manners and social etiquette. The *Almanach Musical* referred to the self-styled musician Jean-Baptiste Davaux, as "cet Amateur plein de goût [qui] séduit sans doute autant par ses manières et son charme naturel que par sa musique".[30]

Activities such as fencing, dancing, and music shared a common vocabulary of gestures and physical appearance that was grounded in politeness and courtly behaviour. For instance, Pierre Rameau advised his dance students to stand with one foot in front of the other, just as deportment treatises had advised for standing in courtly contexts (See Figure 8.3).[31] Later, Nicolas Demeuse described the same posture in detail in his treatise on fencing.[32] This custom, closely associated with the figure of Louis XIV, also related to the earlier fashion for showing off the possession of luxurious shoes, some of which were ornamented with jewellery on the buckles.[33]

This standing posture was also largely cultivated in violin performance. Among his fashion plates and stylish portraits with costume, the engraver Nicolas Arnoult represented a violinist from *Les Vingt-quatre Violons du Roi* – a small string orchestra that played for the court at Versailles from 1626 until mid-eighteenth century (Figure 8.4). The violinist stands with one foot in front

28 Elias, *The Civilising Process*. For a debate on courtliness in French salon sociability, see Brian Cowan, "Public Spaces, Knowledge, and Sociability," in Frank Trentmann, ed., *The Oxford Handbook of the History of Consumption* (Oxford: Oxford University Press, 2012), 251–66.

29 Pierre Rameau, *Le maître à danser* (Paris: Jean Villette, 1725).

30 Quoted in Robert Henri Tissot, "Jean-Baptiste Davaux: un amateur-compositeur", in Robert-Henri Tissot, Camille Bellissant, ed., *Le Concert des Amateurs à l'Hôtel de Soubise (1769–1781)* (Grenoble: MSH-Alpes, 2000), p. 27.

31 Rameau, *Le maître à danser*, p. 3.

32 Nicolas Demeuse, *Le maître d'escrime, ou L'art des armes* (Paris: Chez Delarue, ca.1778), p. 126.

33 Carol Lee, *Ballet in Western Culture: A History of Its Origins and Evolution* (London-New York: Routledge, 2002), p. 81.

FIGURE 8.3 "De la Position du Corps", in Pierre Rameau, *Le maître à danser* (Paris, 1725), plate 3. Engraving by the autor. Bibliothèque nationale de France

FIGURE 8.4 Nicolas Arnoult, "Joueur de violon du roi. Un membre des Vingt-quatre Violons du Roi, en habit" (Paris, 1688). Engraving. Bibliothèque nationale de France

of the other, while displaying a courtly outfit that highlights his laboured visual appearance and courtly demeanour. During the eighteenth century, violin performers were often represented in this posture, and some didactic manuals explicitly recommended it to their violin students up until the nineteenth century.[34] The violin teacher Jean-Baptiste Labadens, for instance, advised his students to carefully examine and learn from the engravings he provided in his manual, including one of a performer standing with one foot ahead.[35]

3 Feminine Touch

The ideal of a "courtly body" of the musical performer comprehended a series of skills, which the musician should learn to master. Music didactic manuals especially stressed features such as grace, naturalness, easiness, and softness, which were crucial in polite sociability and were considered to be more natural to women.[36] Grace was central in treatises on demeanour, and was deemed a feature that students in all physical activities should cultivate throughout the century. Although grace entailed a spiritual dimension – associated with divine grace – it existed in reality through the performance of the body.[37] According to the article "Grâce" by Voltaire in Diderot's and d'Alembert's *Encyclopédie*, grace in a work of art was more than accomplishment: it meant a special attraction, which in the case of human beings, was linked with the command of the body and manners.[38] In musical performance, both female a male performers were taught to cultivate grace through the command of their physical

---

34 Robin Stowell, *Violin Technique and Performance Practice in the Late Eighteenth and Early Nineteenth Centuries* (Cambridge: Cambridge University Press, 1990).

35 Jean Baptiste Labadens, *Nouvelle méthodes pour apprendre à jouer du violon* (Paris, 1772), p. 14.

36 Ursula Rempel, "The Complete Female: Musical 'Accomplishment' in the Late Eighteenth Century", *Man and Nature: Proceedings of the Canadian Society for Eighteenth-Century Studies*, 8 (1989): 39–48. See, also, Julie Candler Hayes, "Sex and Gender, Feeling and Thinking: Imagining Women as Intellectuals", in Daniel Brewer, ed., *The Cambridge Companion to the French Enlightenment* (Cambridge: Cambridge University Press, 2014), pp. 91–104; Anthony J. La Vopa, *The Labor of the Mind: Intellect and Gender in Enlightenment Cultures* (University of Pennsylvania Press, 2017).

37 See Cohen, *Art, Dance, and the Body*.

38 François-Marie Arouet de Voltaire, "Grâce", in Denis Diderot and Jean le Rond d'Alembert, ed., *Encyclopédie, ou dictionnaire raisonné des sciences, des arts et des métiers, par une société de gens de lettres*, vol. 7 vol. 7 (Paris: 1752), p. 805.

body. However, women up to the end of the century were often considered to be naturally endowed with grace.

Deportment manuals advised women to act gracefully in polite sociability, as if their natural purpose was to please an audience.[39] When performing music, moreover, women were to choose musical instruments that enhanced their "graceful" bodies. Beyond the above-mentioned association of musical instruments with female bodies, certain instruments were specifically associated with a notion of femininity through their delicate sounds or the ways in which they displayed the physical attributes of female performers. The fashion for playing musical instruments increased considerably during the second half of the century and was transferred to women from the social elites and learned bourgeoisie from the court. Marie-Antoinette and female members of the Bourbon family allegedly played music and were depicted with musical instruments such as guitars and harps, contributing to the increasing fashion for these instruments as precious commodities.[40] Thereof, such instruments became highly desirable among elite female performers during the last third of the century. A dictionary, created by the historian and lawyer Philippe Macquer, indicated that women playing the guitar allowed them to display the graces with which they were naturally endowed.[41] The posture and "attitude" required by certain musical instruments was especially emphasised in women of marriageable age.[42] This fact was widely acknowledged by authors of teaching manuals. Michel Corrette, the music composer and organist that wrote at least seventeen music manuals for a variety of musical instruments between 1738 and 1784, encouraged women to continue performing even after marriage.[43]

Earlier in the eighteenth century, Couperin claimed that the hands of women performers possessed the flexibility and softness required to play the harpsichord naturally, which responded to women's inner physiological structure:

---

39 Rempel, *The Complete Female*, p. 41.

40 Sylvette Milliot, "La harpe au XVIIIe siècle à travers les documents iconographiques," in Hélène Charnassé, ed., *Instruments et musique instrumentale* (Paris: CNRS, 1986), pp. 141–160.

41 Philippe Macquer, *Dictionnaire raisonné universel des arts et métiers*, vol. 2 (Paris: 1773), p. 621.

42 See François-Alexandre-Pierre de Garsault, *Notionnaire ou Memorial raisonné de ce qu'il y a d'utile et d'interessant dans les connoissances acquises depuis la création du monde jusqu'à présent* (Paris: 1761), p. 54.

43 Michel Corrette, *Le maître de clavecin pour l'accompagnement, méthode theorique et pratique* (Paris: 1753).

FIGURE 8.5 Pierre Pasquier, "Portrait d'une jeune femme jouant de la guitare devant une colonnade" (Paris, 1779). Watercolour, ink and pigments. INV. N°: E.2002.5.1. Photo: Claude Germain © Cité de la musique – Philharmonie de Paris

> J'ai déjà dit, que la souplesse des nerfs contribue beaucoup plus, au bien-jouer, que la force; ma preuve est sensible dans la différence des mains des femmes, à celles des hommes; et de plus, la main gauche des hommes, dont ils se servent moins dans les exercices, est communément la plus souple au clavecin.[44]

This quote reveals the extent to which musical performance followed models of the body that were based on physiological views, gendered and social expectations, as well as practical musical needs. For Couperin, touching was gendered, yet a "feminine" touch constituted an ideal to be achieved by male and female performers alike.[45] This feminine character was persistent in musical learning until the late 1780s, when the emergence of a new model of masculinity opposed to "effeminacy".[46]

Therefore, for most of the century, authors of music manuals advised their male students to practise body posture and appearance thoroughly. In 1786, the composer and harpist Louis-Charles Ragué conceived his manual for harp playing for both women and men, despite the harp was by then one of the most fashionable instruments for elite women. Ragué advised his students to practise in front of a mirror, with the aim of obtaining the graceful and tasteful gestures that were natural to women, and that the figure of the harp contributed to accentuate.[47] Using a mirror for practising both musical instruments and singing, was indeed a common custom for women and men since the beginning of the century. In 1707, the manual written by Jacques-Martin Hotteterre, the most celebrated member of a dynasty of woodwind instrument makers, had listed a number of body gestures and postures necessary to achieving a tasteful and graceful body. Claiming that the body "attitude" should be "pleasurable for the eyes",[48] Hotteterre advised male and female students to follow the posture modelled in an engraving included in his manual, as if they were facing a mirror (Figure 8.6).[49]

44 François Couperin, *L'art de toucher le clavecin* (Paris, 1716), pp. 12–13.

45 See Head, *Sovereign Feminine*.

46 Robert L. Herbert, *David, Voltaire, 'Brutus', and the French Revolution: An Essay in Art and Politics* (New York: Viking, 1972).

47 Louis-Charles Ragué, *Principes de harpe suivis de 18 airs* (Paris: 1786), p. 2.

48 Hotteterre, *Principes de la flute*, p. 3.

49 See Jacques-Martin Hotteterre, *Principes de la flute traversiere ou de la flute a bec* (Paris: 1707), pp. 2–3. Lescat, *Méthodes et traités musicaux*, p. 128.

FIGURE 8.6 Jacques-Martin Hotteterre, *Principes de la flute traversiere ou de la flute a bec*. Engraving by Picart (Paris, 1707). Bibliothèque nationale de France

When explaining the posture of the musical performer, most of the physical prescriptions found in performance manuals described the movement and position of the head and the face, which, except in the case of wind instruments, are generally not involved in the production of sound. Studies of the history of physiognomy have documented an elaborated language of facial expression during the eighteenth century.[50] As Colin Jones has pointed out, smiling was a rather scarce practice in the first half of the eighteenth century, due to a restricted economy of facial expression in the court, epitomised – once again – by Louis XIV.[51] Similarly, in music, grimaces, or any marked expression on the face, were widely condemned. All "awkward" movements were to be avoided, including stooping, placing the hands too close to, or too far from, the body, and any sort of contraction of the face or body.[52] The "courtly body" in musical performance, thus, implied a restricted economy of body movements. This limited repertoire of body and facial motions, however, was partly challenged during the second half of the century in accordance with new physiological knowledge – I will come back to this below.

Grimaces or contortions denoted effort, difficulty, awkwardness, and discomfiture. Musical performers, instead, should pursue "naturalness" in playing, which necessitated concealing any difficulty in playing. In this sense, the pursuit of naturalness followed Castiglione's stress on *sprezzatura* as a principle of courtier behaviour, which was omnipresent in different social, artistic, and literary arenas in the eighteenth century.[53] Yet, naturalness, paradoxically, was the result of hard work obtained through methodical and thorough practice.[54] Although musical performance was acknowledged to be difficult, music methods often stressed that the performer had to appear effortless and show an "air of ease".[55] This quality, again, was more "natural" to female performers, and was also associated with amateur music.

---

50 See Mary Cowling, *The Artist as Anthropologist: The Representation of Type and Character in Victorian Art* (Cambridge: Cambridge University Press, 1989); Anne Vincent-Buffault, *The History of Tears* (London: Macmillan, 1991).

51 Colin Jones, *The Smile Revolution in Eighteenth Century Paris* (Oxford: Oxford University Press, 2017).

52 See Antoine Mahaut, *Méthode pour apprendre en peu de tems à jouer de la flûte traversière* (Amsterdam and Paris: 1759), p. 3.

53 Elena Russo, *Styles of Enlightenment: Taste, Politics, and Authorship in Eighteenth-Century France* (Baltimore: The Johns Hopkins University Press, 2007).

54 Norbert Elias and Michel Foucault stressed the social structures and power relationships underpinning corporeal control and discipline. Elias, *The Civilising Process*; Michel Foucault, *Surveiller et punir: Naissance de la prison* (Paris: Gallimard, 1975).

55 See Couperin, *L'art de toucher le clavecin*, p. 11.

4 Mechanical Bodies

As the taste for instrumental music grew significantly in the second half of the century, so too did the demands placed upon instrumental performers. Musicians were increasingly moving away from courtly patronage and searching for new opportunities beyond the court. These new forms of support – in addition to contemporary claims about the need for an institution for music education coming from mouthpieces of instrumental music such as François-Joseph Gossec – have led historians to identify an increasing "professionalisation" of musicians in this period.[56] This process was paralleled by an intensification of the practice and study required of instrumental performers.

The amount and intensity of practice was key to distinguishing amateur from professional performers.[57] The authors of didactic music manuals believed that practice could be reasoned and systematised, and that the body could be subjected to methodical treatment. This belief was rooted in mechanistic views of the body, as articulated by contemporary anatomists. The words "mechanic" and "mechanical" were constantly repeated in prescriptive literature concerning corporeal performance. This language was used, most notably, by Jean-Philippe Rameau in the 1720s, and became largely extended in mid-century.[58] The ability of fingers to learn and "remember" music, as well as the methodical discipline of the senses required in music, were both regarded as mechanical. Fingers, moreover, were parts of a larger, interconnected machinery which included the whole body of the performer. Teaching manuals increasingly laid stress on the weight of the performer's body, the tension of the strings, and the ways air entered wind instruments, which reveals the extent to which manuals' authors incorporated contemporary language and knowledge coming from the incipient discipline of acoustics. Conversely, musical instruments and musical performers featured in physiological studies of hearing, physics, and human anatomy.[59]

---

56 See Hennebelle, *De Lully à Mozart*; T.C.W. Blanning, *The Triumph of Music: The Rise of Composers, Musicians and Their Art* (Cambridge: Belknap Press of Harvard University Press, 2008).

57 Lescat, *Méthodes et traités musicaux*, p. 99.

58 Jean-Philippe Rameau, *Piéces de clavecin avec une méthode pour la mécanique des doigts* (Paris: 1724; 1731); Herni L. Choquel, *La Musique rendue sensible par la méchanique* (Paris: 1759; 1762).

59 See Jacques Gamelin, *Nouveau recueil d'ostéologie et de myologie dessiné d'après nature* (Toulouse: 1779).

As Simon Schaffer points out, the spread of mechanistic theories about the animal economy was intrinsically connected to the spread of handbooks on physical discipline.[60] The role of physical discipline in musical performance increasingly moved away from being the occasion for moral discipline and preparation for social life, to becoming the occasion for cultivating agility through mechanical processes such as repetition. Consequently, the rise of images of machinery in musical performance was directly related to the rise of virtuoso performers on musical instruments over the same period. From this mechanistic point of view, the body of the virtuoso was an optimal performer: the skilful body-machine working at its best.

In the eighteenth century, the term "virtuose" was used for distinguished musicians, either women or men, singers or instrumentalists.[61] However, the figure of the virtuoso increasingly referred to soloists who excelled at performance on a single instrument, mostly men.[62] Since the foundation of the Concert Spirituel in 1725 – one of the first public concert venues which featured all sorts of musical innovations during religious festivities – the culture of virtuoso performance had been considerably encouraged.[63] In the last decades of the century, virtuoso performers proliferated significantly in relation to the growing culture of leisure and public spectacles, and became public "celebrities".[64] For Paul Metzner, the figure of the virtuoso emerged in different realms of Parisian culture, thanks to widespread interest in skilled performance, as well as increasing technical intensification.[65] The blend of spectacle and technical innovation was conspicuous in new instrumental forms that staged the virtuoso, such as the concerto, which featured one or a small group of soloists accompanied by an orchestra.[66]

---

60 Simon Schaffer, "Enlightened Automata", in Clark, Golinski, Schaffer, ed., *The Sciences in Enlightened Europe* (University of Chicago Press, 1999), 126–67, at p. 141.

61 "Virtuose", in *Dictionnaire de l'Académie française* (Paris: 1762). ARTFL project.

62 For exceptional female virtuoso instrumental performers, see Florence Gétreau, "Une harpiste au Concert Spirituel: Mademoiselle Schencker en 1765", *Musique-Images-Instruments*, 1 (1995): 178–181.

63 Sylvette Milliot, "Le virtuose international: une création du 18e siècle", *Dix-huitième siècle*, 25 (1993): 55–64, p. 59.

64 For the culture of celebrity in this period, see Antoine Lilti, *Figures publiques. L'invention de la célébrité, 1750–1850* (Paris: Fayard, 2014).

65 Paul Metzner, *Crescendo of the Virtuoso* (Berkeley: University of California Press, 1998).

66 Milliot, "Le virtuose international", p. 61; Simon P. Keefe, ed., *The Cambridge Companion to the Concerto* (Cambridge: Cambridge University Press, 2005).

5 Musical Automata

The proliferation of the musical virtuoso as public celebrity and embodiment of the mechanical body was mirrored in the proliferation of musical automata over the same period. The taste for musical automata was intermingled with that for virtuoso performances: both invoked a mechanistic physiology in the context of an emerging culture of spectacle and public display of technique.[67] The proliferation of automata during the eighteenth century has been the subject of a flurry of scholarship over the last two decades. Cultural and social historians of science have shown that android automata served to explain the body, emotions, and politics, while embodying the conjuncture between research on physics and mechanics on the one hand, and the cultures of luxury and consumption on the other.[68] The fact that many of eighteenth-century android automata represented musicians, has received considerably less attention.[69]

Throughout the eighteenth century, the taste for android automata, musical boxes, and mechanical organs, among other sonic objects, was embedded in the increasing consumption of luxury and leisure commodities, and attests to the growing taste for music.[70] Some mechanical instruments doubled as playing, notating, and listening artefacts, and flourished within a demand for accessible enjoyment of music among amateurs. The music resulting from these objects could be reproduced as many times as desired. As such, one might say that musical automata were listening devices that could reproduce music before the age of recording. The great inventiveness in mechanical objects and innovative sound artefacts in this period commodified sound beyond musical practice.

Musical automata lay at the intersection between notions of the body, research into music and sound, and material practices of instrument making.

---

67 Metzner, *Crescendo of the Virtuoso*; Robert M. Isherwood, *Farce and Fantasy: Popular Entertainment in Eighteenth-Century Paris* (Oxford: Oxford University Press, 1989).

68 Schaffer, "Enlightened Automata"; Jessica Riskin, *The Restless Clock: A History of the Centuries-Long Argument Over What Makes Living Things Tick* (Chicago: The University of Chicago Press, 2016).

69 Exceptions include Adelheid Voskuhl, *Androids in the Enlightenment: Mechanics, Artisans, and Cultures of the Self* (Chicago: The University of Chicago Press, 2013); Rebecca Cypess, "'It Would Be without Error': Automated Technology and the Pursuit of Correct Performance in the French Enlightenment", *Journal of the Royal Musical Association*, 142, 1 (2017): 1–29.

70 See Maxine Berg, *Luxury and Pleasure in Eighteenth Century Britain* (Oxford: Oxford University Press USA, 2005); Anne Gerritsen, Giorgio Riello, ed., *The Global Lives of Things: The Material Culture of Connections in the Early Modern World* (London: Routledge, 2015).

The flute player created by Jacques Vaucanson, which was successfully received among Parisian spectators, combined research on the physics of sound, the mechanics of the flute, and the performance techniques of flute players. Vaucanson presented his invention to the Académie Royale des Sciences in 1738, and published a description of his invention that explained the mechanism of sound production at length.[71] It was an object for learning about the physics of sound and body motions of the performer in the same way that his defecating duck was an object for learning about the internal mechanism of food digestion.[72]

Similarly, Marie Antoinette donated her "Joueuse de Tympanon", made by the clock-maker Peter Kintzing and the ebonist David Roentgen, for study and preservation at the Académie Royale des Sciences (Figure 8.7).[73] In a letter that communicated the Queen's bequest to the academicians, the physician François Lassone referred to the automata as an object worthy of examination in the academy, and stressed its "pleasing" physical attributes: "cette figure, dont les traits, les proportions et les ajustements sont fort élégants".[74] Adelheid Voskuhl has associated the automata's physical movements with eighteenth-century "sentimental culture".[75] Yet, behind those physical expressions, there was also the "courtly body" of the performer discussed above. Representing a well-to-do woman, the "Joueuse" embodied the enactment of grace and femininity in its highest degree. In this sense, she was both a model of the mechanical body and a model of courtly demeanour – I shall return to this blend between mechanical, feminine, and courtly bodies below.

Like musical performers, android automata such as the "Joueuse" gathered an audience of listeners.[76] The increasing popularity of mechanical musical instruments was nurtured by the audience's increasing thirst for spectacle and

71 Jacques Vaucanson, *Le mécanisme du fluteur automate, présenté à messieurs de l'Académie royale des sciences* (Paris: 1738), p. 10. See Voskuhl, *Androids in the Enlightenment*, pp. 29–30.

72 Riskin, *The Restless Clock*, pp. 132–45; Emma Spary, *Eating the Enlightenment: Food and the Sciences in Paris, 1670–1760* (Chicago: The University of Chicago Press, 2012), pp. 43–46.

73 For Kintzing and the culture of clockmaking, see Dietrich Fabian, *Kinzing und Roentgen: Uhren aus Neuwied* (Bad Neustadt: Verlag D. Pfaehler, 1984); Voskuhl, *Androids in the Enlightenment*, pp. 110–113; Metzner, *Crescendo of the Virtuoso*.

74 Letter by François Lassone (signed Versailles, 4 March 1785) in Académie des Sciences, Paris, *Pochette de séance 5 mars 1785*.

75 Voskuhl, *Androids in the Enlightenment*.

76 An engraving held at the Betteman Archive (New York) indicates the tickets' prices and performance times of Vaucanson's three automata. See also references to the musical automata made by Robert Richard in Constant Pierre, *Les facteurs d'instruments de musique. Les luthiers et la facture instrumentale. Précis historique* (Paris: Sagot, 1893), p. 98.

FIGURE 8.7 "Joueuse de Tympanon" (The dulcimer player). Made by Pierre Kinzing and David Roentgen (Neuwied, 1784). INV. N°: 07501-0001. © Musée des arts et métiers-Cnam, Paris / Photo J.C. Wetzel

public demonstration of skill. It is in this sense that Paul Metzner links the fashion for automata to the widespread interest in the performance of skill by the "virtuoso".[77] Furthermore, the android automaton itself was a reproduction

77 Metzner, *Crescendo of the Virtuoso*.

of the musical performer, understood as a mechanical body and the provider of an amusing spectacle of skill. The intermingling of human organisms, musical instruments, and machines discussed above was most radically embodied in musical automata. I have already suggested that the representation of a musical instrument as a "person" challenged the boundaries between the agent and the means of sound – the performer and the material object. Both the humanised musical instruments and android automata were not mere means of musical performance, but feeling objects which produced and experienced music in themselves. In android music automata, this blend of bodies and feeling was even more radical: the bodies of the musical instrument and that of the performer were interwoven in the same mechanical fabric, composing a single object. Thus, automata representing musical performers pushed the boundaries between the musical instrument and the musical agent even further. They replaced the performer by situating the production of sound within the material instrument.

6 Expressive Bodies

During the second half of the century, however, there was a reaction against the mechanisation and excessive disciplining of the body. The emergence of the "new sensibility" during this period, emphasised both feeling and sensorial experience, and expressed itself physically: individuals experienced feeling with their bodies, thereby aligning the inner and the outer.[78] Moreover, this new sensibility situated music in a privileged position for the expression of feeling.[79] Throughout the eighteenth century, musical performers were taught to achieve "expression", broadly understood. Yet, from mid-century, the musical performer increasingly moved away from a verbal model of expression to one grounded in the sensible and moving body.

The new sensibility involved a shift in the range of desirable sounds and passions in music, in relationship to the growing taste for Italian music and the association between the Italianate and the feminine. Jean-Jacques Rousseau's defence of Italian music, with his polemic engagement in the *Querelle des Bouffons* (1752–1754), was the cornerstone of new conceptions of both Italian

78 Anne C. Vila, *Enlightenment and Pathology. Sensibility in the Literature and Medicine of Eighteenth-Century France* (Baltimore: Johns Hopkins University Press, 1998); Riskin, *Science in the Age of Sensibility*. For the cult of sensibility, see note 6 above.

79 Downing A. Thomas, *Music and the Origins of Language: Theories from the French Enlightenment* (Cambridge: Cambridge University Press, 1995).

character and music.[80] Whereas French travellers had often remarked upon the sensuality and unruly emotions of Italians, and invoked gender stereotypes – Italy was a "feminine" country that contrasted with the "masculine" England and France – Italians and the Italianate were re-evaluated positively in relation to the cult of sensibility.[81] Furthermore, during the second half of the century, the characterisation of the unequalled "sensibility" of Italians and their "natural" susceptibility to music, was often linked to women's "natural" closeness to feeling and sensibility, and the feminine ideal that was sought in music. Additionally, the association of the new appraisal of music with the feminine was strengthened by the fact that Rousseau's camp in the *Querelle* was called "the queen's corner", and because Marie-Antoinette assisted in the reintroduction of Italian music. With the pursuit of expressive bodies, however, the link between the feminine and Marie Antoinette ultimately became problematic – I will come back to this below.

The new sensibility and the appraisal of the feminine was expressed in the fashion for certain musical instruments – including above-mentioned harps and guitars – and the pursuit of new sonorities. As previously discussed, eighteenth-century writers and musicians conceptualised the relationship between the passions and sound through a notion of musical character. During the second half of the century, Italian musical characters were incorporated into French compositions and manuals, yet musical characters also considerably expanded in number and type.[82] The expansion of characters followed a new demand for expressing the movement and dynamism of human passions in different artistic realms.[83] Crucially, the new stress placed on variety and movement drew upon significant changes within medical and scientific research concerning how the human passions were conceived – namely, a

80 Literature on the *Querelle des Bouffons* is abundant. See Alessandro Di Profio, *La révolution des Bouffons. L'opéra italien au Théâtre de Monsieur 1789–1792* (Paris: CNRS Éditions, 2003); Andrea Fabiano, Sylvie Bouissou, ed., *La 'Querelle des Bouffons' dans la vie culturelle française du XVIIIe siècle* (Paris: CNRS éditions, 2005); Andrea Fabiano, *Histoire de l'opéra italien en France*; David Charlton, *Opera in the Age of Rousseau: Music, Confrontation, Realism* (Cambridge: Cambridge University Press, 2012). Jean-Jacques Rousseau's writings on music include: *Dissertation sur la musique moderne* (Paris: 1743), *Lettre sur la musique française* (Paris: 1753), *Dictionnaire de musique* (Paris: 1768) and *Essai sur l'origine des langues, ou il est parlé de la mélodie et de l'imitation musicale* (Geneva: 1781).

81 Robert Castillo, *The Empire of Stereotypes. Germaine de Staël and the Idea of Italy* (Lonond: Palgrave Macmillan, 2006).

82 Richard Will, *The Characteristic Symphony in the Age of Haydn and Beethoven* (Cambridge: Cambridge University Press, 2012).

83 Melissa Percival, *The Appearance of Character: Physiognomy and Facial Expression in Eighteenth Century France* (Leeds: W.S.Maney & Son LTD, 1999).

departure from fixed notions of the passions, as categorised in the seventeenth century by Descartes and codified in a functional system developed by Charles Le Brun for the use in the visual arts. Instead, by mid-century new medical treatises placed emphasis on both the diversity and flux of human passions. The body-machine model was increasingly replaced by a theory of nervous function and structure that explained both the passions and bodily movements. The vitalist body, for example, was represented as inherently dynamic and moving.[84] Additionally, the increasing emphasis on sensibility and irritability of the nerves and muscles gave place to the rise of the model of nervous disorders – that is, that any perturbation or alteration in the nerves could be the cause of illness.[85] These different models of the body and nervous action, were essential to explain both the functioning of music in the internal body and its effects upon the bodies of listeners.

Melissa Percival explains how these physiological theories penetrated theatre performance and the visual arts, as characters moved from fixed to permeable notions in the 1750s.[86] Similarly, teaching manuals for musical performance during the second half of the century not only embraced a wider variety of passions, but remarked that passions could frequently become mixed. Moving away from a fixed repertoire of characters, this approach to musical performance coincides with the progressive distancing of music from the principle of mimesis which scholars have traditionally situated in this period, and the embrace, instead, of a broader notion of expression.[87]

The physiological emphasis on movement and flux of the passions was mirrored in the expanding range of expressive resources, mainly as provided by the development of instrumental techniques. Historians of music have written of a considerable flowering of instrumental technique during this period, as seen in the growth of orchestras and the symphonic form, among other developments relating to instrumental music. The orchestra of Mannheim has been considered one of the main drivers of these developments, which laid emphasis on expressive elements such as dynamics (volume marks) and their

84 Elizabeth A. Williams, *A Cultural History of Medical Vitalism in Enlightenment Montpellier* (London: Routledge, 2003); Roselyne Rey, *Naissance et développement du vitalisme en France de la deuxième moitié du 18e siècle à la fin du Premier Empire* (Liverpool: Voltaire Foundation/Liverpool University Press, 2000).

85 See George Rousseau, "Discourses of the Nerve", in Frederick Amrine, ed., *Literature and Science as Modes of Expression* (Boston: Boston Studies in the Philosophy of Science Book Series, 1989), pp. 29–60.

86 Percival, *The Appearance of Character*.

87 John Neubauer, *The Emancipation of Music from Language: Departure from Mimesis in Eighteenth-Century Aesthetics* (New Haven, CT: Yale University Press, 1986).

transition through the techniques of *crescendo* and *diminuendo* – processes of growing or diminishing in intensity, respectively.[88]

Accordingly, new "expressive" musical instruments were invented and old instruments were improved. As expression became increasingly associated with variability and transition from one passion to another, instrument makers sought to expand the potential characters and expressive possibilities of their instruments. Instruments that would allow performers to achieve greater levels of nuance and volume dynamics were preferable, including such instruments linked to women and the feminine. Harps, lutes, and guitars produced primarily subtle and soft sounds. Indeed, the stress placed on sensibility encouraged the use of delicate sounds and the appraisal of low-volume instruments under the rubric of intimate feeling. This appraisal was expressed in German lands through the fashion for clavichords, which produced a rich range of low-volume sounds and nuances.[89]

It is striking to notice the importance of timbral nuance and widening volume dynamics in the last third of the century, as the expressive elements *sine qua non* in music. This phenomenon famously epitomised in debates about the dynamic possibilities of harpsichords and the introduction of the forte-piano in the last decades of the century. Although harpsichords were seen as achieving delicacy and grace better than other instruments, they were criticised for their inability to offer an extensive dynamic range; the forte-piano, which used a system of hammers instead of plucking the strings like the harpsichord, extended the range of volume in both extremes – *forte* (loud) and *piano* (soft). This criticism, thus, reveals how qualities associated to the "courtly body" model, became contentious in light of the new requirements for expressivity. Yet, attesting to how old and new rubrics of the body coexisted, new technologies were created in France to improve the dynamic range of the harpsichords, which were still highly fashionable instruments.[90] For instance, a harpsichord made by Sebastian Érard in 1783 which allowed for dynamic gradations, was praised for expressing performers' sensibilities and their varied, changing, and

88 See e.g. Emily I. Dolan, *The Orchestral Revolution: Haydn and the Technologies of Timbre* (Cambridge: Cambridge University Press, 2013).

89 Bernard Brauchli, *The Clavichord* (Cambridge: Cambridge University Press, 1998), p. 137; Robert Marshall, *Eighteenth-Century Keyboard Music* (London: Routledge, 2004), pp. 12–14.

90 See Michael Latcham, "The Combination of the Harpsichord and the Piano in the Eighteenth Century", in Thomas Steiner, ed., *Instruments à claviers – expressivité et flexibilité sonore/Keyboard Instruments – Flexibility of Sound and Expression* (Bern: Peter Lang, 2004).

nuanced passions.[91] Mirroring the new expressive human body was, thus, the newly improved expressive body of the musical instrument.

7 Bodies of Dispute

Performers had at their disposal material technologies which gave expression to a new emotional palette, which they expressed through both their playing and their own bodies. Similarly, historians of dance, theatre, and body language more generally have claimed that the new sensibility witnessed the pursuit of expressive bodies which could communicate feeling beyond words.[92] Writers on dance and theatre carried out processes of reform which placed the body at the centre of the artistic aim, as a highly meaningful signifier of the passions. In dance, this reform was led by Jean-Georges Noverre, who sought a form of primitive expression of the passions which did not obey geometrical laws but was attributable to an inner state and feeling.[93] The pursuit of new expressive bodies, was thus, strongly opposed to the existent stress on mechanization and cultivation of skill.

Although the principles of taste and grace were deemed positive throughout the century, excessive stress on the corporeal and manners came to be seen as affectation, mannerism, and posturing. In musical performance, these critiques extended to virtuoso performers, who epitomised the excessive display of skill and speed at the expense of expression. Objections to "excess" are evident in music manuals from the beginnings of the century. Geminiani, deemed a talented performer himself, strongly condemned the exhibition of manual skills and "tricks" by violin performers on his violin method.[94] Virtuoso performers were later criticised within a more general perception of manual labour as being inferior to intellectual capacity.[95] The Chevalier J. J. O. de Meude-Monpas, a composer of music who published a musical dictionary

91 "Clavecin d'une nouvelle construction, inventé par M. Erard, attaché à Madame la Duchesse de Villeroi", *Almanach Musical* (Paris: Ruault, 1783), p. 54.

92 Cecilia Feilla, *The Sentimental Theater of the French Revolution* (Aldershot: Ashgate, 2013); Edward Nye, *Mime, Music and Drama on the Eighteenth-Century Stage: The Ballet d'Action* (Cambridge: Cambridge University Press, 2011); Percival, *The Appearance of Character*; Sophia Rosenfeld, *A Revolution in Language. The Problem of Signs in Late Eighteenth-Century France* (Stanford: Stanford University Press, 2001).

93 Jean-Georges Noverre, *Lettres sur la danse, et sur les ballets* (Lyon: Aimé Delaroche, 1760).

94 Francesco Geminiani, *The Art of Playing on The Violin* (London: 1751).

95 See Lissa Roberts, Simon Schaffer, Peter Dear, ed., *The Mindful Hand: Inquiry and Invention from Late Reinassance to Industrialisation* (Baltimore: The Johns Hopkins Univiversity Press, 2007).

in 1787, stressed this "manual" character of virtuoso performers, calling them "ouvriers des notes".[96] Consequently, they were derogatively associated to automata. Noverre depicted the appearance of a successful virtuoso violinist in a Parisian stage, yet referred to him as "l'homme-machine et sans tête", void of both intellect and feeling.[97]

Hence the critique to virtuosi performers intertwined social, gendered, and political issues. The fashionable pleasure arising from virtuosi was relegated to the street fairs, boulevard entertainment, where "tricks", according to elites, had a well-established place.[98] Furthermore, the critique to virtuoso performers also channelled a broader criticism to the "artificiality" of courtly life and aristocratic bodies, in the same ways scholars have studied for the case of android automata.[99] Aristocratic women performers, specifically, represented the mechanical body *par excellence*, as exemplified in the popularity of female android automata. Meude-Monpas criticised the excessive training of the musical body and compared it to a female courtier's "toilette", which became a target for criticism of courtly excesses during the Revolutionary period.[100] Consequently, the excess of attention on the body and the feminine ideal in music, became increasingly associated with effeminacy and courtly adornment.

Conversely, the ideal expressive body was devoid of ornament, deprived of any accessory; almost nude, the new "naturalness" of the body was unmediated and uncultivated.[101] The dancer Marie Salle – Noverre's teacher – triggered a transformation of the costumes worn by female dancers by appearing on stage wearing simpler clothing and without masks, thus showing her hair and face. Later, in the Revolutionary period, the theatre reformer François-Joseph Talma appeared wearing a Roman toga in his first performance, which, drawing on neoclassic associations between nature, authenticity, and antiquity,

---

96 Chevalier J. J. O. de Meude-Monpas, *Dictionnaire de musique* (Paris: Knapen et fils, 1787), p. 212.

97 Noverre, *Lettres sur la danse*, pp. 273–4.

98 Le Guin, *Boccherini's Body*, p. 138.

99 Outram, *The Body and the French Revolution*; Schaffer, "Enlightened Automata"; Voskuhl, *Androids in the Enlightenment.*

100 Meude-Monpas, *Dictionnaire de musique*, p. 211. See Kimberly Chrisman-Campbell, "Dressing to Impress: The Morning Toilette and the Fabrication of Femininity", in Charissa Bremer-David, ed., *Paris: Life & Luxury in the Eighteenth Century* (Los Angeles: Getty Publications, 2011), pp. 53–74.

101 For a similar discussion on gardens, see Rémy Gilbert Saisselin, "The French Garden in the Eighteenth Century: From Belle Nature to the Landscape of Time", *The Journal of Garden History*, 5, 3 (1985): 284–297; Emma Spary, *Utopia's Garden: French Natural History from Old Regime to Revolution* (Chicago: The University of Chicago Press, 2000).

placed a sharp contrast with theatre's pompous costumes.[102] This reform of theatre, scholars have argued, was paralleled in other art forms and expressed the emergence of a new ideal citizen, which was aligned with a new masculine ideal of virtue.[103] For Dorinda Outram, this reform followed an increasing differentiation of the sexes, and the consequent displacement of women from political culture.[104] This shift was paralleled by the growing discredit of female "sovereignty" in musical practice, as discussed by Mathew Head.[105] After the French Revolution, the "feminine" was ultimately excluded from the body politic.[106] In a similar way, the new demands on authenticity and classical virtue, verbalised by reformers in theatre, dance, and the visual arts, increasingly excluded the "feminine" from the ideal musical body.

8 Conclusions

This essay argued that the body of the musical performer was cultivated and signified according to changing models of the body, which comprehended notions of feeling, human physiology, gender, materiality, and social and political concerns. Models of the performer's body were negotiated through musical instruments and new material technologies such a musical automata and the development of new instrumental techniques in the case of the expressive body.

The study of musical performers illustrates that different models of the body coexisted in the last half of the eighteenth century, including the courtly, mechanical, and expressive body. For much of the century, performing methods of musical instruments stipulated a series of requirements on body posture, gestures, and appearance that drew upon courtly codes of behaviour. In this context, the ideal performing body was physically, morally, and sonically disciplined according to the aims of polite sociability and gendered values. Similarly, musical performance followed an ideal of femininity, which spanned from a courtly model that stressed soft and graceful gestures – which contemporaries deemed more natural to women – towards the embrace of a certain sensibility and "naturalness" of feeling that were linked to women.

---

102 Marie-Joseph de Chénier, *Charles IX, ou l'école des rois* (Paris: 1790), pp. 34–5.

103 Herbert, *David, Voltaire, "Brutus"*; Feilla, *The Sentimental Theater*.

104 Outram, *The Body and the French Revolution*.

105 Head, *Sovereign Feminine*.

106 Lynn Hunt, "Pourquoi la République est-elle une femme? La symbolique républicaine et l'opposition des genres, 1792–99", in Michel Vovelle, ed., *Révolution et République. L'exception française* (Paris: Kimé, 1994).

Yet the body of the musical performer also became a site of contestation of those same standards. Growing demands for technical skill and virtuosic playing, together with new mechanistic views of human anatomy, further emphasised the importance of physical discipline and repetitive training. Nevertheless, during the last decades of the century, musical automata and virtuoso performers were jointly criticised as part of a wider critique of the artificiality of courtly behaviour. Opposition to artificiality and courtly constraint targeted the very concepts of naturalness and feminine performance long cultivated by the musical performer. The association of women with automata and the Ancien Régime ultimately undermined the deep-rooted relationship between musical bodies, courtliness, and a "feminine touch".

References

Académie Royale des Sciences, *Machines et inventions approuvées par l'Académie royale des sciences* (Paris: 1735–1777).

Ancelet, *Observations sur la musique, les musiciens, et les instrumens* (Amsterdam [i.e. Paris]: 1757).

Anon., "Clavecin d'une nouvelle construction, inventé par M. Erard, attaché à Madame la Duchesse de Villeroi", *Almanach Musical* (Paris: Ruault, 1783), 54.

Berg, Maxine, *Luxury and Pleasure in Eighteenth Century Britain* (Oxford: Oxford University Press USA, 2005).

Blanning, T.C.W., *The Triumph of Music: The Rise of Composers, Musicians and Their Art* (Cambridge: Belknap Press of Harvard University Press, 2008).

Boüin, François, *La Vielleuse habile, ou nouvelle méthode courte, très facile et très sure pour apprendre à jouer de la vielle* (Paris, 1761).

Bouissou, Sylvie, Andrea Fabiano, ed., *La 'Querelle des Bouffons' dans la vie culturelle française du XVIIIe siècle* (Paris: CNRS éditions, 2005).

Brauchli, Bernard, *The Clavichord* (Cambridge: Cambridge University Press, 1998).

Brenet, Michel, *Les concerts en France sous l'Ancien Régime* (Paris: Libraire Fischbacher, 1900).

Brévan, Bruno, *Les changements de la vie musicale parisienne de 1774 à 1799* (Paris: Presses Universitaires de France, 1980).

Burke, Peter, *The Fabrication of Louis XIV* (New Heaven, CT: Yale University Press, 1992).

Carpentier, Joseph, *Méthode distribuée par leçons pour apprendre en peu de temps à jouer de l'instrument appellé cytre où guitthare allemande* (Paris : Joseph Carpentier, 1771).

Castiglione, Baldassare, *Il cortegiano* (Venice: 1528).

Castillo, Robert, *The Empire of Stereotypes. Germaine de Staël and the Idea of Italy* (Lonond: Palgrave Macmillan, 2006).

Charles-Dominique, Luc, "*'Jouer', 'sonner', 'toucher'* Une taxinomie française historique et dualiste du geste musical", *Cahiers d'ethnomusicologie*, 2001, 14:111–123.

Charlton, David, *Opera in the Age of Rousseau: Music, Confrontation, Realism* (Cambridge: Cambridge University Press, 2012).

Chénier, Marie-Joseph, *Charles IX, ou l'école des rois* (Paris: 1790).

Choquel, Herni L., *La Musique rendue sensible par la méchanique* (Paris: 1759; 1762).

Chrisman-Campbell, Kimberly, "Dressing to Impress: The Morning Toilette and the Fabrication of Femininity", in Charissa Bremer-David, ed., *Paris: Life & Luxury in the Eighteenth Century* (Los Angeles: Getty Publications, 2011).

Cohen, Sarah R. *Art, Dance, and the Body in French Culture of the Ancien Régime* (Cambridge: Cambridge University Press, 2000).

Corrette, Michel, *Le maître de clavecin pour l'accompagnement, méthode theorique et pratique* (Paris: 1753).

Couperin, François, *L'art de toucher le clavecin*; Francesco Geminiani, *L'art de jouer le violon: contenant les règles nécessaires à la perfection de cet instrument* (Paris: 1752).

Courtin, Antoine de, *Nouveau traité de la civilité qui se pratique en France parmi les honnestes gens* (Amsterdam: 1708).

Cowan, Brian, "Public Spaces, Knowledge, and Sociability," in Frank Trentmann, ed., *The Oxford Handbook of the History of Consumption* (Oxford: Oxford University Press, 2012), 251–66.

Cowling, Mary, *The Artist as Anthropologist: The Representation of Type and Character in Victorian Art*, (Cambridge: Cambridge University Press, 1989).

Cypess, Rebecca, *Curious and Modern Inventions: Instrumental Music as Discovery in Galileo's Italy* (Chicago: University of Chicago Press, 2016).

Cypess, Rebecca, "'It Would Be without Error': Automated Technology and the Pursuit of Correct Performance in the French Enlightenment", *Journal of the Royal Musical Association*, 142, 1 (2017): 1–29.

Darlow, Mark, *Dissonance in the Republic of Letters: The Querelle des Gluckistes et de Piccinnistes* (London: Routledge, 2013).

Darnton, Robert, *The Great Cat Massacre: And Other Episodes in French Cultural History* (New York: Basic Books, 2009).

Demeuse, Nicolas, *Le maître d'escrime, ou L'art des armes* (Paris: Chez Delarue, ca. 1778).

Dena Goodman, *The Republic of Letters: A Cultural History of the French Enlightenment* (Ithaca, NY: Cornell University Press, 1996).

Di Profio, Alessandro, *La révolution des Bouffons. L'opéra italien au Théâtre de Monsieur 1789–1792* (Paris: CNRS Éditions, 2003).

Dictionnaire de l'Académie française (Paris: 1694): http://artfl-project.uchicago.edu.

Diderot, Denis and Jean le Rond d'Alembert, *Encyclopédie, ou dictionnaire raisonné des sciences, des arts et des métiers, par une société de gens de lettres*, eds. Denis Diderot and Jean le Rond d'Alembert, vol. 8 (Paris: 1765). ARTFL Encyclopédie Project, University of Chicago: http://encyclopedie.uchicago.edu.

Dolan, Emily I., *The Orchestral Revolution: Haydn and the Technologies of Timbre* (Cambridge: Cambridge University Press, 2013).

Elias, Norbert, *The Civilising Process*, tr. Edmunf Jephcott (Oxford: Blackwell Publishing, 2000).

Erlmann, Veit, *Reason and Resonance: A History of Modern Aurality* (New York: Zone Books, 2010).

Fabian, Dietrich, *Kinzing und Roentgen: Uhren aus Neuwied* (Bad Neustadt: Verlag D. Pfaehler, 1984).

Fabiano, Andrea, *Histoire de l'opéra italien en France (1752–1815). Héros et héroînes d'un roman théâtral* (Paris: CNRS Editions, 2006).

Feilla, Cecilia, *The Sentimental Theater of the French Revolution* (Aldershot: Ashgate, 2013).

Foucault, Michel, *Surveiller et punir: Naissance de la prison* (Paris: Gallimard, 1975).

Gamelin, Jacques, *Nouveau recueil d'ostéologie et de myologie dessiné d'après nature* (Toulouse: 1779).

Garsault, François-Alexandre-Pierre de, *Notionnaire ou Memorial raisonné de ce qu'il y a d'utile et d'interessant dans les connoissances acquises depuis la création du monde jusqu'à présent* (Paris: 1761).

Geminiani, Francesco, *The Art of Playing on The Violin.* (London: 1751).

Gerritsen, Anne, Giorgio Riello, ed., *The Global Lives of Things: The Material Culture of Connections in the Early Modern World* (London: Routledge, 2015).

Gervasio, Giovanni Battista, *Méthode très facile pour apprendre à jouer de la mandoline à quatre cordes, instrument fait pour les dames* (Paris: Boüin, ca. 1767).

Gétreau, Florence, "Corps, mains et visages de musiciens sous les crayons de Watteau", in *Antoine Watteau (1684–1721). La leçon de musique*, ed. Florence Raymond (Bruxelles: Bozar Books/Hannibal, 2013).

Gétreau, Florence, "Une harpiste au Concert Spirituel: Mademoiselle Schencker en 1765", *Musique-Images-Instruments*, 1 (1995): 178–181.

Gétreau, Florence, *Aux origines du Musée de la Musique: les collections instrumentales du Conservatoire de Paris. 1793–1993* (Paris: Klincksieck, 1996).

Goehr, Lydia, *The Imaginary Museum of Musical Works: An Essay in the Philosophy of Music* (Oxford: Oxford University Press, 1992).

Gordon, Daniel, *Citizens without Sovereignty: Equality and Sociability in French Thought, 1670–1789* (Princeton: Princeton University Press, 1994).

Gouk, Penelope and Ingrid Sykes, "Hearing Science in the Eighteenth Century", *Journal of the History of Medicine and Allied Sciences*, 66, 4 (2011): 507–545.

Gouk, Penelope, *Music, Science, and Natural Magic in Seventeenth-Century England* (New Haven, CT: Yale University Press, 1999).

Gritten, Anthony, Elaine King, ed., *Music and Gesture* (London and New York: Routledge, 2006).

Guichard, Charlotte, *Les Amateurs d'art à Paris au XVIIIe siècle* (Seyssel: Champ Vallon, 2008).

Hayes, Julie Candler, "Sex and Gender, Feeling and Thinking: Imagining Women as Intellectuals", in Daniel Brewer, ed., *The Cambridge Companion to the French Enlightenment* (Cambridge: Cambridge University Press, 2014), pp. 91–104.

Head, Matthew, *Sovereign Feminine: Music and Gender in Eighteenth-Century Germany* (Berkeley: University of California Press, 2013).

Hennebelle, David, *De Lully à Mozart. Aristocratie, musique et musiciens à Paris (XVIIe–XVIIIe siècles)* (Seyssel: Champ Vallon, 2009).

Herbert, Robert L., *David, Voltaire, "Brutus", and the French Revolution: An Essay in Art and Politics* (New York: Viking, 1972).

Hotteterre, Jacques-Martin, *Principes de la flute traversiere ou de la flute a bec* (Paris: 1707).

Hunt, Lynn, "Pourquoi la République est-elle une femme? La symbolique républicaine et l'opposition des genres, 1792–99," in Michel Vovelle, ed., *Révolution et République. L'exception française* (Paris: Kimé, 1994).

Isherwood, Robert M., *Farce and Fantasy: Popular Entertainment in Eighteenth-Century Paris* (Oxford: Oxford University Press, 1989).

Jones, Colin, *The Smile Revolution in Eighteenth Century Paris* (Oxford: Oxford University Press, 2017).

Keele, Simon P., ed., *The Cambridge Companion to the Concerto* (Cambridge: Cambridge University Press, 2005).

Kennaway, James, *Bad Vibrations. The History of the Idea of Music as a Cause of Disease* (Aldershot: Ashgate, 2012).

La Mettrie, Julien Offray de, *L'Homme-machine* (Leyden: Elie Luzac, 1748).

La Vopa, Anthony, *The Labor of the Mind: Intellect and Gender in Enlightenment Cultures* (University of Pennsylvania Press, 2017).

Latcham, Michael, "The Combination of the Harpsichord and the Piano in the Eighteenth Century", in Thomas Steiner, ed., *Instruments à claviers – expressivité et flexibilité sonore/ Keyboard Instruments – Flexibility of Sound and Expression* (Bern: Peter Lang, 2004).

Le Guin, Elisabeth, *Boccherini's Body. An Essay in Carnal Musicology* (Berkeley: University of California Press, 2006).

Lee, Carol, *Ballet in Western Culture: A History of Its Origins and Evolution* (London-New York: Routledge, 2002).

Leppert, Richard, *The Sight of Sound. Music, Representation and History of the Body* (Berkeley: University of California Press, 1995).

Lescat, Philippe, *Méthodes et traités musicaux en France 1660–1800, réflexions sur l'écriture de la pédagogie musicale en France* (Paris: Institut de pédagogie musicale et chorégraphique-la Villette, 1991).

Lilti, Antoine, *Figures publiques. L'invention de la célébrité, 1750–1850* (Paris: Fayard, 2014).

Macquer, Philippe, *Dictionnaire raisonné universel des arts et métiers*, vol. 2 (Paris: 1773).

Mahaut, Antoine, *Méthode pour apprendre en peu de tems à jouer de la flûte traversière* (Amsterdam and Paris: 1759).

Marshall, Robert, *Eighteenth-Century Keyboard Music* (London: Routledge, 2004).

Metzner, Paul, *Crescendo of the Virtuoso* (Berkeley: University of California Press, 1998).

Meude-Monpas, (Chevalier) J. J. O. de, *Dictionnaire de musique* (Paris: Knapen et fils, 1787).

Milliot, Sylvette, "Le virtuose international: une création du 18e siècle", *Dix-huitième siècle*, 25 (1993): 55–64.

Milliot, Sylvette, La harpe au XVIIIe siècle à travers les documents iconographiques , in Hélène Charnassé, ed., *Instruments et musique instrumentale* (Paris: CNRS, 1986), 141–160.

Montagu, Jeremy, *The World of Baroque and Classical Musical Instruments* (Woodstock, NY: Overlook Press, 1979).

Morvan, Jean-Baptiste, *Réflexions sur la politesse des mœurs* (Paris: 1698).

Neubauer, John, *The Emancipation of Music from Language: Departure from Mimesis in Eighteenth-Century Aesthetics* (New Haven, CT: Yale University Press, 1986).

Noverre, Jean-Georges, *Lettres sur la danse, et sur les ballets* (Lyon: Aimé Delaroche, 1760).

Nye, Edward, *Mime, Music and Drama on the Eighteenth-Century Stage: The Ballet d'Action* (Cambridge: Cambridge University Press, 2011).

Outram, Dorinda, *The Body and the French Revolution. Sex, Class, and Political Culture* (New Haven, CT: Yale University Press, 1989).

Percival, Melissa, *The Appearance of Character: Physiognomy and Facial Expression in Eighteenth Century France* (Leeds: W.S.Maney & Son LTD, 1999).

Pierre, Constant, *Les facteurs d'instruments de musique. Les luthiers et la facture instrumentale. Précis historique* (Paris: Sagot, 1893).

Ragué, Louis-Charles, *Principes de harpe suivis de 18 airs* (Paris: 1786).

Rameau, Jean-Philippe, *Piéces de clavessin avec une méthode pour la mécanique des doigts* (Paris: 1724; 1731).

Rameau, Pierre, *Le maître à danser* (Paris: Jean Villette, 1725).

Rempel, Ursula, "The Complete Female: Musical 'Accomplishment' in the Late Eighteenth Century", *Man and Nature: Proceedings of the Canadian Society for Eighteenth-Century Studies*, 8 (1989): 39–48.

Rey, Roselyne, *Naissance et développement du vitalisme en France de la deuxième moitié du 18e siècle à la fin du Premier Empire* (Liverpool: Voltaire Foundation/Liverpool University Press, 2000).

Riskin, Jessica, *Science in the Age of Sensibility: The Sentimental Empiricists of the French Enlightenment* (Chicago: The University of Chicago Press, 2002).

Riskin, Jessica, *The Restless Clock: A History of the Centuries-Long Argument Over What Makes Living Things Tick* (Chicago: The University of Chicago Press, 2016).

Roberts, Lissa, Simon Schaffer, Peter Dear, ed., *The Mindful Hand: Inquiry and Invention from Late Reinassance to Industrialisation* (Baltimore: The Johns Hopkins Univiversity Press, 2007).

Rosenfeld, Sophia, *A Revolution in Language. The Problem of Signs in Late Eighteenth-Century France* (Stanford: Stanford University Press, 2001).

Rousseau, George, "Discourses of the Nerve", in Frederick Amrine, ed., *Literature and Science as Modes of Expression* (Boston: Boston Studies in the Philosophy of Science Book Series, 1989), 29–60.

Russo, Elena, *Styles of Enlightenment: Taste, Politics, and Authorship in Eighteenth-Century France* (Baltimore: The Johns Hopkins Univiversity Press, 2007).

Saisselin, Rémy Gilbert, *Taste in Eighteenth Century France. Critical Reflections on the Origins of Aesthetics; or, an Apology for Amateurs* (Syracuse, NY: Syracuse University Press, 1965).

Saisselin, Rémy Gilbert, "The French Garden in the Eighteenth Century: From Belle Nature to the Landscape of Time," *The Journal of Garden History* 5, 3 (1985): 284–297.

Sarrasin Robichaud, Philippe, *L'Homme-clavecin, une analogie diderotienne* (Paris: Classiques Garnier, 2017).

Schaffer, Simon, "Enlightened Automata", in William Clark, Jan Golinski, Simon Schaffer, ed., *The Sciences in Enlightened Europe* (Chicago: The University of Chicago Press, 1999), pp. 126–67.

Spary, Emma, *Eating the Enlightenment: Food and the Sciences in Paris, 1670–1760* (Chicago: The University of Chicago Press, 2012).

Spary, Emma, *Utopia's Garden. French Natural History from Old Regime to Revolution* (Chicago: The University of Chicago Press, 2000).

Stowell, Robin, *Violin Technique and Performance Practice in the Late Eighteenth and Early Nineteenth Centuries* (Cambridge: Cambridge University Press, 1990).

Thomas, Downing A., *Music and the Origins of Language: Theories from the French Enlightement* (Cambridge: Cambridge University Press, 1995).

Tissot, Robert-Henri, Camille Bellissant, ed., *Le Concert des Amateurs à l'Hôtel de Soubise (1769–1781). Une institution musicale parisienne en marge de la Cour* (Grenoble: MSH-Alpes, 2000).

Van Orden, Kate, *Music, Discipline, and Arms in Early Modern France* (Chicago: The University of Chicago Press, 2005).

Vaucanson, Jacques, *Le mécanisme du fluteur automate, présenté à messieurs de l'Académie royale des sciences* (Paris: 1738).

Vila, Anne C., *Enlightenment and Pathology. Sensibility in the Literature and Medicine of Eighteenth-Century France* (Baltimore: Johns Hopkins University Press, 1998).

Vincent-Buffault, Anne, *The History of Tears* (London: Macmillan, 1991).

Voltaire, François-Marie Arouet de, "Grâce", in Denis Diderot and Jean le Rond d'Alembert, ed., *Encyclopédie, ou dictionnaire raisonné des sciences, des arts et des métiers, par une société de gens de lettres*, vol. 7 (Paris: 1752): http://encyclopedie.uchicago.edu.

Voskuhl, Adelheid, *Androids in the Enlightenment: Mechanics, Artisans, and Cultures of the Self* (Chicago: The University of Chicago Press, 2013).

Will, Richard, *The Characteristic Symphony in the Age of Haydn and Beethoven* (Cambridge: Cambridge University Press, 2012).

Williams, Elizabeth A., *A Cultural History of Medical Vitalism in Enlightenment Montpellier* (London: Routledge, 2003).

Postface

On Hands, Feelings, and a Nose: Bodies Beyond Gender as Transdisciplinary Tools in Science

Paola Govoni

A variety of quantitative studies I will discuss below confirm that the integrated or transdisciplinary approaches long called-for by practitioners in science and technology studies are increasingly being adopted. From nanotechnology to neuroscience, from artificial intelligence to climate, cancer and "one health" studies, a growing number of research groups are moving beyond speaking only their own specialist dialect. Efforts are being made to understand, if not actually to speak, the languages of neighboring fields, in order to draw from these fields both inspiration and data useful for rounding out their own understanding. There is an increasingly shared awareness that complex problems require multiple tools to be tackled: indeed, this notion has become so often repeated that some argue it is becoming a risky cliché.[1] Exchanges between different fields, some even quite distant, provide confirmation of an insight science studies began to develop long ago: that interactions between different contents and contexts play a crucial role in the process of building innovative (natural) knowledge.[2] From climate change to the spread of viral epidemics, from the increase in average life expectancy to the rise in cognitive abilities, every phenomenon involving humanity is at the intersections of natural, technological, and social spheres. In order to be understood, these phenomena must often be studied using both the natural sciences and the social sciences: the kind of integrated approach typical of studies on gender and science.

From the beginning, gender studies adopted research strategies that were simultaneously biological, historical and sociological, questioning the

1 Ellie Bothweel, "Is Interdisciplinary Research Really the Best way to Tackle Global Challenges?", *The Times Higher Education*, February 13, 2020, https://www.timeshighereducation.com/features/interdisciplinary-research-really-best-way-tackle-global-challenges. But a few days later, punctually: Editorial, "Economists and Scientists: Solve Big Societal Problems by Working Together," *Nature*, 27 February 2020: 489, doi: 10.1038/d41586-020-00532-4.

2 See at least: Bruno Latour, *Science in Action: How to Follow Scientists and Engineers Through Society* (Milton Keynes: Open University Press, 1987); Jan Golinski, *Making Natural Knowledge: Constructivism and the History of Science* (Chicago: Chicago University Press, 2005^2); Steven Shapin, *Never Pure: Historical Studies of Science as if it was Produced by People with Bodies, Situated in Time, Space, Culture, and Society, and Struggling for Credibility and Authority* (Baltimore: Johns Hopkins University Press, 2010).

 | DOI:10.1163/9789004512610_011

boundaries that allegedly divide the natural dimensions from the cultural ones.[3] Boundaries which, as many scholars have noted, have long been patrolled by society as well as by science. This line of research has helped to unravel the political origins of the image of science as a body of knowledge which is supposed to be somehow immune from social and cultural values, including gender-based values.[4] In fact, this research line – and this has received less attention – has also helped us to understand how techno-scientific creativity works by calling attention to the deep face-to-face relationship that is established between the individual and nature in the natural sciences. In the following pages I shall limit myself, through some brief snapshots, to addressing some of the best-known authors – women as well as men – who, working in different periods, disciplines and contexts, have shown awareness of the role of their bodies in their deep relationship with nature, often going beyond a gendered perspective. These examples will show just how meaningless the nature vs. nurture debate has become in light of the knowledge provided by biology and genetics, as well as by primatology and gender studies. Or at least how meaningless it is to engage with that debate as it has been framed by many thinkers since the Nineteenth century.[5] With a simple example, primatologist Hans Kummer (1930–2013) has suggested that asking how much is produced by nature and how much by nurture, how much by genes and how much by the environment, makes just as much sense in some ways as asking if the music we might hear in the distance is coming from the instruments or the people playing them.[6]

---

3 For an outline of this early phrase of research by biologist and gender studies scholar see: Lynda Birke, "'Life' as We Have Known It: Feminism and the Biology of Gender," in Marina Benjamin, ed., *Sense and Sensibility: Gender and Scientific Enquiry, 1780–1945* (Oxford: Basil Blackwell, 1991), pp. 243–263. For further sources, I would cite at least the work of Anne Fausto-Sterling, biologist as well as STS scholar: *Myths of Gender: Biological Theories about Women and Men* (New York: Basic Books, 1985); *Sexing the Body: Gender Politics and the Construction of Sexuality* (New York: Basic Books, 2000).

4 The literature on these topics is obviously very extensive. To cite just a few: Carolyn Merchant, *The Death of Nature: Women, Ecology and Scientific Revolution* (New York: Harper and Row 1980); Evelyn Fox Keller, *Reflections on Gender and Science* (New Haven: Yale University Press, 1985). In addition, one other key source is Katharine Park, "Women, Gender, and Utopia: *The Death of Nature* and the Historiography of Early Modern Science," *Isis*, 97 (2006): 487–495.

5 For a brilliant synthesis of that long-term debate and in light of today's science, see: Evelyn Fox Keller, *The Mirage of a Space between Nature and Nurture* ([Durham:] Duke University Press, 2010).

6 Of course, Kummer continues, "we can very well discuss whether two recorded performances sound different because of a new drummer or a new instrument." Hans Kummer, *Primate Societies: Group Techniques of Ecological Adaptation* (New Brunswick, NJ: Aldine Transaction,

Keeping in mind these broad themes, and in the context of a book on the material and visual, modern and early modern history of science in relation to touch – whether gendered or not – in this chapter I suggest a long-durée – and scandalously shallow – journey through some well-known cases in which touch played a significant role. It is a journey that starts in today's laboratories, in order to address the question: are there or are there not sex/gender differences in touch? The journey continues with a leap back into an early modern anatomy lab, then back to the Twentieth century. The aim is to highlight the results that I believe the recent dialogue around gender issues that has sprung up between the natural and social sciences has produced in science, technology and medicine. Ironically, while professional women scientists are still struggling with discrimination *ratione sexus*[7] in more or less every country in the world, the concept of gender has successfully made its way into many cutting-edge labs, breaking new ground in otherwise well-established, traditional research areas.

1 Gender in the Lab and in the History of Science

In a volume focused on cases drawn from the early modern age, the rapid journey through the centuries by an historian of contemporary and recent science is meant to think about some of the tools that in recent years have helped renew the field of gender studies, both in the natural sciences and in science studies.

Late in 2019, the science weekly *Nature* (which released its first issue on November 4, 1869) published a series of articles to commemorate its 150th year of activity. One interesting article presents the results of a broad, long-term analysis of the bibliographic references mentioned in the papers published in *Nature*. The survey shows that, since the foundation of the journal, the papers show a growing awareness of interdisciplinary research; a trend that in recent

---

2006), p. 12. The example is evoked by Frans de Waal, *The Ape and the Sushi Master: Cultural Reflections of a Primatologist* (New York: Basic, 2002), p. 7; Fox Keller, *The Mirage of a Space between Nature and Nurture*, p. 7.

7 This was the expression used to define the limitations imposed on Laura Bassi (1711–1778) in 1732 when she secured a paid academic position. Paula Findlen, "Science as a Career in Enlightenment Italy: The Strategies of Laura Bassi," *Isis*, 84 (1993): 441–469. Special thanks to Paula Findlen for the exchanges we had on Bassi and several issues discussed in this chapter over the years.

times has come to involve the social sciences as well.[8] The data provide additional evidence of the findings of a study published in 2009 in *PLOS One*. An analysis of the clickstream produced by the readers of 35,000 specialist digital publications showed that scientists drew inspiration from research in the social sciences, including literature and art much more frequently than the publications' bibliographic references would display.[9] Quantitative studies such as these seem to corroborate some long-term trends that the history of science, using its own tools and languages, has highlighted in recent decades. In examining the biographies of several natural philosophers and scientists – both men and women – we have discovered that, in many cases, the habit of haunting far-flung and ostensibly "non-scientific" fields that has led to some of the most fascinating discoveries or innovative leaps. In evolutionism, it is well known, Darwin drew inspiration and data from many areas of naturalistic research as well as from demography and economics. In quantum physics, scientists have often crossed the boundaries between mathematics, physics, philosophy, art and politics. The work of geneticist Barbara McClintock (1902–1992), which I will return to below, merged scientific practices traditional in the field with other, quite bizarre approaches.[10]

Today, it is increasingly common to find scientists – female and male – bringing the social sciences and the concept of gender into their laboratory work. For some time now, a so-called transdisciplinary approach has been generating new research fields: on gender *and* medicine, gender *and* genomics, as well as new ways of practicing pharmacology, the neurosciences, and evolutionary psychology etc.[11] These approaches, more recently, have led scientists to think about and occasionally change their daily lab practices, for

8 Albert-László Barabási, et al., "A Network of Science: 150 years of Nature papers," *Nature*, 2019, 7 November, doi: 10.1038/d41586-019-03325-6 https://www.nature.com/articles/d41586-019-03325-6.

9 Johan Bollen, et al., "Clickstream Data Yields High-Resolution Maps of Science," *PLoSONE*, 2009:4 www.plosone.org/article/info:doi/10.1371/journal.pone.0004803.

10 Evelleen Richards, *Darwin and the Making of Sexual Selection* (Chicago: University of Chicago Press, 2017); Evelyn Fox Keller, *A Feeling for the Organism: The Life and Work of Barbara McClintock* (San Francisco: W.H. Freeman, 1983); David Kaiser, *Quantum Legacies: Dispatches from an Uncertain World* (Chicago: University of Chicago Press, 2020).

11 Gillian Einstein, "Situated Neuroscience: Elucidating a Biology of Diversity," in R. Bluhm, A. Jaap Jacobson, and H. Lene Maibom, ed., *Neurofeminism: Issues at the Intersection of Feminist Theory and Cognitive Science* (New York: Palgrave Macmillan, 2012), pp. 145–174; Gillian Einstein, "Sex and Gender in Health: The World Writes on the Body," in Marianne J. Legato, Marek Glezerman, ed., *The International Society for Gender Medicine: History and Highlights* (London: Elsevier, Academic Press, 2017), pp. 45–55.

instance by considering the importance of indicating in their papers the sex or gender of the researcher collecting data in order to monitor replicability; and declaring whether they have worked on male or female cells, and with male or female mice.[12] This mindfulness is of course the result of complex social as well as scientific attitudes. Practitioners more attentive to social questions have brought a new awareness of multiple social questions into the laboratories; an awareness that includes sex/gender issues, as they are addressed both today and in historical perspective. Awareness of the importance of considering such issues inside the lab has also permeated research institutions thanks to a close dialogue between biologists and scholars such as Evelyn Fox Keller, Anne Fausto-Sterling, Patricia Adair Gowaty, Lynda Birke, Isabelle Stengers and many others: scholars who have been simultaneously philosophers, sociologists and historians trained and often still practicing as scientists. Scholars delving into the history of European women in Eighteenth-century natural philosophy have also made an important contribution in this direction. Years ago, Londa Schiebinger – a professional historian of science – launched a simultaneously scientific and political project called gendered innovations in science, technology and medicine.[13]

Interactions between these diverse approaches may have a positive boomerang effect on historians who investigate specific questions such as: Can different ways of knowing be gendered? Were practitioners and scientists of the past aware of their own (gendered) touch? To explore questions such as these, a broad approach that puts science into dialogue with the social sciences enriched by long-durée historical awareness can help free us from some stereotypes that seem to affect also touch, in both a literal and broader

12 Cara Tannenbaum, Robert P. Ellis, Friederike Eyssel et al., "Sex and Gender Analysis Improves Science and Engineering," *Nature* 575, (2019): 137–146. See also: Colin D. Chapman, Christian Benedict, Helgi B. Schiöth, "Experimenter gender and replicability in science," *Science Advances*, 4, 1 (2018); Annaliese K. Beery, Irving Zucker, "Sex bias in neuroscience and biomedical research," *Neuroscience & Biobehavioral Reviews*, 35, 3 (2011), pp. 565–72; Brian J. Prendergast, Kennet G. Onishi, Irving Zucker, "Female Mice Liberated for Inclusion," *Neuroscience & Biobehavioral Reviews*, 40 (2014), DOI: 10.1016/j.neubiorev.2014.01.001. The Canadian Institutes of Health Research (http://www.cihr-irsc.gc.ca/e/50836.html) and National Institutes of Health (https://orwh.od.nih.gov/research/sex-gender/methods-and-techniques) offer guidelines.

13 Londa Schiebinger, *Has Feminism Changed Science?* (Cambridge: Harvard University Press, 1999); Londa Schiebinger, ed., *Gendered Innovations in Science and Engineering* (Stanford: Stanford University Press, 2008).

sense.[14] The results achieved by some historians of science who have adopted this approach in order to re-recount the lives and works of several women practicing natural philosophy or mathematics in the age of Enlightenment are important and well-known.

Many female Eighteenth-century natural philosophers and mathematicians that historians have written about in recent decades were already mentioned in various dictionaries and biographical repertoires in the Nineteenth and Twentieth centuries. The innovative move in historiography observed since the 1980s has lain not so much in (re)discovering the exemplary figures, but rather in granting a proper place to those same women by contextualizing their lives and works, putting those women back into dialogue with the men who occasionally took them seriously, and positioning them within scientific institutions in dialogue with the politics, religion and society of their times. In studying the networks of different people, powers, and places, practices, cultures and roles, historians working on female natural philosophers such as Laura Bassi (1711–1778) – the first woman in Europe, as Paula Findlen has documented, to be given a paid academic position in 1732 – have helped rewrite entire chapters of the history of science and culture, going beyond gender issues.[15] An historiographical approach open to science and technology studies has contributed to interpreting the work of Bassi and many other coeval women in science in ways that have proven inspirational for today's female scientists, as well as contemporary historians. Indeed, modern-age cases in context by scholars such as Marta Cavazza, Paula Findlen, Rebecca Messbarger, Massimo Mazzotti, and Londa Schiebinger have suggested new perspectives from which it is hoped that we will consider the strategies involved in the relations among individuals – men and women – in contemporary labs. By offering new, fresh reconstructions showing how European women in science, such as Bassi or Maria Gaetana Agnesi (1718–1799) – women who most people of the time intended should remain *exceptions* – have displayed the spaces of personal autonomy which since the Eighteenth century were leading to new

14 Autobiographical accounts of this can be found in Londa Schiebinger, "Following the Story: From *The Mind Has No Sex?* to *Gendered Innovations*," in Paola Govoni, Zelda A. Franceschi, ed., *Writing about Lives in Science: (Auto)biography, Gender and Genre* (Göttingen: V&R, 2014), pp. 43–54; Paula Findlen, "Listening to the Archives: Searching for the Eighteenth-Century Women of Science," ibid., pp. 87–115.

15 On Bassi, besides Findlen, "Science as a Career in Enlightenment Italy," see Beate Ceranski, *"Und sie fürchtet sich vor niemandem": Die Physikerin Laura Bassi (1711–1778)* (Frankfurt-New York: Campus, 1996); Marta Cavazza, *Laura Bassi: Donne, genere e scienza nell'Italia del Settecento* (Milan: Editrice Bibliografica, 2020).

female subjectivities, albeit always within male strategies of power.[16] These new historiographical approaches, together with the exchanges between gender scholars and natural scientists rooted in 1970s feminist studies, have helped us to better understand what gender is and why it matters.

Those studies have shown that gender is indeterminate or fluid, and that is why it is such a wonderful tool for researchers to employ in today's labs as well as for historians and social scientists to explore what science is and how it works. In discussing gender, we touch – literally, thanks also to our scientist colleagues – what we have meant in the last few decades when debating interactions between science and society, between nature and culture. As we shape the natural and social world around us, to some extent the "world writes on the body":[17] an insight offered by Gillian Einstein, a neuroscientist who set off from studies, including historical research, on women and gender studies and ended up bringing gender into the laboratory by interweaving psychological, endocrinological and neuroscientific research. The intertwining among those different approaches, including historical investigations into the professional challenges facing women in science,[18] have helped to show that not only are health and life expectancy affected by gendered systems, but the structure and functions of our plastic brains also respond in different ways to contexts in which women and men are assigned different tasks.[19] The interactions involved in this process are are profound and quantifiable, and they have been rendered perceptible – for example, by using the new brain imaging technologies. As more and more scientists become aware of how much cultural and social issues, by becoming "natural," affect our ways of knowing through our plastic brains and fingers, we are better able to understand issues of gender difference that manifest themselves, in different cognitive mathematical abilities.[20] Long-term historical accounts have helped us to understand such phenomena. Indeed, the neuroscientist Gina Rippon envisions a social

16 Marta Cavazza, *Between Modesty and Spectacle: Women and Science in Eighteenth-Century Italy*, in Paula Findlen, Wendy Wassyng Roworth, Catherine Sama, ed., *Italy's Eighteenth-Century: Gender and Culture in the Age of the Grand Tour* (Stanford: Stanford University Press, 2009), pp. 275–302. Besides the scholars mentioned, early modern historians such as Lorraine Daston, Katharine Park, Gianna Pomata, Joan Scott have also provided important historiographical considerations on the issues discussed here.

17 Gillian Einstein, *Sex and Gender in Health*, p. 46.

18 Many female scientists mention the history of women in science in their texts, including Fausto-Sterling, Birke, Einstein and Rippon in the publication cited above.

19 See Gina Rippon, *The Gendered Brain: The New Neuroscience that Shatters the Myth of the Female Brain* (London: The Bodley Head, 2019).

20 Paola Govoni, "Hearsay, Not-So-Big Data, and Choice: On Understanding Science and Maths by Looking at Men Who Supported Women," in Eva Kaufholz-Soldat, Nicola

cognitive neuroscience engaged in "[investigating] how a gendered world can produce a gendered brain."[21]

Such new developments of the old nature vs. nurture controversy encompass the issue of human touch as well.[22] For centuries, indeed millennia, touch was considered linked to sex: an all-female physical "sensitivity" was traditionally opposed to male "strength", and even men's vaunted logical and rational skills. What do the most recent neurosciences say on such issues? In 2009, a study offering some temporarily conclusive answers found that

> tactile perception improves with decreasing finger size, and that this correlation fully explains the better perception of women, who on average have smaller fingers than men. Indeed, when sex and finger size are both considered in statistical analyses, only finger size predicts tactile acuity. Thus, a man and a woman with fingers of equal size will, on average, enjoy equal tactile acuity. [This research] further show[s] that sweat pores, and presumably the Merkel receptors beneath them, are packed more densely in smaller fingers.[23]

It seems that people operating in various technological settings in early modern Bologna had already grasped that the question of women's superior manual sensitivity was more complex than it might appear from centuries of gendered-biased stereotypes on the subject. The owners of the textile mills in early modern Bologna, known as the more technologically advanced mills in Europe in those days, did not rely on sex alone when choosing the workers employed in winding silk thread, a job that required particularly sensitive fingers.[24] Similarly, Anna Morandi Manzolini (1714–1774), a professional anatomist in papal Eighteenth-century Bologna, also seems to have understood the point quite well.

Oswald, ed., *Against all Odds: Women's Ways to Mathematical Research Since 1800* (Cham: Springer, 2020), pp. 281–314.

21 Rippon, *The Gendered Brain*, p. 348.

22 For a wide-ranging and long-term overview of studies on touch that includes but also extends beyond scientific work and offers a rich bibliography, see Claudio Pogliano, *Senso lato: Il tatto e la cultura occidentale* (Rome: Carocci, 2015).

23 Ryan M. Peters et al., "Diminutive Digits Discern Delicate Details: Fingertip Size and the Sex Difference in Tactile Spatial Acuity," *Journal of Neuroscience*, 29, 50 (16 December 2009), 15756–15761.

24 Space limitations prevent me from exploring this highly interesting topic in depth, but see Carlo Poni, *La seta in Italia: Una grande industria prima della rivoluzione industriale* (Bologna: il Mulino, 2009).

2 Jumping through Centuries and Genes

In a 250-page manuscript notebook, anatomical sculptor Anna Morandi Manzolini wrote on the sense of touch that: "All of the identified nerves extend to the summit and apex of the fingers and here they multiply in most copious and minute ramifications … and for this reason … sensation is rendered more acute and delicate than anywhere else."[25] These remarks are surprising if read in the light of what we know today, thanks to the sophisticated technologies used by neuroscientists, in contrast with the primitive resources available to Morandi Manzolini. Her remarks do not include any explicit mention of the supposed greater tactile sensitivity attributed to women. What Morandi Manzolini's extraordinary wax models, writings and other accounts do display is the ability to delve imaginatively into those "minute ramifications". And she sought to communicate this cognitive and artistic ability with her impressive self-portraits.

Morandi Manzolini's wax models include also a famous portrait of Giovanni Manzolini (1700–1755), her husband and first anatomical teacher, portrayed while anatomizing a human heart. In the self-portrait, already internationally famous during her times, Morandi Manzolini represents herself intent on dissecting a brain. As part of Morandi Manzolini's dialogue across space and time with the great male masters of anatomy, Rebecca Messbarger argues that her self-representation conveyed an intent to emphasize "femininity" through the exaggerated display of crochet and lacework: details meant to celebrate Morandi Manzolini's own feminine hands, adorned with flashy rings, holding a scalpel and pliers in an authoritative grip. The staging thus communicates a powerful message: instead of the hand to be studied, as in the famous portrait of Giulio Cesare Casserio (but also the one of Vesalius), Morandi puts a brain, and she dissects it according to the new Enlightenment order stressing the hand-brain relationship underlying all knowledge.[26] Morandi Manzolini thus allows us to see the advantages in being a woman studying anatomy by re-signifying the stereotype of women's manual dexterity often used in the *querelle des femmes* to keep women in their place, that is, on the margins of knowledge.

---

25 Quoted in Rebecca Messbarger, *The Lady Anatomist: The Life and Work of Anna Morandi Manzolini* (Chicago: Chicago University Press, 2010), p. 130; Rebecca Messbarger, "Waxing Poeting: Anna Morandi Manzolini's Anatomical Stulptures," *Configurations*, 9, 1 (2001): 65–97.

26 My thanks to Rebecca Messbarger for our discussions on this issue. On the interpretation of the self-portrait, in addition to Messbarger's texts cited above, see also Marta Cavazza, "Women's Dialectics, or the Thinking Uterus: An Eighteenth-century Controversy on Gender and Education," in Lorraine Daston, Gianna Pomata, ed., *The Faces of Nature in Enlightenment Europe* (Berlin: Berliner Wissenschafts-Verlag, 2003), pp. 237–257.

FIGURE P.1 Giulio Cesare Casserio, approximately 1552–1616, line engraving by G. van Veen. Wellcome Collection

FIGURE P.2 Anna Morandi Manzolini, wax self-portrait with brain. Courtesy of Museo di Palazzo Poggi, Sistema Museale di Ateneo, Università di Bologna, photo by Fulvio Simoni

In both scientific and historical research, there are obvious physical reasons why we can never really take on that "view from nowhere" that – we are sometimes still tempted to think – would allow us to forget about our gendered bodies and minds in order to produce research free of conditioning. With this awareness, several natural philosophers and scientists such as Morandi

Manzolini have managed to transform those bodily conditionings into a formidable tool for understanding nature. Moving from papal Bologna to the Twentieth century, let me briefly outline some well-known cases of women and men struggling with their "touch beyond gender".

Let us jump from Morandi Manzolini's Eighteenth-century laboratory to Twentieth-century American science labs where, particularly after the First World War, competition between women and men in science often bordered on psychological violence.[27] In those contexts, similarly deliberate, powerful choices about issues of "gendered touch" often led in opposite directions from those of Morandi Manzolini, as shown by the well-known case of Barbara McClintock.

McClintock spoke of a "feeling for the organism" that she consciously used when observing and 'meeting' the corn plants in her field one by one, or when 'observing' her jumping genes. The people who knew her often called her way of knowing 'mystic,' meaning strange. A mysticism not religious but stemming from the concentration with which McClintock – since childhood – had developed a particular way of using her body when studying nature.[28] McClintock's practices were consistent with her interest in Buddhism and biofeedback techniques, interests which were actually fairly widespread among scientists, especially physicists. Other interests pursued by Nobel prize winner McClintock were unquestionably less common, such as her curiosity for UFOs, poltergeists or parapsychology.[29] Today, as in the past, such interests represent a serious challenge for those who prefer to cling to an idea of science as a culture of certainty.[30] The fact that McClintock also investigated these phenomena brings us face to face with issues which are not scientific in the traditional sense, but which – as in this case – we cannot ignore if we want to understand the creativity of complex figures like hers. McClintock was a great scientist for whom the body – and touch – played a central role in science practices. In her work, she sought to push such speculation beyond any form of consolidated culture so as to plunge that very culture into crisis in the quest for new

27 By now the literature on this issue is extraordinarily extensive, but see Margaret W. Rossiter's three volumes: *Women Scientists in America: Struggles and Strategies to 1940* (Baltimore: Johns Hopkins University Press, 1982); *Before Affirmative Action, 1940–1972* (1995); *Forging a New World since 1972* (2012).

28 Evelyn Fox Keller, *A Feeling for the Organism: The Life and Work of Barbara McClintock* (San Francisco: W.H. Freeman, 1983).

29 For a discussion of these aspects see Nathaniel C. Comfort, *The Tangled Field: Barbara McClintock's Search for the Patterns of Genetic Control* (Cambridge: Harvard University Press, 2001), pp. 151–155.

30 Comfort, *The Tangled Field*, pp. 151–155; Michael D. Gordin, *On the Fringe: Where Science Meets Pseudoscience* (Oxford: Oxford University Press, 2021).

FIGURE P.3 Barbara McClintock (1902–1992), US geneticist, in a maize field taking notes. Photographed in around 1960. ©American Philosophical Society/Science Photo Library

answers. Just as Morandi Manzolini in the Eighteenth century emphasized a female touch to support women in anatomy and its institutions, so McClintock pursued the same – 'political' – objective by rejecting some slippery concept of femininity and her own situation as a woman. While denying the label of feminist, she spoke up on multiple occasions to denounce the discrimination suffered by women, as well as by Jews and African Americans in the academic field. Deliberately ignoring questions of sex/gender, she demanded to be taken as only a mind working on nature through the use of an un-gendered body: "The body was something you dragged around [...] I always wished that I could be an objective observer, and not be what is known as 'me' to other people."[31] Now let us take another bold leap to Frans de Waal, class of 1948 and one of the most interesting among today's primatologists by virtue of his ability to combine a naturalistic gaze with a sociological one. In his many scientific and popular publications, de Waal offers some highly interesting considerations about the debate on "nurture as nature". Like McClintock, de Waal also seems wholly indifferent to disciplinary and other borders. As early as fifteen years ago he declared: "In about three decades, the general public in the West has

31 McClintock quoted in Keller, *A Feeling for the Organism*, p. 36.

moved from fear of biology to fascination. Now let's see if the social sciences will follow by putting more evolution into their thinking and curriculum. It is bound to happen."[32] Through a cross-disciplinary lens and meditating on his own way of knowing, de Waal talks about "bodily connection, usually called emotional contagion", a process that is, he argues, "easily demonstrable in lots of animals." After clarifying that he is not religious, de Waal speaks of an "experience of transcendence" among scientists:

> We [scientists] have a connection to the larger universe through our science. So that's part of the transcendence that we feel. [...] most of us [primatologists] were completely fascinated the first time we looked into the eyes of an ape. We felt an immediate connection between apes and humans. We feel this connection at a very visceral level.[33]

This move to go beyond the cultural and social self in order to be simply mammals, capable of establishing a feeling with nature, is not only for anatomists, geneticists or primatologists. It can also involve physicists and chemists.[34] With these latter in mind, it is with a nose that I want to conclude these reflections. A special nose, that of the internationally recognized writer and professional chemist Primo Levi (1919–1987).

As is well known, Levi was a holocaust survivor and writer with interests in art, the natural sciences, and technoscience, all of which he pursued while working his entire professional life as a chemist in industrial research. In 1984, reflecting on his own way of knowing through the body, Levi wrote:

> It is clear that from practical point of view a machine-made analysis is more convenient. But manual analysis, like all manual work, has a formative value; it is too similar to our origins as *mammals* to be neglected. We should after all know how to use our hands, our eyes, our nose. I'm very glad that I educated my nose: I'm able to identify by smell certain

32 "Q&A: Frans de Waal", *Current Biology*, 16, 21 (2006): R903–R904, R904.

33 From an interview with Frans de Waal in Steve Paulson, "Empathy, Morality, Community, Culture – Apes Have It All: Primatologist Frans de Waal takes exception with human exceptionalism," *Nautilus*, March 28, 2019, http://nautil.us//issue/70/variables/empathy-morality-community-cultureapes-have-it-all.

34 For an interesting example in physics, see the sensory deprivation practices attempted by members of the Fundamental Fysiks Group. See David Kaiser, *How the Hippies Saved Physics: Science, Counterculture, and the Quantum Revival* (New York: Norton&Company, 2011).

> functional groups more quickly than the infrared spectrometer and the gas chromatographer. In this way the physicist's trade integrates you in your function as a complete person who does not neglect any of *his* potential. Whether as a student or a varnish technician, I made a continuous effort to exploit fully the help that eyes, fingers, and nose can afford.[35]

This quote is from the English translation of a book featuring a lengthy dialogue between Levi and Tullio Regge (1931–2014), a theoretical physicist. The original Italian text – published in 1984, shortly before Levi's death – does not use 'his.' As Levi was known to be secular, I like to think that he – who worked side by side with the chemist Giovanna Balzaretti in the lab – avoided gender markers precisely to emphasize that 'mammal' point of view. Translated literally, the sentence sounds something like this: "In this [sense] the job of chemist integrates you in your function as a complete person, none of whose possible faculties are neglected."[36] Levi, who today we would call anti-speciesist, also incorporated traces of this approach throughout much of his literary production.[37] The issue alluded to by Levi is not so different from those explored by Morandi Manzolini when she wondered how humans understand the properties of things, emphasizing the direct relationship between the hands and the brain; or from the work of McClintock, who understood how to tune into jumping genes through a particular use of the body and practices which were markedly unusual for a professional scientist; or a primatologist such as de Waal, who seems capable of conducting a dialogue among equals with the apes he studies by taking on observational practices innovatively developed by women primatologists starting from the 1960s.[38] All of this has to do with issues like touch that I believe historians also deal with.

35 Primo Levi, Tullio Regge, *Conversations*, tr. Raymond Rosenthal (London: I.B. Tauris &Co, 1989), p. 62. Italics are mine.

36 Primo Levi, Tullio Regge, *Dialogo*, ed. by Ernesto Ferrero (Torino: Einaudi, 2018), p. 63.

37 On the vast literature on these issues see: Ian Thompson, *Primo Levi: A Life* (New York: Picador, 2002); Roberta Mori, Domenico Scarpa, ed., *Album Primo Levi* (Torino: Einaudi, 2017), pp. 233–279.

38 See Donna Haraway, *Primate Visions: Gender, Race, and Nature in the World of Modern Science* (New York and London: Routledge, 1989); ZuleymaTang-Martínez, "The history and impact of women in animal behaviour and the ABS: a North American perspective," *Animal Behaviour* (2020), https://doi.org/10.1016/j.anbehav.2019.12.011.

3 Conclusion: Beyond Gender

In these pages, I have offered a bird's-eye view of a few well-known cases of men and women from different periods and fields who have struggled with, or used with success, their touch, bodies and feelings in connection with their objects of investigation. Knowing how to position oneself from a mammal standpoint has emerged as the equivalent of the physicists' effort to position herself from the standpoint of an electron when dealing with relativity. The feelings we have been dealing with are both natural and cultural and they apply in anatomy or when identifying jumping genes, understanding quanta or apes, or smelling chemical substances. If we trust historians, we can confidently say that similar feelings are also important when working on archival documents and lives from the past.[39] Historians' sources are likewise material – that is, natural as well as cultural – whether we are dealing with wax models, naturalistic collections, instruments or institutions, written documents or oral accounts.[40] Such sources produce in the historian feelings which, as Levi would say, must be educated to include "any possible potential": what at the beginning of this chapter I have called an integrated approach open to different practices and languages. From the standpoint of science, just like from that of history and science studies, I think we should definitely shelve the supposed fracture between nature and culture that has dominated and distressed both academic and popular thought for so long. Our culture – quite obviously – is part of our nature, and vice versa. We should regard nature as implied in the works of Morandi Manzolini and McClintock, de Waal or Levi, consciously using our bodies as knowing tools, sometimes emphasizing gender or perhaps more often denying it.

In the ways of knowing practiced by perceptive historians, gendered touch may play an important role, but only in that it serves to engage those social and political dimensions of knowledge-building that historians contextualize through the lives of the people they want to understand. Any skilled historian will know how to hide that instrument away backstage. As in Primo Levi's writings, backgrounding the nose will allow the intricacies of nature and culture to fully emerge, in chemistry as elsewhere.

---

39 See the sources cited in note 14 and Paola Govoni, "Drawing Nature, Building Knowledge: Between Beauty and 'Gut Feelings' in the Sciences," in Giorgia Grilli, ed., *Non-Fiction Picture Books: Sharing Knowledge as an Aesthetic Experience* (Pisa: ETS, 2020), pp. 202–219.

40 Of the extensive literature on this point see Simon Schaffer, Adriana Craciun, ed., *The Material Culture of Enlightenment Arts and Sciences* (London: Palgrave Macmillan, 2016); Ivan Gaskell, Sarah Anne Carter, ed., *The Oxford Handbook of History and Material Culture* (Oxford: Oxford University Press, 2020).

Acknowledgements

In addition to the editors of this volume, I would like to thank the participants in the Bologna June 2019 workshop on 'Gendered Touch', for the interesting insights they offered with their papers and the lively exchanges that followed. Special thanks are due to the anonymous referees for their generous comments on this chapter. Finally, this is the occasion for a special note of thanks to my lifelong friends Marta Cavazza, Paula Findlen and Rebecca Messbarger, from whom I have learned so much about women, science, and society in modern-age Bologna, and much more.

References

"Economists and Scientists: Solve Big Societal Problems by Working Together," *Nature*, 27 February 2020: 489. doi: 10.1038/d41586-020-00532-4.

"Q&A: Frans de Waal", *Current Biology*, 16, 21 (2006).

Barabási, Albert-László, et al., "A Network of Science: 150 years of Nature papers," *Nature*, 2019, 7 November, doi: 10.1038/d41586-019-03325-6.

Beery, Annaliese K., Irving Zucker, "Sex Bias in Neuroscience and Biomedical Research," *Neuroscience & Biobehavioral Reviews*, 35, 3 (2011): 565–572.

Birke, Lynda, "'Life' as We Have Known It: Feminism and the Biology of Gender," in Marina Benjamin, ed., *Sense and Sensibility: Gender and Scientific Enquiry, 1780–1945*, (Oxford: Basil Blackwell, 1991), 243–263.

Bollen, Johan, et al., "Clickstream Data Yields High-Resolution Maps of Science," *PLoSONE*, 2009:4 www.plosone.org/article/info:doi/10.1371/journal.pone.0004803.

Bothweel, Ellie, "Is Interdisciplinary Research Really the Best way to Tackle Global Challenges?" *The Times Higher Education*, February 13, 2020, https://www.timeshighereducation.com/features/interdisciplinary-research-really-best-way-tackle-global-challenges.

Cavazza, Marta, "Women's Dialectics, or the Thinking Uterus: An Eighteenth-century Controversy on Gender and Education," in Lorraine Daston and Gianna Pomata, ed., *The Faces of Nature in Enlightenment Europe* (Berlin: Berliner Wissenschafts-Verlag: 2003), 237–257.

Cavazza, Marta, *Between Modesty and Spectacle: Women and Science in Eighteenth-Century Italy*, in Paula Findlen, Wendy Roworth, Wassyng, and Catherine Sama, ed., *Italy's Eighteenth-Century: Gender and Culture in the Age of the Grand Tour* (Stanford: Stanford University Press, 2009), 275–302.

Cavazza, Marta, *Laura Bassi. Donne, genere e scienza nell'Italia del Settecento* (Milan: Editrice Bibliografica, 2020).

Ceranski, Beate, *"Und sie fürchtet sich vor niemandem": Die Physikerin Laura Bassi (1711–1778)* (Frankfurt-New York: Campus, 1996).

Chapman, Colin D., Christian Benedict, and Helgi B. Schiöth, "Experimenter Gender and Replicability in Science," *Science Advances*, 4, 1 (2018).

Comfort, Nathaniel C., *The Tangled Field: Barbara McClintock's Search for the Patterns of Genetic Control* (Cambridge: Harvard University Press, 2001).

de Waal, Frans, *The Ape and the Sushi Master: Cultural Reflections of a Primatologist* (New York: Basic, 2002).

Einstein, Gillian, "Sex and Gender in Health: The World Writes on the Body," in Marianne J. Legato, Marek Glezerman, ed., *The International Society for Gender Medicine: History and Highlights* (London: Elsevier, Academic Press, 2017), 45–55.

Einstein, Gillian, "Situated Neuroscience: Elucidating a Biology of Diversity," in R. Bluhm, A. Jaap Jacobson, H. Lene Maibom, ed., *Neurofeminism: Issues at the Intersection of Feminist Theory and Cognitive Science* (New York: Palgrave Macmillan, 2012), 145–174.

Fausto-Sterling, Anne, *Myths of Gender: Biological Theories about Women and Men* (New York: Basic Books, 1985).

Fausto-Sterling, Anne, *Sexing the Body: Gender Politics and the Construction of Sexuality* (New York: Basic Books, 2000).

Findlen, Paula, "Listening to the Archives: Searching for the Eighteenth-Century Women of Science," in Paola Govoni, Zelda A. Franceschi, ed., *Writing about Lives in Science: (Auto)biography, Gender and Genre* (Göttingen: V&R, 2014), 87–115.

Findlen, Paula, "Science as a Career in Enlightenment Italy: The Strategies of Laura Bassi," *Isis*, 84 (1993): 441–469.

Fox Keller, Evelyn, *A Feeling for the Organism: The Life and Work of Barbara McClintock* (San Francisco: W.H. Freeman, 1983).

Fox Keller, Evelyn, *Reflections on Gender and Science* (New Haven: Yale University Press, 1985).

Fox Keller, Evelyn, *The Mirage of a Space between Nature and Nurture* ([Durham:] Duke University Press, 2010).

Gaskell, Ivan, Sarah Anne Carter, ed., *The Oxford Handbook of History and Material Culture* (Oxford: Oxford University Press, 2020).

Golinski, Jan, *Making Natural Knowledge: Constructivism and the History of Science* (Chicago: Chicago University Press, 2005).

Gordin, Michael D., *On the Fringe: Where Science Meets Pseudoscience* (Oxford: Oxford University Press, 2021).

Govoni, Paola, "Drawing Nature, Building Knowledge: Between Beauty and 'Gut Feelings' in the Sciences," in Giorgia Grilli, ed., *Non-Fiction Picture Books: Sharing Knowledge as an Aesthetic Experience* (Pisa: ETS, 2020), 202–219.

Govoni, Paola, "Hearsay, Not-So-Big Data, and Choice: On Understanding Science and Maths by Looking at Men Who Supported Women," in Eva Kaufholz-Soldat, Nicola Oswald, ed., *Against all Odds: Women's Ways to Mathematical Research Since 1800* (Cham: Springer, 2020), 281–314.

Haraway, Donna, *Primate Visions: Gender, Race, and Nature in the World of Modern Science* (New York and London: Routledge, 1989).

Kaiser, David, *How the Hippies Saved Physics: Science, Counterculture, and the Quantum Revival* (New York: Norton&Company, 2011).

Kaiser, David, *Quantum Legacies: Dispatches from an Uncertain World* (Chicago: University of Chicago Press, 2020).

Kummer, Hans, *Primate Societies: Group Techniques of Ecological Adaptation* (New Brunswick, NJ: Aldine Transaction, 2006).

Latour, Bruno, *Science in Action: How to Follow Scientists and Engineers Through Society* (Milton Keynes: Open University Press, 1987).

Levi, Primo, Tullio Regge, *Conversations*, tr. Raymond Rosenthal (London: I.B. Tauris&Co, 1989).

Levi, Primo, Tullio Regge, *Dialogo* (Torino: Einaudi, 2018).

Merchant, Carolyn, *The Death of Nature: Women, Ecology and Scientific Revolution* (New York: Harper and Row 1980).

Messbarger, Rebecca, "Waxing Poetiıg: Anna Morandi Manzolini's Anatomical Stulptures," *Configurations*, 9, 1 (2001): 65–97.

Messbarger, Rebecca, *The Lady Anatomist: The Life and Work of Anna Morandi Manzolini* (Chicago: Chicago University Press, 2010).

Mori, Roberta, Domenico Scarpa, ed., *Album Primo Levi* (Torino: Einaudi, 2017).

Park, Katharine, "Women, Gender, and Utopia: *The Death of Nature* and the Historiography of Early Modern Science," *Isis*, 97 (2006): 487–495.

Paulson, Steve, "Empathy, Morality, Community, Culture – Apes Have It All: Primatologist Frans de Waal takes exception with human exceptionalism," *Nautilus*, March 28, 2019, http://nautil.us//issue/70/variables/empathy-morality-community-cultureapes-have-it-all.

Peters, Ryan M. et al., "Diminutive Digits Discern Delicate Details: Fingertip Size and the Sex Difference in Tactile Spatial Acuity," *Journal of Neuroscience*, 29, 50 (2009): 15756–15761.

Pogliano, Claudio, *Senso lato: Il tatto e la cultura occidentale* (Rome: Carocci, 2015).

Poni, Carlo, *La seta in Italia: Una grande industria prima della rivoluzione industriale* (Bologna: il Mulino, 2009).

Prendergast, Brian J., Kennet G. Onishi, Irving Zucker, "Female Mice Liberated for Inclusion," *Neuroscience & Biobehavioral Reviews*, 40 (2014), DOI: 10.1016/j.neubiorev.2014.01.001.

Richards, Evelleen, *Darwin and the Making of Sexual Selection* (Chicago: University of Chicago Press, 2017).

Rippon, Gina *The Gendered Brain: The New Neuroscience that Shatters the Myth of the Female Brain* (London: The Bodley Head, 2019).

Rossiter, Margaret W., *Women Scientists in America: Struggles and Strategies to 1940* (Baltimore: Johns Hopkins University Press, 1982).

Schaffer, Simon, Adriana Craciun, ed., *The Material Culture of Enlightenment Arts and Sciences* (London: Palgrave Macmillan, 2016).

Schiebinger, Londa, "Following the Story: From *The Mind Has No Sex?* to *Gendered Innovations*," in Paola Govoni, Zelda A. Franceschi, ed., *Writing about Lives in Science: (Auto)biography, Gender and Genre* (Göttingen: V&R, 2014), 43–54.

Schiebinger, Londa, ed., *Gendered Innovations in Science and Engineering* (Stanford: Stanford University Press, 2008).

Schiebinger, Londa, *Has Feminism Changed Science?* (Cambridge, Mass.: Harvard University Press, 1999).

Shapin, Steven, *Never Pure: Historical Studies of Science as if it was Produced by People with Bodies, Situated in Time, Space, Culture, and Society, and Struggling for Credibility and Authority* (Baltimore: Johns Hopkins University Press, 2010).

Tang-Martínez, Zuleyma, "The history and impact of women in animal behaviour and the ABS: a North American perspective," *Animal Behaviour* (2020), https://doi.org/10.1016/j.anbehav.2019.12.011.

Tannenbaum, Cara, Robert P. Ellis, Friederike Eyssel et al., "Sex and Gender Analysis Improves Science and Engineering," *Nature* 575 (2019), 137–146.

Thompson, Ian, *Primo Levi: A Life* (New York: Picador, 2002).

Index

www.ingramcontent.com/pod-product-compliance
Lightning Source LLC
LaVergne TN
LVHW010544300926
849162LV00002B/226
* 9 7 8 9 0 0 4 5 1 2 6 0 3 *

www.ingramcontent.com/pod-product-compliance
Lightning Source LLC
LaVergne TN
LVHW010544300926
849162LV00002B/226